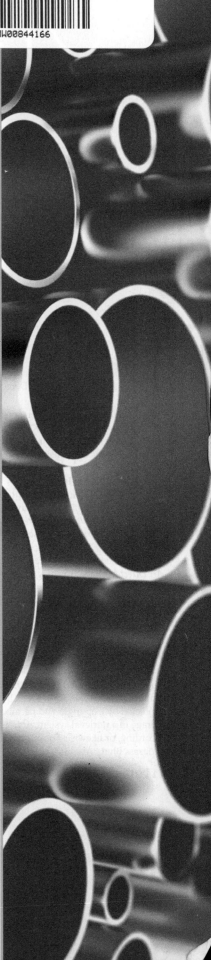

Plumbing
Level Four

Trainee Guide
Fourth Edition

PEARSON

Boston   Columbus   Indianapolis   New York   San Francisco   Upper Saddle River
Amsterdam   CapeTown   Dubai   London   Madrid   Milan   Munich   Paris   Montreal   Toronto
Delhi   Mexico City   São Paulo   Sydney   Hong Kong   Seoul   Singapore   Taipei   Tokyo

NCCER

President: Don Whyte
Director of Product Development: Daniele Dixon
Plumbing Project Manager: Chris Wilson
Senior Manager: Tim Davis
Quality Assurance Coordinator: Debie Hicks

Desktop Publishing Coordinator: James McKay
Permissions Specialists: Megan Casey / Adrienne Payne
Production Assistant: Adrienne Payne
Editor: Tanner Yea

Writing and development services provided by S4Carlisle Publishing Services, Dubuque, IA

Project Manager: Michael B. Kopf
Writers: Paul Lagasse and Jack Klasey
Art Development: S4Carlisle Publishing Services

Permissions Specialists: Kim Schmidt, Karyn Morrison
Media Specialist: Genevieve Brand
Copy Editor: Michael H. Toporek

Pearson Education, Inc.

Director, Global Employability Solutions: Jonell Sanchez
Head of Associations: Andrew Taylor
Editorial Assistant: Douglas Greive
Program Manager: Alexandrina B. Wolf
Project Manager: Janet Portisch
Operations Supervisor: Deidra M. Skahill
Art Director: Diane Ernsberger
Directors of Marketing: David Gesell, Margaret Waples
Field Marketers: Brian Hoehl, Stacey Martinez

Composition: NCCER
Printer/Binder: LSC Communications
Cover Printer: LSC Communications
Text Fonts: Palatino and Univers

Credits and acknowledgments for content borrowed from other sources and reproduced, with permission, in this textbook appear at the end of each module.

11 2019

PEARSON

ISBN-13: 978-0-13-382422-3
ISBN-10:   0-13-382422-5

Preface

To the Trainee

Most people are familiar with plumbers who come to their home to unclog a drain or install an appliance. In addition to these activities, however, plumbers install, maintain, and repair many different types of pipe systems. For example, some systems move water to a municipal water treatment plant and then to residential, commercial, and public buildings. Other systems dispose of waste, provide gas to stoves and furnaces, or supply air conditioning. Pipe systems in generation plants carry the steam that powers turbines. Pipes are also used in manufacturing plants, such as wineries, to move material through production processes.

Plumbers and their associated trades constitute one of the largest construction occupations, holding about 420,000 jobs. The occupation continues to grow, and nearly 229,000 job openings are expected to be available by 2020. Plumbers are also among the highest paid construction occupations.

New with *Plumbing Level Four*

This fourth edition of *Plumbing Level Four* presents a new instructional design and features a streamlined teaching order to better prepare you for your career as a plumber. In addition, *Plumbing Level Four* is now produced in full color, and contains updated artwork, detailed figures, and the latest tools and technology available to the plumbing trade. Throughout this title, "Going Green" features emphasize sustainability and environmentally constructive procedures.

This edition of *Plumbing Level Four* has a new module. *Introduction to Medical Gas and Vacuum Systems* provides information on these systems used in health care facilities, as outlined in NFPA 99, *Health Care Facilities Code*, and the professional qualifications as outlined in ASSE/IAPMO/ANSI Series 6000, *Professional Standard for Medical Gas Systems Personnel*.

Your instructor also has been provided with a new set of tools to better ensure your success as an apprentice plumber. As you continue with your third year of training in the plumbing craft, we wish you the best and hope that you'll continue your training beyond this curriculum. As many of the craftspeople employed in this trade will tell you, there are numerous opportunities awaiting those with the skills and the desire to move forward in the plumbing profession.

We invite you to visit the NCCER website at **www.nccer.org** for information on the latest product releases and training, as well as online versions of the *Cornerstone* magazine and Pearson's NCCER product catalog.

Your feedback is welcome. You may email your comments to **curriculum@nccer.org** or send general comments and inquiries to **info@nccer.org**.

NCCER Standardized Curricula

NCCER is a not-for-profit 501(c)(3) education foundation established in 1996 by the world's largest and most progressive construction companies and national construction associations. It was founded to address the severe workforce shortage facing the industry and to develop a standardized training process and curricula. Today, NCCER is supported by hundreds of leading construction and maintenance companies, manufacturers, and national associations. The NCCER Standardized Curricula was developed by NCCER in partnership with Pearson, the world's largest educational publisher.

Some features of the NCCER Standardized Curricula are as follows:

- An industry-proven record of success
- Curricula developed by the industry for the industry
- National standardization providing portability of learned job skills and educational credits
- Compliance with the Office of Apprenticeship requirements for related classroom training (*CFR 29:29*)
- Well-illustrated, up-to-date, and practical information

NCCER also maintains a Registry that provides transcripts, certificates, and wallet cards to individuals who have successfully completed a level of training within a craft in NCCER's Curricula. *Training programs must be delivered by an NCCER Accredited Training Sponsor in order to receive these credentials.*

Special Features

In an effort to provide a comprehensive, user-friendly training resource, we have incorporated many different features for your use. Whether you are a visual or hands-on learner, this book will provide you with the proper tools to get started in the plumbing trade.

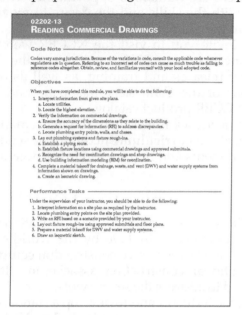

Introduction

This page is found at the beginning of each module and lists the Objectives, Performance Tasks, Trade Terms, and Required Trainee Materials for that module. The Objectives list the skills and knowledge you will need in order to complete the module successfully. The Performance Tasks give you an opportunity to apply your knowledge to the real-world duties that heavy equipment operators perform. The list of Trade Terms identifies important terms you will need to know by the end of the module. Required Trainee Materials list the materials and supplies needed for the module.

Special Features

Features provide a head start for those entering the Plumbing field by presenting technical tips and professional practices from operators in various disciplines. These features often include real-life scenarios similar to those you might encounter on the job site.

A Museum for Architecture

The Athenaeum of Philadelphia is a museum of American architecture. The museum, which has collected the work of about 1,000 American architects, has 150,000 drawings, 50,000 photographs, and many architectural documents.

Color Illustrations and Photographs

Full-color illustrations and photographs are used throughout each module to provide vivid detail. These figures highlight important concepts from the text and provide clarity for complex instructions. Each figure reference is denoted in the text in *italics* for easy reference.

Figure 13 Examples of BIM drawings.

Notes, Cautions, and Warnings

Safety features are set off from the main text in highlighted boxes and are organized into three categories based on the potential danger of the issue being addressed. Notes simply provide additional information on the topic area. Cautions alert you of a danger that does not present potential injury but may cause damage to equipment. Warnings stress a potentially dangerous situation that may cause injury to you or a co-worker.

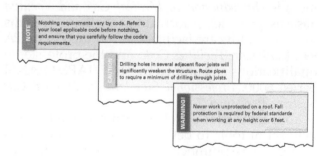

Going Green

Going Green looks at ways to preserve the environment, save energy, and make good choices regarding the health of the planet. Through the introduction of new construction practices and products, you will see how the "greening of America" has already taken root.

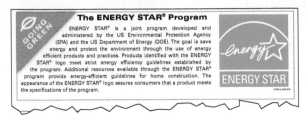

The ENERGY STAR® Program

ENERGY STAR® is a joint program developed and administered by the US Environmental Protection Agency (EPA) and the US Department of Energy (DOE). The goal is save energy and protect the environment through the use of energy efficient products and practices. Products identified with the ENERGY STAR® logo meet strict energy efficiency guidelines established by the program. Additional resources available through the ENERGY STAR® program provide energy-efficient guidelines for home construction. The appearance of the ENERGY STAR® logo assures consumers that a product meets the specifications of the program.

Did You Know?

The Did You Know? features offer hints, tips, and other helpful bits of information from the trade.

Did You Know?

Copper is one of the most plentiful metals. It has been in use for thousands of years. In fact, a piece of copper pipe used by the Egyptians more than 5,000 years ago is still in good condition. When the first Europeans arrived in the new world, they found the Native Americans using copper for jewelry and decoration. Much of this copper was from the region around Lake Superior, which separates northern Michigan and Canada.

One of the first copper mines in North America was established in 1664 in Massachusetts. During

Step-by-Step Instructions

Step-by-step instructions are used throughout to guide you through technical procedures and tasks from start to finish. These steps show you not only how to perform a task but also how to do it safely and efficiently.

Step 1	Drive a stake on one side of the excavation trench.
Step 2	Drive a second stake on the opposite side of the trench, in line with the first stake. If a length of pipe has already been positioned in the trench, drive the stakes beside the hub end of the pipe.
Step 3	Place the batter board in position, and level it using a builder's level.
Step 4	For ease in calculating grade, locate the batter board at a convenient height above the pipe. General-purpose levels are suitable for this task, though the results will not be as accurate

Trade Terms

Each module presents a list of Trade Terms that are discussed within the text and defined in the Glossary at the end of the module. These terms are denoted in the text with bold, blue type upon their first occurrence. To make searches for key information easier, a comprehensive Glossary of Trade Terms from all modules is located at the back of this book.

sic parts and in a wide variety of styles. You can interchange these basic parts and produce an almost endless variety of drains. The basic parts are the drain body, grate (dome), deck clamp, clamping ring or flashing clamp, drain receiver (bearing pan), and extensions.

The basic component of all drains, the drain body, funnels the water into the piping system. Usually made of cast iron or plastic, it is designed to connect to the roof, floor, or foundation. Special installations may require stainless steel or porcelain-enameled cast iron. Examples of floor drains are shown in *Figure 1*.

Floor and area drains often contain a separate sediment bucket to prevent solids from entering the piping (see *Figure 2*). The drain body connects

Review Questions

Review Questions reinforce the knowledge you have gained and are a useful tool for measuring what you have learned.

Review Questions

1. The abbreviation DWV stands for _____.
   a. digital water valve
   b. drain, water, and vent
   c. drain, waste, and vent
   d. double water volume

2. Which of the following statements is true?
   a. It is more difficult to route DWV piping around water supply piping than vice-versa.
   b. The grade used for water supply piping is not important.
   c. The grade used for DWV piping is not important.
   d. It is more difficult to route water supply piping around DWV piping than vice-versa.

3. To calculate the load factor for a DWV system, you can use information contained in the _____.
   a. *Plumber's Handbook*
   b. applicable local code
   c. floor plans
   d. materials takeoff

4. Drainage and waste piping systems must be installed _____.
   a. below the frost line
   b. at a slope greater than 30°
   c. at a specified, constant grade
   d. by a master plumber

5. A general-purpose level can be used to provide _____.
   a. the same accuracy as a plumb bob or laser level
   b. a precise calculation of grade
   c. an accurate estimate of distance between level points
   d. a very rough approximation of grade

6. How can a general-purpose level be modified to make it more accurate for grade measurements?
   a. Extend its length with a wood lath.
   b. Attach a suitably sized wood block to one end.

7. What can a transit level do that a builder's level cannot?
   a. Tilt upward to measure angles from the vertical
   b. Provide digital readouts
   c. Rotate horizontally to measure angles
   d. Be properly leveled on a sloping site

8. A vernier scale, used to make more precise readings, is found on a _____.
   a. laser level
   b. stadia rod
   c. steel tape
   d. builder's level

9. When setting up a builder's level, the leveling screws should be _____.
   a. backed out fully, then adjusted individually
   b. turned in opposite directions
   c. rotated counterclockwise
   d. tightened as far as possible, then backed off as necessary

10. Convert 8⅞" to decimals of a foot.
    a. 0.872
    b. 0.688
    c. 0.693
    d. 0.698

11. Which of the following statements about the cold beam laser level is *not* true?
    a. The light beam is perfectly straight.
    b. It projects a beam of ultraviolet light.
    c. The light beam does not expand appreciably.
    d. You should not look directly into the light beam, even though it is considered harmless.

12. To protect a pipe that will pass through a concrete footing, install a(n) _____.
    a. insulation boot
    b. sleeve
    c. buffer
    d. expansion joint

NCCER Standardized Curricula

NCCER's training programs comprise more than 80 construction, maintenance, pipeline, and utility areas and include skills assessments, safety training, and management education.

Boilermaking
Cabinetmaking
Carpentry
Concrete Finishing
Construction Craft Laborer
Construction Technology
Core Curriculum:
　Introductory Craft Skills
Drywall
Electrical
Electronic Systems Technician
Heating, Ventilating, and
　Air Conditioning
Heavy Equipment Operations
Highway/Heavy Construction
Hydroblasting
Industrial Coating and Lining
　Application Specialist
Industrial Maintenance
　Electrical and Instrumentation
　Technician
Industrial Maintenance
　Mechanic
Instrumentation
Insulating
Ironworking
Masonry
Millwright
Mobile Crane Operations
Painting
Painting, Industrial
Pipefitting
Pipelayer
Plumbing
Reinforcing Ironwork
Rigging
Scaffolding
Sheet Metal
Signal Person
Site Layout
Sprinkler Fitting
Tower Crane Operator
Welding

Maritime

Maritime Industry Fundamentals
Maritime Pipefitting
Maritime Structural Fitter

Green/Sustainable Construction

Building Auditor
Fundamentals of Weatherization
Introduction to Weatherization
Sustainable Construction
　Supervisor
Weatherization Crew Chief
Weatherization Technician
Your Role in the Green
　Environment

Energy

Alternative Energy
Introduction to the Power
　Industry
Introduction to Solar
　Photovoltaics
Introduction to Wind Energy
Power Industry Fundamentals
Power Generation Maintenance
　Electrician
Power Generation I&C
　Maintenance Technician
Power Generation Maintenance
　Mechanic
Power Line Worker
Power Line Worker: Distribution
Power Line Worker: Substation
Power Line Worker:
　Transmission
Solar Photovoltaic Systems
　Installer
Wind Turbine Maintenance
　Technician

Pipeline

Control Center Operations,
　Liquid
Corrosion Control
Electrical and Instrumentation
Field Operations, Liquid
Field Operations, Gas
Maintenance
Mechanical

Safety

Field Safety
Safety Orientation
Safety Technology

Management

Fundamentals of Crew
　Leadership
Project Management
Project Supervision

Supplemental Titles

Applied Construction Math
Tools for Success

Spanish Translations

Basic Rigging
　(Principios Básicos de
　Maniobras)
Carpentry Fundamentals
　(Introducción a la
　Carpintería, Nivel Uno)
Carpentry Forms
　(Formas para Carpintería,
　Nivel Trés)
Concrete Finishing, Level One
　(Acabado de Concreto,
　Nivel Uno)
Core Curriculum:
　Introductory Craft Skills
　(Currículo Básico:
　Habilidades Introductorias del
　Oficio)
Drywall, Level One
　(Paneles de Yeso, Nivel Uno)
Electrical, Level One
　(Electricidad, Nivel Uno)
Field Safety
　(Seguridad de Campo)
Insulating, Level One
　(Aislamiento, Nivel Uno)
Ironworking, Level One
　(Herrería, Nivel Uno)
Masonry, Level One
　(Albañilería, Nivel Uno)
Pipefitting, Level One
　(Instalación de Tubería
　Industrial, Nivel Uno)
Reinforcing Ironwork, Level One
　(Herrería de Refuerzo,
　Nivel Uno)
Safety Orientation
　(Orientación de Seguridad)
Scaffolding
　(Andamios)
Sprinkler Fitting, Level One
　(Instalación de Rociadores,
　Nivel Uno)

Acknowledgments

This curriculum was revised as a result of the farsightedness and leadership of the following sponsors:

ABC Southern California Chapter
Ivey Mechanical Company, LLC
Lake Mechanical Contractors, Inc.
Lee Company

Putnam Career and Technical Center
Sundt
Wat-Kem Mechanical, Inc.
Worth & Company, Inc.

This curriculum would not exist were it not for the dedication and unselfish energy of those volunteers who served on the Authoring Team. A sincere thanks is extended to the following:

Doug Allen
Jonathan Byrd
Stanley Cordova
Steve Guy

David Hoover
Jan Prakke
Harry Rimbey

Brad Sims
Brent Thompson
Ray Thornton

NCCER Partners

American Fire Sprinkler Association
Associated Builders and Contractors, Inc.
Associated General Contractors of America
Association for Career and Technical Education
Association for Skilled and Technical Sciences
Carolinas AGC, Inc.
Carolinas Electrical Contractors Association
Center for the Improvement of Construction Management and Processes
Construction Industry Institute
Construction Users Roundtable
Construction Workforce Development Center
Design Build Institute of America
GSSC – Gulf States Shipbuilders Consortium
Manufacturing Institute
Mason Contractors Association of America
Merit Contractors Association of Canada
NACE International
National Association of Minority Contractors
National Association of Women in Construction
National Insulation Association
National Ready Mixed Concrete Association
National Technical Honor Society
National Utility Contractors Association

NAWIC Education Foundation
North American Technician Excellence
Painting & Decorating Contractors of America
Portland Cement Association
SkillsUSA®
Steel Erectors Association of America
U.S. Army Corps of Engineers
University of Florida, M. E. Rinker School of Building Construction
Women Construction Owners & Executives, USA

Contents

Module One
Business Principles for Plumbers

Introduces trainees to concepts and practices that are essential for competitive, successful plumbing businesses. The module covers basic business accounting and project estimating, as well as techniques for cost control and task organization. (Module ID 02401-14; 15 Hours)

Module Two
Fundamentals of Crew Leadership

While this module has been designed to assist the recently promoted crew leader, it is beneficial for anyone in management. The course covers basic leadership skills and explains different leadership styles, communication, delegating, and problem solving. Job-site safety and the crew leader's role in safety are discussed, as well as project planning, scheduling, and estimating. Includes performance tasks to assist the learning process. (Module ID 46101-11; 20 Hours)

Module Three
Water Pressure Booster and Recirculation Systems

Builds on trainees' previous experience with pumps, storage tanks, controls, and pipes and fittings by teaching them to assemble those components into systems that boost water pressure and provide hot water. (Module ID 02403-14; 12.5 Hours)

Module Four
Indirect and Special Waste

Describes the code requirements and installation procedures for systems that protect against contamination from indirect and special wastes. (Module ID 02404-14; 17.5 Hours)

Module Five
Hydronic and Solar Heating Systems

Introduces the basic types of hydronic and solar heating systems and their components. The module reviews hydronic and solar heating system layout, installation, testing, and balancing, and also discusses methods that inhibit corrosion in hydronic and solar heating systems. (Module ID 02405-14; 17.5 Hours)

Module Six
Codes

Discusses the different codes used by plumbers across the country and explains how those codes are written, adopted, modified, and implemented. (Module ID 02406-14; 12.5 Hours)

Module Seven
Private Water Supply Well Systems

Describes the operation of pumps and well components. Reviews the qualities of good wells and how to assemble and disassemble pumps and components. (Module ID 02408-14; 10 Hours)

Module Eight
Private Waste Disposal Systems

Describes the types of private waste disposal systems, discusses the maintenance and installation of these systems, and explains how to determine the local code requirements for these systems. Covers percolation tests and sewage system planning and layout. (Module ID 02409-14; 10 Hours)

Module Nine

Swimming Pools and Hot Tubs

Introduces trainees to plumbing systems in swimming pools, hot tubs, and spas. (Module ID 02410-14; 7.5 Hours)

Module Ten

Plumbing for Mobile Homes and Travel Trailer Parks

Describes the location and layout of plumbing systems for mobile homes and travel trailer parks. The module reviews how to design and lay out a system, how to connect water and sewer lines to a mobile home, and how to estimate materials for the park. (Module ID 02411-14; 7.5 Hours)

Module Eleven

Introduction to Medical Gas and Vacuum Systems

Provides an introduction to the various types of medical gas and vacuum systems used in health care facilities today. Covers the system requirements and professional qualifications required by code, describes common types of medical gas and vacuum systems, and introduces trainees to the safety requirements that must be observed when installing, testing, and servicing these systems. (Module ID 02412-14; 15 Hours)

Glossary

Index

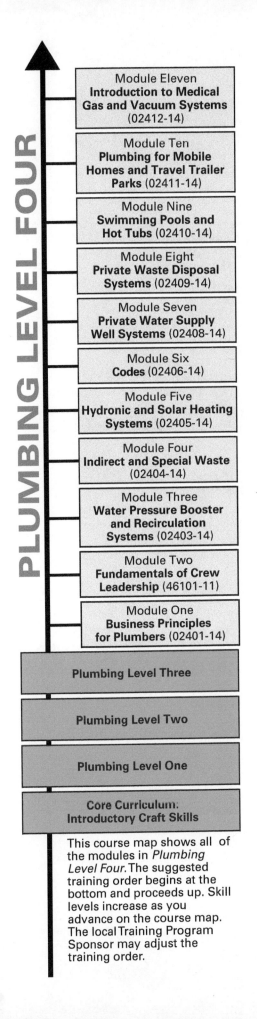

PLUMBING LEVEL FOUR

Module Eleven
Introduction to Medical Gas and Vacuum Systems
(02412-14)

Module Ten
Plumbing for Mobile Homes and Travel Trailer Parks (02411-14)

Module Nine
Swimming Pools and Hot Tubs (02410-14)

Module Eight
Private Waste Disposal Systems (02409-14)

Module Seven
Private Water Supply Well Systems (02408-14)

Module Six
Codes (02406-14)

Module Five
Hydronic and Solar Heating Systems (02405-14)

Module Four
Indirect and Special Waste (02404-14)

Module Three
Water Pressure Booster and Recirculation Systems (02403-14)

Module Two
Fundamentals of Crew Leadership (46101-11)

Module One
Business Principles for Plumbers (02401-14)

Plumbing Level Three

Plumbing Level Two

Plumbing Level One

Core Curriculum: Introductory Craft Skills

This course map shows all of the modules in *Plumbing Level Four*. The suggested training order begins at the bottom and proceeds up. Skill levels increase as you advance on the course map. The local Training Program Sponsor may adjust the training order.

02401-14

Business Principles for Plumbers

Overview

A successful business must make a profit, compete successfully for work, keep costs under control, and work efficiently. Plumbers should understand basic business accounting, and practice effective techniques for bidding jobs, controlling costs, and organizing projects. Project management involves keeping control of orders, purchases, and deliveries; safe use and storage of materials, tools, and equipment; and issuing change orders. Task-planning skills enable plumbers to complete work on time and within budget.

Module One

Trainees with successful module completions may be eligible for credentialing through the NCCER Registry. To learn more, go to **www.nccer.org** or contact us at **1.888.622.3720**. Our website has information on the latest product releases and training, as well as online versions of our *Cornerstone* magazine and Pearson's product catalog.

Your feedback is welcome. You may email your comments to **curriculum@nccer.org**, send general comments and inquiries to **info@nccer.org**, or fill in the User Update form at the back of this module.

This information is general in nature and intended for training purposes only. Actual performance of activities described in this manual requires compliance with all applicable operating, service, maintenance, and safety procedures under the direction of qualified personnel. References in this manual to patented or proprietary devices do not constitute a recommendation of their use.

Objectives

When you have completed this module, you will be able to do the following:

1. Identify cost control measures.
   a. Interpret a balance sheet and a profit and loss statement.
   b. Interpret how cost control measures affect the profit and loss statement.
   c. Interpret how on-the-job task organization affects profit and loss.
2. Identify the information required to prepare a material takeoff.
   a. Identify the information required for insurance and liability.
   b. Identify the information required for the estimating process.
   c. Identify the information required for updates and revisions.

Performance Task

Under the supervision of your instructor, you should be able to do the following:

1. Prepare a material takeoff as part of an estimate.

Trade Terms

Accounting	Cradle-to-grave	Gross income	Production figure
Accounts receivable	responsibility	Indirect cost	Profit
Asset	Critical activities	Invoice	Profit and loss statement
Balance	Critical path management	Liability	Purchase order
Balance sheet	Daily work report	Mobilization	Recapitulation sheet
Bid	Debt	Net income	Rework
Bond	Direct cost	Overhead	Sole source
Change order	Equity	Premium	Task planning
Contract	Estimating	Preplanning	Unit cost

Industry-Recognized Credentials

If you're training through an NCCER-accredited sponsor, you may be eligible for credentials from NCCER's Registry. The ID number for this module is 02401-14. Note that this module may have been used in other NCCER curricula and may apply to other level completions. Contact NCCER's Registry at 888.622.3720 or go to **www.nccer.org** for more information.

Code Note

Codes vary among jurisdictions. Because of the variations in code, consult the applicable code whenever regulations are in question. Referring to an incorrect set of codes can cause as much trouble as failing to reference codes altogether. Obtain, review, and familiarize yourself with your local adopted code.

Contents

Topics to be presented in this module include:

Figures

1.0.0 COST CONTROL MEASURES

Objective

Identify cost control measures.
  a. Interpret a balance sheet and a profit and loss statement.
  b. Interpret how cost control measures affect the profit and loss statement.
  c. Interpret how on-the-job task organization affects profit and loss.

Trade Terms

Accounting: The process of compiling and analyzing financial records.

Accounts receivable: Money that is owed to a business but has not been paid.

Asset: Any type of property owned by a business.

Balance: The condition of an account when assets and liabilities are equal.

Balance sheet: A form used to track assets and liabilities.

Bid: A formal offer to do work according to an estimate.

Change order: An agreement to perform work in addition to what was originally agreed to in a contract.

Critical activities: Tasks in the critical path management process that could cause significant delay if not completed on time.

Critical path management (CPM): A task-planning process in which diagrams show the relationships between tasks and estimated completion times.

Daily work report: A report written by a job superintendent describing activities at a job site.

Debt: Money owed by a business for payment of direct costs and indirect costs.

Equity: The value of the business if all debts are paid.

Gross income: All money earned before expenses are subtracted.

Invoice: An itemized list of materials for purchase or rent.

Liability: In accounting, money owed by a business. In insurance, a legal responsibility.

Mobilization: The final stage of preplanning, during which all items required to start construction are moved to the job site, arrangements are made with the other trades to coordinate job-site activities, and all job-site details are organized in preparation for construction.

Net income: Another term for profit.

Preplanning: The phase immediately following the awarding of a contract but before the start of construction, during which subcontractors are selected, purchase orders are issued, construction schedules are developed, and estimates are reviewed in depth.

Profit: Money left over after expenses are subtracted from income.

Profit and loss statement: A list of all income and expenses over a given period of time.

Purchase order: An order form submitted to a supplier.

Rework: To redo a previously completed task.

Task planning: The process of organizing activities and materials so that tasks are completed in the right order and on time.

All successful businesses start out with a set of clear goals and objectives. For plumbing businesses, the most important goal is to provide the highest-quality work possible at a reasonable cost to the client. Keeping clients happy is an important part of being in business, because happy clients will hire you again and recommend you to others. However, a business cannot run simply on goodwill. A successful business must also make a profit, compete successfully for work, keep costs under control, and work efficiently. That's a tall order for any business, but good businesses succeed because their employees understand how essential these goals are.

In this section, you will learn techniques for ensuring the success of your plumbing business through the use of cost control measures. You will review the basic principles of accounting to learn how companies stay in business. You will learn how to calculate and submit offers for new work, and you will examine effective ways to control costs and organize projects. All of these techniques are essential for ensuring that a plumbing business can meet all its goals and be successful.

1.1.0 Interpreting the Balance Sheet and the Profit and Loss Statement

Like many people, you probably take time once or twice a month to balance your checkbook. You understand the importance of having enough money in the bank to cover your expenses. You also usually try to end up with some money left over. The extra money goes to things that improve the quality of your life, like movies, vacations, or a new car. Businesses do the same thing but on a larger scale. Money comes in from projects and is used to pay for expenses, including salaries, equipment, and materials. The money that remains is called profit.

Without profit, a business will not survive. In the construction trades, like any other business, owners track profits by compiling and analyzing financial records. This process is called accounting. Many people feel intimidated by accounting, but you don't need a degree in accounting to understand how it works. Basic accounting is a good skill for a plumber to have. Like a toolbox, you can take the knowledge with you wherever you decide to work. If you learn the basics of accounting, you will be able to take on positions of greater responsibility in the plumbing trade.

The two basic tools of accounting are the balance sheet and the profit and loss statement. A balance sheet tracks what a business has and what it owes others, while a profit and loss statement lists all income and expenses over a period of time. Each of these basic tools is described in more detail in the following sections. There are also many books available on accounting for small businesses and the construction trades. When on the job, ask your employer how their accounting system works. Every system is a little different, but they all follow the same basic rules.

1.1.1 The Balance Sheet

Anything that a business owns is called an asset, while anything that the business owes is called a liability. Assets include the following:

- Cash
- Inventory
- Property
- Real estate holdings

Liabilities consist of debt and equity. Debt is money that the business owes to banks, other businesses, and individuals. Mortgages, taxes, loan repayments, and salaries are examples of debt. Equity is the value of the business if all the debts are paid. A business needs to have enough

assets to pay its liabilities. This concept can be expressed as a simple equation:

$$\text{Assets} \geq \text{Liabilities}$$

When the records show that the assets are at least equal to the liabilities, the business can pay all of its debt. This is the goal of accounting. When assets and liabilities are equal, the account is in balance. That is why one of the basic tools of accounting is called the balance sheet. Think of a balance sheet as a snapshot of the business's finances at one moment in time.

Here is a simple example of how a balance sheet works. Imagine that you are the owner of Bay City Plumbing and Heating, and you want to determine your company's assets, debt, and equity. Begin by looking at your records. Your bank statement shows $19,000 in the bank. You are awaiting $1,200 in payment from one client and $700 from another. Money that is owed to you but that has not yet been paid is called accounts receivable. Bay City has $3,900 worth of plumbing supplies in stock. Adding up the value of all the equipment, tools, and vehicles, you find they are all worth $50,000. Now add up the various amounts to determine Bay City's assets. *Figure 1* shows that the total assets equal $74,800.

Now determine Bay City's debts. Looking over the bills, you see that Bay City owes $1,300 for a new set of tools. The company has $4,400 left to pay on a business development loan. The company will have to pay $1,100 in taxes for the year. Insurance payments are $2,800 for the year. Salary totals $44,000 for the year. Now add up the debts. *Figure 2* shows that the total debts equal $53,600.

To determine the equity, start with the basic formula you just learned:

$$\text{Assets} \geq \text{Liabilities}$$

Remember that liabilities consist of debt and equity, which translates as:

$$\text{Assets} \geq \text{Debt} + \text{Equity}$$

	A	B	C
1	Cash	$19,000.00	
2	Accounts Receivable	$1,900.00	
3	Plumbing Supplies	$3,900.00	
4	Equipment, Tools, Vehicles	$50,000.00	
5	**TOTAL ASSETS**	$74,800.00	
6			
7			

02401-14_F01.EPS

Figure 1 Sample asset calculation for balance sheet.

	A	B	C
1	Accounts Owned	$1,300.00	
2	Loan	$4,400.00	
3	Taxes	$1,100.00	
4	Insurance	$2,800.00	
5	Salary	$44,000.00	
6	**TOTAL DEBT**	$53,600.00	
7			

02401-14_F02.EPS

Figure 2 Sample debt calculation for balance sheet.

This means that you can calculate both debt and equity by using simple substitution. Then, you can apply this formula to find the equity of Bay City Plumbing and Heating:

Equity = Assets – Debt
Equity = $74,800 – Debt
Equity = $74,800 – $53,600
Equity = $21,200

Of course, most real-life cases are more complicated, but you can use the same basic formula to find the assets, debt, and equity of any company.

1.1.2 The Profit and Loss Statement

The other basic tool of accounting is the profit and loss statement. This is a summary of income and expenses over a given period. If a balance sheet is like a snapshot, then a profit and loss statement is like a short film.

The profit and loss statement indicates how well the business has done over time. It does this by tracking the business's income and expenses, each of which are on separate accounts. The profit and loss statement tells how much profit the business made during a month, a quarter, or a year.

Profit is the money left after all expenses have been paid. The relationship can be expressed as another simple equation

Profit = Income – Expenses

Profit, also called net income, is what allows a business to keep operating. Contractors use profit to buy new equipment, attract investors, and stay in business between projects. If a project goes over its budget, a contractor will have to make up the difference from its profit. If that happens too often, the contractor could go out of business. It is a delicate balance.

Consider this example: imagine again that you are the owner of Bay City Plumbing and Heating. Looking over the paperwork for April, you find that the company was paid $30,300 for five separate projects. This is called the gross income.

Gross income is all the money made before related expenses are subtracted.

Next, add up the expenses. The company spent $10,120 on materials and equipment for the projects. Records show that the company paid $3,300 in salary. The company made its yearly tax payment of $1,100. The monthly loan payment was $175. To determine the profit, subtract the expenses from the gross income of $30,300. The net income therefore equals $15,605.

Balance sheets and profit and loss statements show how well a business is doing. This information does more than help you pay taxes and suppliers; it helps you plan for the future of the business. If you see that expenses in a certain area are higher than you expected, you can check whether there is a problem. If the profit margin

The Chart of Accounts

Businesses usually set up a separate account for each type of transaction. The accounts are grouped together in a logical order. Each account is given a number. The following example is a simplified chart of accounts:

00	Assets
	110 Current Assets
	120 Long-Term Assets
	130 Other Assets
200	Liabilities
	210 Current Liabilities
	220 Long-Term Liabilities
	230 Accounts Payable
300	Net Worth
	310 Capital Stock
	320 Retained Earnings
	330 Equity Account
400	Income
	410 Completed Contracts
	420 Sales Returned
	430 Discounts on Sales
500	Direct Costs
	510 Materials Purchased
	520 Direct Labor
	530 Freight and Delivery

Take time to learn how the chart of accounts is structured where you work. It will help you understand how the business is organized.

	A	B	C
1	Gross Income	$30,300.00	
2	Materials	$10,120.00	
3	Salary	$3,300.00	
4	Taxes	$1,100.00	
5	Loan Payment	$175.00	
6	**NET INCOME**	$15,605.00	
7			

02401-14_F03.EPS

Figure 3 Sample profit and loss statement.

is lower than you would like, consider advertising for more business or raising your rates. Unless you look at the money coming in and going out, your business will eventually fail. Keeping track of the flow of money will help your company stay healthy and grow over time.

1.2.0 Understanding the Impact of Cost Control on the Profit and Loss Statement

Imagine that you are part of a crew of four plumbers working at a construction site. Your crew's task is to install soil pipe. Everything is going well until suddenly the team runs out of wyes. The nearest plumbing supply store is about 20 minutes away. The team leader instructs a plumber to go and buy enough wyes to finish the job. The plumber returns in an hour with the wyes. The team finishes installing the soil pipe and moves on to the next task.

The story seems innocent enough, but look at it through the eyes of a project manager. Because no one made sure that there were enough wyes to complete the job, the following unnecessary costs were added to the project budget:

- Four man-hours of work
- Four hours' pay
- Mileage costs on a company vehicle
- Potential delays for other trades

In addition, the project schedule was pushed back an hour. That could mean your team starts its next task late, or it could mean that there will be less time to deal with other issues in the future. Project budgets are designed to allow workers to do the best job in the shortest practical time, for the lowest cost to the client, and with a profit that allows the contractor to stay in business. Plumbers on the team are responsible for ensuring that the project is kept on schedule and within budget. Costs can quickly spiral out of control if any of the following happen:

- Materials are ordered late or incorrectly.
- Materials are stored incorrectly, insecurely, or too far away.
- Tools and equipment are used improperly or damaged.
- Company vehicles are damaged or lost because of unsafe operation or negligence.
- Work that is not specified in the contract is done.

Plumbers are responsible for preventing these problems. If you see a problem, take steps to correct it. You will learn how to start and finish tasks in the right order and on time later in this module. The following sections discuss how to keep a project within budget.

1.2.1 Preplanning

Once a company has been awarded the contract for a project, the company must complete a number of steps before construction can actually begin. These steps are called preplanning. Preplanning steps are required to ensure that cost control measures are in place from the very beginning of a project, in order to prevent cost overruns that can eat into the company's profits. During this phase, subcontractors are selected, purchase orders are issued, construction schedules are developed, and the estimate is reviewed in depth. The project team holds discussions on staffing, equipment, and cost control, and then mobilization begins. Each of these steps is discussed in detail in the following.

Selecting Subcontractors

A bid is a formal offer to do the work according to the estimate. Just as the general contractor obtains a contract by submitting a bid to the project owner, a subcontractor obtains work by submitting a bid to the general contractor. Using a description of the work to be done, the plumbing subcontractor develops an estimate based on takeoffs, prices obtained from suppliers, vendors, the company's cost reports, and labor rates for the crew as well as workers who will be needed to support the crew's work.

Taxes

Plumbing tests nationwide often feature several questions about taxes. Before taking the test, be sure to learn how your local tax system works. Talk to an accountant to learn more about how businesses pay taxes. A little preparation on this topic can help improve your test scores.

When the general contractor receives bids from subcontractors, the bids are reviewed carefully, making certain they are complete. Items usually reviewed include the following:

- Scope of work
- Cost
- Staffing resources
- Schedule

The general contractor then awards the contract to the subcontractor that proposes the most cost-effective mix of those items.

Issuing Purchase Orders

The contractor orders most materials for a project using purchase orders. A purchase order, often called a PO, is an order form submitted to a sup-

plier. Purchase orders list the materials being ordered, the number needed, and the unit and total prices. Follow the contractor's procedures when issuing a purchase order (*Figure 4*). Distribute copies to the project file and to the superintendent. Purchase orders ensure that bills are paid only after the materials have been received.

Developing a Schedule

The project schedule should be developed with cooperation and input from the following contributors:

- Members of the company's project team
- Major subcontractors
- Key suppliers
- Others who can contribute to a realistic schedule

Smartphone Apps for Construction

Although the craft of plumbing has been around for centuries, not every tool that a plumber uses is old. Smartphones are becoming an increasingly popular form of communication, and also offer a great deal of versatility for craftworkers. Smartphone cameras can be used to document on-the-job activities or potential safety violations. Best practices can be communicated to crew members using video clips. Construction calculators can be downloaded, providing craftworkers with the same (or even greater) capabilities than a handheld calculator.

The smartphone version of Calculated Industries, Inc.'s popular Construction Master® Pro, for example, lets craftworkers calculate and convert dimensions, plot right-angle conversions, find area and volume, and determine measures such as board feet, cost per unit, and even the angles of an equal-sided polygon. Calculated Industries, Inc., also makes an app version of its ConcreteCalc™ Pro (shown in the figure), which masons can use to calculate stair dimensions, the length and weight of rebar, and the number of brick loads and mortar bags needed for a job.

Note that company and job-site policies may not allow workers to carry smartphones while on the job. Cell-phone use can distract you from paying attention to safety hazards. Camera-equipped phones may raise privacy concerns on job sites where photography is not allowed. Be sure to familiarize yourself with the applicable policies and procedures before bringing a smartphone or other cell phone to the job site.

02401-14_SA01.EPS

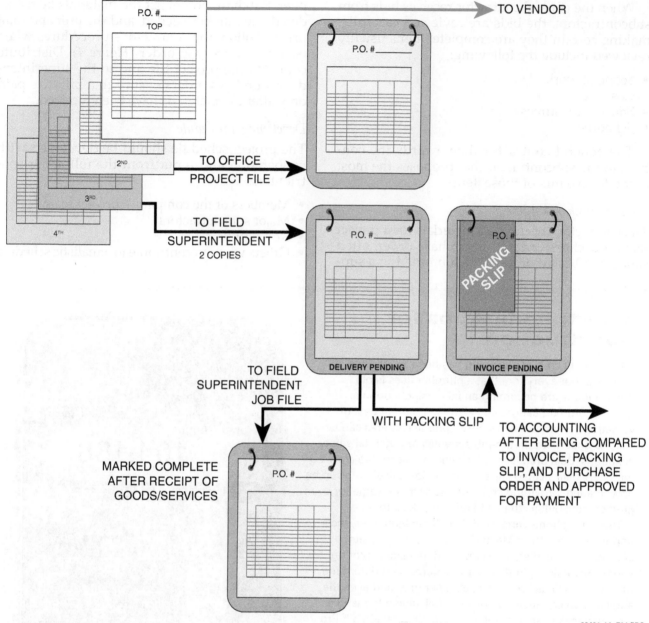

TO VENDOR

P.O. #_____
TOP

2ND

3RD

4TH

TO OFFICE
PROJECT FILE

P.O. #_____

TO FIELD
SUPERINTENDENT
2 COPIES

P.O. #_____

P.O. #_____

PACKING SLIP

DELIVERY PENDING

INVOICE PENDING

TO FIELD
SUPERINTENDENT
JOB FILE

WITH PACKING SLIP

TO ACCOUNTING
AFTER BEING COMPARED
TO INVOICE, PACKING
SLIP, AND PURCHASE
ORDER AND APPROVED
FOR PAYMENT

MARKED COMPLETE
AFTER RECEIPT OF
GOODS/SERVICES

P.O. #_____

02401-14_F04.EPS

Figure 4 Typical work flow for a purchase order.

Throughout the scheduling process, the highest priority should be on the following issues:

- Long lead times
- Critical operations
- Sequencing
- Duration of activities
- Equipment deliveries and installation
- Staffing resources

Reviewing the Estimate

The preplanning review of the estimate is not a search for errors but a review to be sure all costs have been identified. Many times, errors both in the contractor's favor and against it are uncovered during the review. Even if no errors are found, a thorough project-team review ensures that everyone on the project has a clear understanding of the project and the costs involved.

Many contractors add an additional step that requires the construction team to prepare a construction estimate to compare against the original bid. This helps identify any lapses or doubling up of items that were in the original estimate.

Identifying Staffing Resources and Equipment Requirements

Labor and construction equipment requirements for all construction phases should be identified during review of the estimate and preparation of the schedule.

Determining the number of craftworkers from the various trades required at each stage of work gives the project team a basis for:

- Determining the availability of personnel
- Identifying possible worker shortages and surpluses
- Controlling labor cash flow

Reviewing equipment requirements ensures that cranes, lifts, generators, and other construction equipment are available when needed, used wisely and efficiently on site, and removed from the job site when no longer required.

Establishing a Cost Control System

Every step of the project, from beginning to end, is based on cost management. This makes a cost control system essential to all aspects of the proper management of any project.

- The owner considers cost when developing the project concept.
- The architect or engineer considers cost throughout the design phase.
- Cost is the basis for accepting and rejecting bids.
- Cost plays a central role in whether the project is profitable for those involved.

During the preplanning meeting, a review of the cost system, the extent of control, and the need for control must be established and understood.

Mobilizing

Mobilization includes the following activities:

- Moving all items required to start construction to the job site
- Coordinating job-site arrangements with the other trades on the job
- Organizing all job-site details in preparation for construction

- Setting up office and storage trailers
- Locating temporary facilities
- Establishing storage areas
- Assigning specific responsibilities to members of the project team

At the beginning of the project, find out when and from where you will receive the materials for each task. These factors will affect how you schedule your time. If some materials are scheduled to be delivered at the start of the project, begin with the tasks that use those materials. Schedule the other tasks to take place when the necessary materials arrive. Find out if plumbing suppliers in your area have orders delivered to them. Ask them when an order must be placed for it to arrive on schedule and when you can expect to pick it up.

Contractors often order materials from more than one supplier. Choose suppliers based on the location of the project and the price of the materials. Suppliers that have their own warehouses are more likely to have the materials you need in stock. Otherwise, they have to order supplies for the project, potentially causing delays if the supplier has problems getting the materials on time. Large orders can be delivered directly to the job site. Before arranging for a delivery, ensure that the project budget allows for the cost of delivery.

Special requests and emergencies can eat into a budget quickly. Before a project starts, establish procedures for dealing with unexpected delays. Establish an account with a nearby supplier to meet emergency supply needs. Note the different suppliers for all the materials in the project schedule so that you can contact suppliers quickly in case of problems.

Establishing Storage Areas

In addition to the steps listed above, another important preplanning step is to establish areas where materials will be stored. Project costs are affected by the location of the stored plumbing materials on the job site. Whenever practical, have

Lean and LEED Certifications

Don't confuse the terms *Lean* and *LEED*. They are complementary, but they are not the same. LEED (Leadership in Energy and Environmental Design) is a set of standards that building designers, builders, owners, and operators follow to ensure that a building has been constructed using environmentally sound practices. The LEED standard was developed in 1998 by the US Green Building Council (USGBC). The Lean approach to business management focuses on identifying and eliminating wasteful practices and processes that cost time and money. The principles of Lean manufacturing are based on the Toyota Production System (TPS) developed by the car manufacturer in the late 1940s and now widely used in industries around the world. Construction professionals can obtain certifications as LEED and Lean professionals from approved accreditation organizations.

materials placed near where they will be needed. This step will help reduce labor costs during the installation.

Materials stored on site must be secured as they can be stolen. Store smaller items in a company vehicle or in the field office (*Figure 5*). The cost of replacing stolen materials and the time lost waiting for the replacements can eat away at a project budget. Store materials in the building only after doors with locks are installed. Account for materials on a regular basis. Depending on the project, this can be monthly, weekly, or daily. Track materials using an inventory form (*Figure 6*). Add materials to the inventory form when they are delivered and subtract them when they are used.

1.2.2 Ordering and Taking Delivery of Materials

Time spent standing in line at a supplier to place an order and then waiting to collect the materials from storage could be better spent on the job. Consider ordering supplies by telephone. Learn the style of your company; as a company representative, you must present a good image when talking with suppliers and clients on the phone.

Ensure that telephone orders are clear and complete. Before you place the order, make a list of the materials to be ordered and the questions you have. Speak slowly and clearly with a friendly voice. Include the quantities, sizes, and grades of the materials. Ask the supplier about the availability of each of the items on the list. Finally, ask the supplier to read back the order. This will cut down on errors.

02401-14_F05.EPS

Figure 5 A field office.

If the person you are trying to reach is not available, leave your phone number and a brief message. If you cannot leave a message, find out when the person will be available. Keep a phone log of important calls, and note all business calls in a job diary. Whenever possible, try to get a written confirmation of an order, as this will help eliminate mistakes and misunderstandings.

Most suppliers accept emailed and faxed orders as well. Some suppliers may have a standard order form on their website that should be used instead of sending a regular email. The form will prompt you to provide all the information that the supplier needs, and will ensure that the order is sent to the correct person. When faxing an order, be sure to write clearly and use white paper. Some colored papers, especially yellow, do not fax well.

When a supplier sends a bill for the delivery of materials, the contractor checks the bill against a copy of the invoice for that shipment. An invoice is an itemized list of materials specifying prices and the terms of the agreement (*Figure 7*). An invoice is the only record that a contractor has to confirm that the bill can be paid. Also, the contractor will not know if the materials have been delivered on time. Without this information, the project could fall behind schedule, causing costs to escalate quickly. If you receive a shipment of materials, follow the correct procedures for handling the invoice.

Ensure that the shipment contains the correct amount and type of materials, and that they are all in good condition. With fixtures, open each box to ensure that the fixtures are not damaged. Do not sign for the delivery until you are sure that everything that was ordered has been delivered in good condition. After signing the invoice, give a copy to your supervisor. If you purchased the materials with cash, obtain a receipt. The contractor will use the receipt to reimburse you, and receipts are also used by contractors to control the project budget.

1.2.3 Proper Use of Tools and Equipment

Always use the proper tools and equipment on the job. This is not only a good habit but also sound business practice. Using the correct pipe and fitting the first time means you will not have to waste time and money in rework. Using the correct tools will save you time and energy. Finally, use tools properly. A damaged tool can cause injury and slow down the work. Keep records of tools and equipment to ensure that they are accounted for (see *Figure 8*).

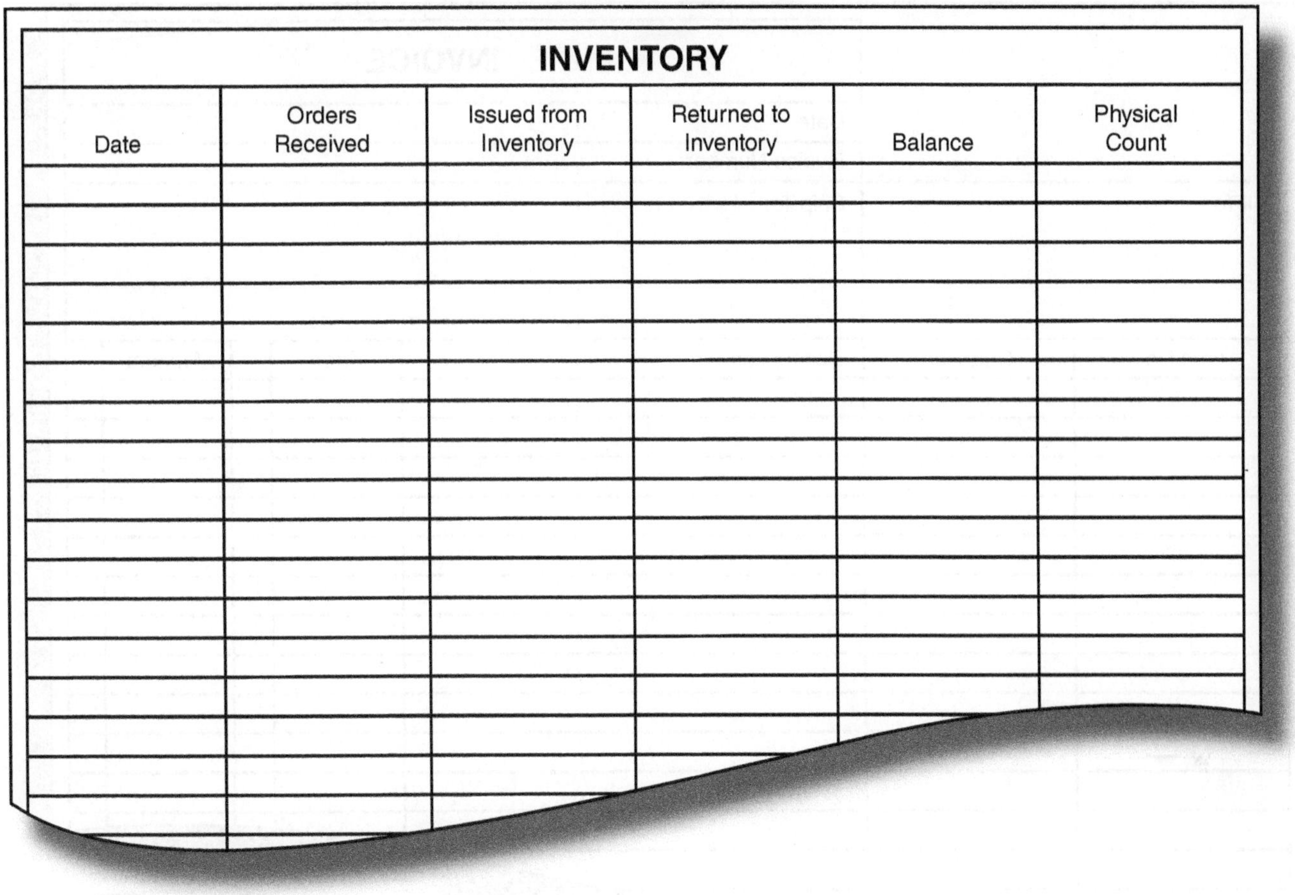

INVENTORY					
Date	Orders Received	Issued from Inventory	Returned to Inventory	Balance	Physical Count

02401-14_F06.EPS

Figure 6 Sample inventory form.

Clean, dry, and oil tools before storing them. Proper maintenance prevents rust from developing and ensures that the tool stays in proper working order. Repair or replace damaged tools promptly.

Establish a storage area for tools and equipment. Return every tool to its proper place when you are finished using it. Store your own tools in a toolbox, and organize it so that you can find tools quickly.

1.2.4 Company Vehicles

Operate the company's vehicles safely and courteously. Remember that as the driver, you are advertising the company. This is especially true if the vehicle has the company's name on it. Reckless or discourteous driving will attract attention, but it will not be the kind that the company likes to get. Lock the vehicle when it is parked, and ensure that all toolboxes are secure. If you are responsible for a vehicle overnight, park it in a safe place. This can substantially reduce the risk of theft.

Use a checklist to account for all items in the vehicle that should include all tools, materials, and equipment. A checklist also serves as a record of the materials that have been used. Consult the checklist when resupplying the vehicle.

Keep the vehicle clean both inside and out. If you are responsible for maintaining the vehicle, follow the company schedule, and keep records of all oil changes and tune-ups. Take the vehicle in for regular service when scheduled, and keep track of the mileage. If you are using a company credit card, save all the sales slips, submitting them to your supervisor regularly. Find out what the contractor expects you to do if the vehicle breaks down, and whom to contact and what to do in case of an accident.

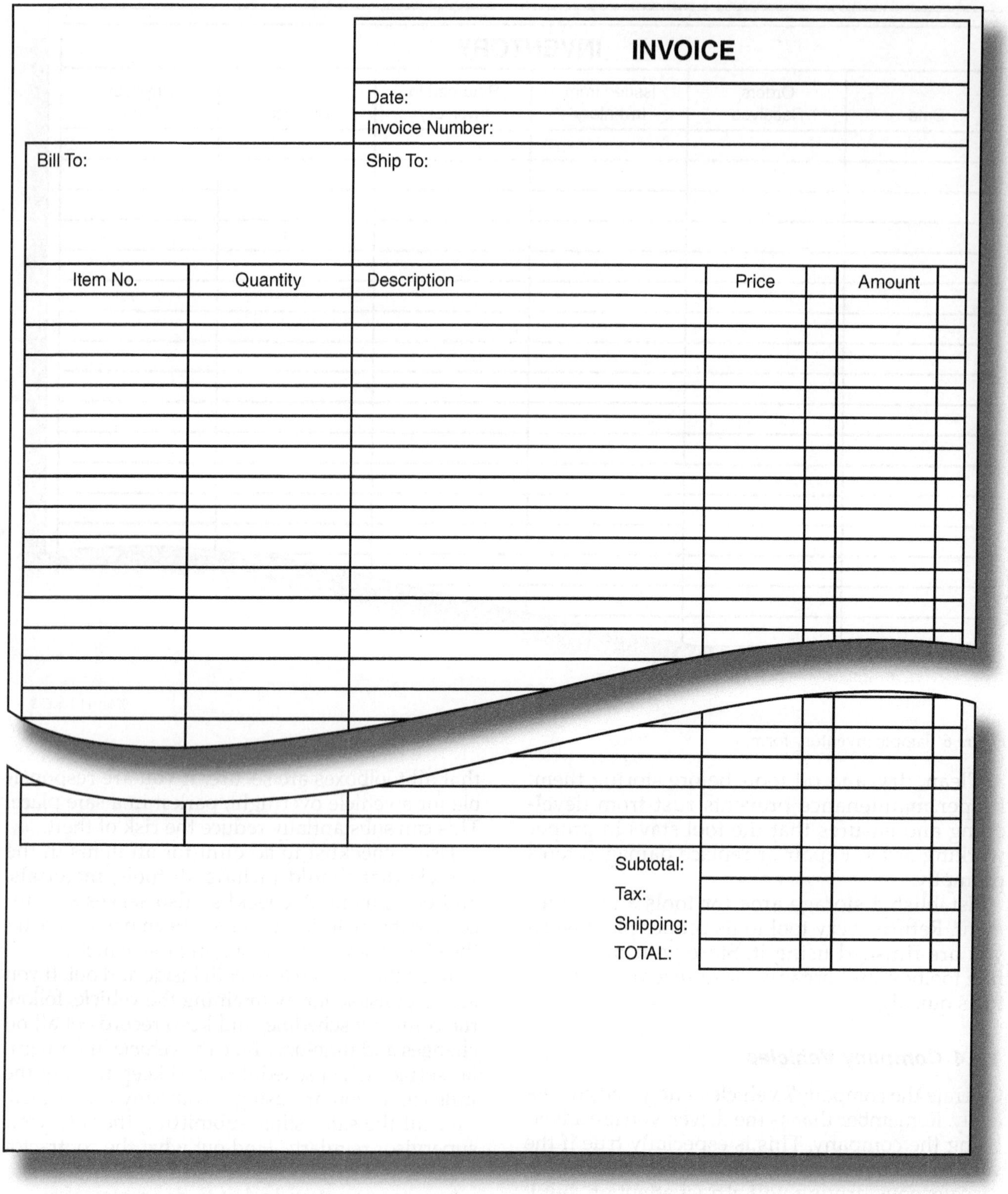

INVOICE

| Date: |
| Invoice Number: |

| Bill To: | Ship To: |

Item No.	Quantity	Description	Price		Amount	
			Subtotal:			
			Tax:			
			Shipping:			
			TOTAL:			

02401-14_F07.EPS

Figure 7 Sample invoice.

EQUIPMENT RECORD

Description:				
Identification Number:		Classification:		
Size:	Model:	Style:		Engine Number:
New/Used:		Serial Number:		
Purchased From:				
Date of Purchase:		Terms of Purchase:		

Price:
Tax:
Delivery Charges:
Installation Charges:
Other Charges:
Total:

Date Sold:		Sold To:

Gross Sale Price:
Tax:
Total Sale Price:

02401-14_F08.EPS

Figure 8 Sample equipment record for a project.

1.2.5 Change Orders

The success of the project depends on plumbers performing their assigned tasks. Do not take on additional tasks outside the scope of what you are required to do. If you do, the company may lose money on the project. If additional work needs to be done, the client issues a change order. Change orders (*Figure 9*) are agreements to perform additional work outside the scope of the original contract. Change orders are supplements to the contract and are issued for the following:

- Additional tasks not foreseen in the original contract
- Tasks to be deleted from the contract
- Changes to tasks outlined in the contract
- Work to correct problems discovered during the course of the job

Sometimes, a change order requires the team to redo a task that it has already completed. For example, a change order could require moving a bathroom fixture group because the client changed the floor plan. This type of activity is called rework. Rework can be frustrating because it requires the team to tear out something that took time to install. Do not take this type of work personally; it is not a reflection on the quality of the team's workmanship. Do not let the need for change cause the team to lower its standards of quality. Perform the rework as if it were a brand-new task. Keeping a positive attitude shows people that you take pride in your work and care about your job.

1.3.0 Implementing On-the-Job Task Organization

The division of project responsibilities is an essential part of on-the-job task organization. Project responsibilities are typically divided among a contractor's staff (*Figure 10*). Common divisions of responsibility are as follows:

- Either the owner of the contracting business or the project manager is responsible for ensuring that the entire project stays on schedule and within budget and that it meets the client's needs.
- The job superintendent is responsible for coordinating the work at the job site.
- The plumbing foreman is responsible for coordinating the installation of the plumbing if the project involves more than one trade (such as a new building); the foreman is also responsible for assigning work to each member of the plumbing crew.
- The plumber is responsible for specific tasks, such as locating the main waste stack or installing bathroom fixtures.
- The apprentice is expected to serve on field crews for several years (typically three to four years) in order to obtain sufficient on-the-job experience to meet an apprentice's educational requirements, and also to learn the trade by working alongside more experienced co-workers.

This division of responsibilities is part of the process of task planning. Task planning allows plumbers to undertake and complete tasks in the right order and on time.

Electronic Aids for Estimating

In the field, plumbers use specialized calculators to estimate materials. In addition to the standard calculator functions, these calculators can perform mathematical operations on dimensional measurements. Dimensional measurements can be easily converted into other English or metric units. Estimating calculators such as the Pipe Trades Pro™ by Calculated Industries include specialized functions designed to perform many typical calculations used in estimating:

- An area, given the length and width
- A volume, given the length, width, and height
- Cost estimates and per-unit costs
- The weight of materials given the volume, or the volume for a given weight
- The square-up or diagonal length of a rectangle, given the length and width

02401-14_SA02.EPS

Check Your Math

Electronic aids such as calculators, smartphone apps, and online estimators increase the accuracy and speed of the estimating process. However, it is still important to double-check all calculations. Calculators and computer programs can still produce incorrect estimates if the information entered is incorrect. As computer programmers say, "Garbage in, garbage out."

Estimating Software

Back in the office, plumbers can also use special computer software such as AccuBid by Trimble® to develop estimates for a project. The AccuBid system includes a built-in database of materials with unit prices, labor estimates, and cost tracking codes for each item. Plumbers can use this software to prepare takeoffs on the fly. Like a spreadsheet, the estimating software reflects changes in real time so that the plumber can see the effects of changes immediately without having to recalculate by hand.

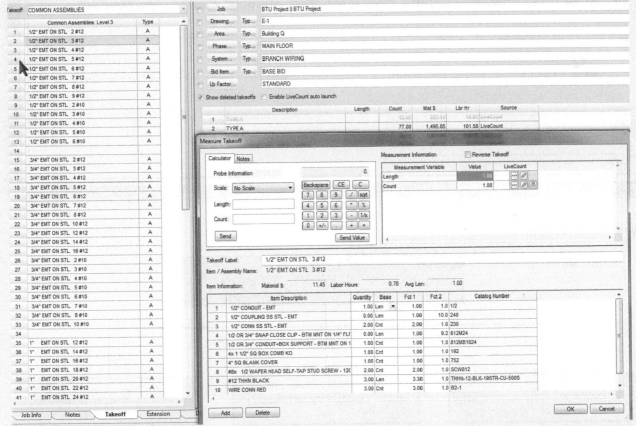

02401-14_SA03.EPS

CHANGE ORDER

This CHANGE ORDER is an integral part of the CONTRACT dated _____ between _____ (contractor) and _____ (client) for the following work: _____ to be performed at the following location: _____. The following changes are the only changes allowed to be made under this CHANGE ORDER. This CHANGE ORDER signifies that the changes outlined below shall henceforth become a part of the original CONTRACT and may not be altered again without written authorization and consent of the undersigned parties.

CHANGES ARE AS FOLLOWS:

_____
_____
_____
_____

These changes will: _____ increase _____ decrease the original contract amount. The terms of payment for these changes are as follows: _____. The amount of change in the contract price will be as follows: _____($). The new total contract price shall be as follows: _____ ($).

The undersigned parties hereby agree that the changes outlined in this CHANGE ORDER are the only changes allowed to be made. Verbal agreements shall not be considered valid. Further alterations to the CONTRACT shall not be allowed unless additional written authorization is first obtained. Written authorizations must be agreed to and signed by all parties.

This CHANGE ORDER constitutes the complete and entire agreement between all parties to alter the original CONTRACT.

_____   _____
Client   Contractor

_____   _____
Date   Date

_____
Client

_____
Date

02401-14_F09.EPS

Figure 9 Sample change order.

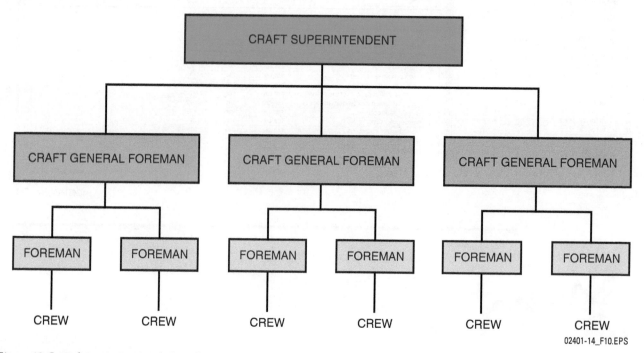

02401-14_F10.EPS

Figure 10 Sample organizational chart for a project.

Every task has a logical position in the overall project schedule. Each task must occur before, after, or at the same time as other tasks. Critical path management (CPM) is a popular way to visualize how tasks are arranged. CPM uses diagrams that show the relationships between tasks and the estimated time that each will take (see *Figure 11*).

Using CPM, activities that could significantly delay the project if not completed on time are labeled critical activities. They are placed along the main schedule line, or critical path, in the diagram. Other, less critical, activities are shown in parallel to the critical activities. They intersect the critical path at the point where they must be performed for the project to stay on schedule. By adding up the durations of all the activities along the critical path, you can determine the duration of the entire project.

Before beginning each task, ensure that you understand what you are expected to do and what is required to accomplish the task. This includes knowing where your work is to begin and end and what types of materials are to be installed. Divide the task into individual steps, and decide the sequence in which you will perform the work. Teams need to know the following before starting a task:

- The start and end dates of the project
- The team size for each task
- The dates that materials will be delivered
- The schedules of the other trades
- Potential bottlenecks

Contractors rely on daily work reports to keep track of the project's schedule (*Figure 12*). The job superintendent writes up a daily work report at the end of each day's work. The daily report shows the contractor how much work was completed that day. If there were problems or delays, the contractor can decide what steps to take to keep the project on schedule. Consult the daily work reports to spot delays that could affect your team.

Obtain all the materials, tools, equipment, and supplies that you will need to perform the task. Perform the task according to plans, speci-

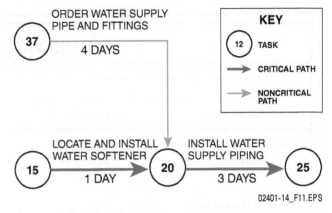

Figure 11 Sample critical path management (CPM) diagram.

fications, and code requirements, referring to the drawings and specifications, if needed. If the installation includes fixture rough-ins, ensure that you have the correct manuals. Confirm the finished elevations of drains installed in the ground.

One of the most important parts of task planning is to think ahead. At least once a day, stop and ask yourself the following questions:

- What materials, equipment, and information will I need during the next couple of days?
- Are all of the supplies, special tools, and equipment available?
- Will I need help to perform these tasks?
- Are there any potential conflicts between my work and that of other trades?
- Do I need the cooperation of other trades to perform my work efficiently?
- What other work could I do if the task is interrupted?

If supplies, tools, and equipment are not available, inform your supervisor so that the supervisor can get them to you quickly. Look ahead to ensure that you collect all the information you will need. Plan your task so that you can work around other trades. If their work interferes with what you want to do, turn to other tasks waiting to be done. Good task planning ensures that all time is used wisely.

DAILY WORK REPORT

Project: _____   Weather: _____

Job number: _____   Temperature: _____

Superintendent: _____   Wind: _____

Work force/subcontractors:

_____ Foreman   _____ Plumbers

_____ Pipefitters   _____ (Other)

Equipment on the job: _____

Remarks: _____
_____

Visitors:

Time:	Name:	Representing:	Remarks:
_____	_____	_____	_____
_____	_____	_____	_____
_____	_____	_____	_____
_____	_____	_____	_____
_____	_____	_____	_____

Equipment needs, rentals, problems: _____
_____

Work completed (describe): _____
_____

Work in progress (describe): _____
_____

Remarks: _____
_____

Superintendent's signature: _____   Date: _____

02401-14_F12.EPS

Figure 12 Sample daily work report for a project.

Effective Communication

A successful project depends on clear communication. Always speak with precision and provide the necessary information. When someone asks you a question, listen closely and answer promptly. These may seem like basic skills, but they are worth practicing all the time.

Getting along with your co-workers is an important part of communication. You will be working with other plumbers, workers in other trades, supervisors, plumbing suppliers, and inspectors. All of these people deserve your respect. Cooperation leads to productivity, and the more productive the crew, the better the work. Address personal problems or conflicts outside of regular work hours. Otherwise, they will decrease your productivity and slow down your colleagues. Do not let a problem or misunderstanding on the job go unresolved, because it will harden into resentment and mistrust. Ask a supervisor for advice in solving the problem if necessary.

Additional Resources

Construction Accounting & Financial Management, Third Edition. 2012. Steven Peterson. New York: Prentice Hall.

Financial Management and Accounting for the Construction Industry. 1988. New York: Construction Financial Management Association.

Financial Management in Construction Contracting. 2013. Andrew Ross and Peter Williams. Ames, IA: Wiley-Blackwell.

Small Business Accounting Simplified, Fifth Edition. 2010. Daniel Sitarz. Carbondale, IL: Nova Publishing Company.

1.0.0 Section Review

1. Profit is _____.
   a. income plus expenses
   b. income minus expenses
   c. assets plus debt
   d. assets minus debt

2. To account for all items in a company vehicle, _____.
   a. leave all items where you found them
   b. store them in a secure locker
   c. take photos of the items
   d. use a checklist

3. Typically, apprentices are expected to serve on field crews for _____.
   a. one to two years
   b. three to four years
   c. five to six years
   d. seven to eight years

2.0.0 PREPARING A MATERIAL TAKEOFF

Objective

Identify the information required to prepare a material takeoff.

a. Identify the information required for insurance and liability.
b. Identify the information required for the estimating process.
c. Identify the information required for updates and revisions.

Performance Task

Prepare a material takeoff as part of an estimate.

Trade Terms

Bond: A financial guarantee that work will be completed according to the terms of a contract.

Contract: A legal agreement between a contractor and a client for work to be done.

Cradle-to-grave responsibility: A contractor's legal responsibility for the management of hazardous wastes from creation to disposal.

Direct cost: An expense for an activity or an item related directly to a project.

Estimating: Calculating the costs to complete a specific project, which are submitted as part of a bid.

Indirect cost: An expense for an activity or an item that supports the administrative and overhead activities of a business.

Overhead: The costs of maintaining a business that are not applicable to a specific job. Office and administrative expenses are common examples of overhead.

Premium: What the contractor pays the insurance company to ensure that the project is covered.

Production figure: An estimate of the time required for a crew to perform a task.

Recapitulation sheet: A summary of material, labor, and equipment costs for each task in a project. Also called a recap sheet.

Sole source: A type of bid in which a project is offered to a contractor of the client's choice.

Unit cost: In estimating materials used in a job, the total material cost divided by the number of units used.

Many commercial and municipal projects choose plumbing companies through a competitive process called bidding. In the bidding process, each contractor estimates how much time and money it will take to complete the project. The contractors then submit these estimates to the client as bids. When the client settles on the winning bid, the contractor and the client sign a contract. This is a legal agreement for the work to be done on the project.

Public and commercial projects do not solicit bids in the same way. Public projects are bid under formal rules. They must be advertised and open to all qualified bidders. Commercial projects can be offered to a few selected bidders. They can also be offered to a single contractor of the owner's choosing. This is called a sole source bid.

Contractors calculate costs for a bid through a process called estimating. People often use the word estimating to mean guessing. An estimate is a forecast of something that has not happened yet, but it is more than a guess; plumbing estimates are careful calculations. They are based on information provided by the client and on the contractor's past experience. They take into account a wide variety of factors:

- Direct costs such as:
  – Materials
  – Labor
  – Tools
  – Equipment
  – Worker insurance

- Indirect costs such as:
  – Salaries and benefits
  – Office rent
  – Permits
  – Utilities
  – Insurance
  – Office supplies
  – Advertising
  – Property and equipment depreciation

- Profit

Direct costs apply to items and activities that are directly related to the project. Indirect costs are the items and activities that support the contractor's business and are sometimes called overhead. Payroll taxes are an important overhead expense. Payroll taxes are money withheld from employee paychecks to cover health insurance, Social Security, and other individual obligations.

Profit, as discussed in the section *The Profit and Loss Statement*, is the amount of money left after all the expenses, including overhead, have been paid. Profit allows the contractor to stay in business. If estimates are not accurate and realistic, the contractor will not make a profit.

In this section, you will learn how to estimate and bid on a plumbing job. To prepare an accurate estimate, you will use skills that are already familiar to you. Estimating involves reading blueprints and creating material takeoffs, which you have learned how to do. You will use your basic math skills to calculate the time, materials, and labor involved. You will also use effective writing skills to clearly communicate the cost estimate. The process can be intimidating at first, but with practice you will become familiar with the steps involved in estimating and bidding on a wide variety of jobs.

2.1.0 Preparing Insurance and Liability Information

You probably know that insurance protects your car, home, and health in case of an accident. As a contractor, you will need insurance too. Contractors are responsible for the safety of their employees. They are also responsible for ensuring that the client's property is not damaged while the plumbing work is being performed. These are only two of the many things that a contractor is responsible for on the job. The legal term for such responsibility is liability. With insurance, the term *liability* has a different meaning than it does with accounting.

Insurance is one of the most important indirect costs on a project. Find out early on what the insurance requirements are for the project. Ensure that you have the proper coverage. For the estimate, you will need to determine the premium. This is what you pay the insurance company to cover the project. You can get this information from your insurance specialist. If you do not have all the necessary coverage, find out if it will be possible to get the coverage before the project's start date. Some clients prefer that the contractor have the proper insurance before submitting a bid for the work. Be sure to read the client's bid information carefully.

Clients often require contractors to guarantee that all work will be completed correctly, on time, and within budget. The guarantee usually requires that the contractor pay a penalty to the client if the work is poor, late, or more expensive than agreed upon. Such financial guarantees are called bonds. Performance bonds and payment bonds are common. A performance bond guarantees that the work will be done according to the terms of the contract. A payment bond guarantees that the contractor will pay subcontractors brought in to do specific tasks. Banks issue bonds to contractors, so a contractor needs to have a good reputation in order to secure a bond.

Some projects result in the creation of hazardous wastes. Pollution liability coverage covers the management of hazardous wastes. Hazardous wastes include the following:

- Acids and alkyds
- Adhesives, epoxies, and solvents
- Cleaners and degreasers
- Fuels, motor oil, and hydraulic fluids
- Insulation and asbestos
- Primers, paints, and thinners
- Scrap metal and soiled rags

Federal rules say that contractors have cradle-to-grave responsibility for hazardous waste. This means that the contractor is responsible for the wastes from their creation to their disposal. Ensure that the estimate takes into account the cost of an approved waste hauler, who must dispose of the waste using proper methods. Ensure that the work crew is trained to work with hazardous materials.

2.2.0 Preparing Estimates

Begin an estimate by reviewing records of previous jobs. Most contractors keep detailed records of past work, for reference and accounting. Records of past projects should contain the following information:

- Material, equipment, and labor costs
- Labor productivity rates by task
- Effective crew sizes and combinations of supervisors, apprentices, journeymen, and laborers

Use this information as a guide when developing a new estimate. Remember that the material, equipment, and labor costs in old project records may be out of date. Be sure to use the most recent figures.

Legal and contract requirements may affect how estimates are developed. Use the following steps as a guideline. Note that the steps will vary depending on the contractor.

Step 1 Using the project's drawings and specifications, perform a takeoff of all the materials needed for the job. Write this information on a takeoff worksheet (*Figure 13*).

Step 2 Determine the size of the crew needed for the job. Calculate how long it will take the crew to perform each task. These calculations are known as production figures. Refer to figures from past jobs. Published estimating manuals are also available, but should be used only if figures from past jobs are not available.

Step 3 Multiply the labor hours for each task by the hourly labor rates. For subcontractor labor, have the subcontractor calculate the costs per task.

Step 4 Calculate the equipment costs. If you have to rent the equipment, be sure to use up-to-date rental costs.

Step 5 Transfer the total material quantities, material costs, labor costs, and equipment costs to a summary sheet (*Figure 14*). Include the unit cost for the materials used. This is the total cost divided by the number of units of material.

Step 6 Transfer the material, labor, and equipment costs for each activity to a recapitulation sheet, also called a recap sheet (*Figure 15*). A recapitulation sheet serves as a task-by-task summary of an entire project.

Step 7 Calculate the indirect costs, including insurance, taxes, overhead, and profit. Consult an accounting specialist for this information. Add the indirect costs to the direct costs on the recapitulation sheet. The resulting figure is the estimated total job cost.

Step 8 When the estimate is complete, submit it to the client as a bid.

The two most common mistakes that plumbers make when performing an estimate are omitting all or part of an item and making an arithmetical error.

Always recheck the material takeoffs against the construction drawings. Check all the arithmetic at each stage. Review the summary worksheet against the individual item worksheets, then review the recapitulation sheet against the summary sheet. This way, you have a better chance of catching errors.

Takeoff By:	WORKSHEET										PAGE #

DATE _____   PROJECT _____

SHEET _____ of _____   ARCHITECT _____

REF.	DESCRIPTION	DIMENSIONS				EXTENSION	QUANTITY	UNIT	TOTAL		REMARKS
		NO.	LENGTH	WIDTH	HEIGHT				QUANTITY	UNIT	

02401-14_F13.EPS

Figure 13 Sample takeoff worksheet.

| By: | | | SUMMARY SHEET | | | | | | | | | | PAGE # |

DATE _____ PROJECT _____

SHEET _____ of _____ TITLE: _____ WORK ORDER # _____

	DESCRIPTION	QUANTITY		MATERIAL COST		LABOR MAN HOURS FACTORS					LABOR COST		ITEM COST	
		TOTAL	UT	PER UNIT	TOTAL	CRAFT	PR UNIT	TOTAL	RATE	COST PR	PER	TOTAL	TOTAL	PER UNIT
					MATERIAL							LABOR	TOTAL	

02401-14_F14.EPS

Figure 14 Sample summary sheet.

2.3.0 Preparing Updates and Revisions

Project schedules and budgets are continually updated throughout the life of a project. Updates allow the contractor to ensure that the costs, materials, and schedule are being followed as the project moves along. It is important to monitor the progress of a project. If expenses are not in line with the progress of the job, the contractor may need to adjust the cost structure and budget. If a task is taking too long, the contractor may need to find ways to speed it up. For these reasons, project managers in the field continually track the actual costs against the estimated costs. Doing so allows the contractor to determine if the project is staying within budget. Otherwise, the contractor may find out about problems too late to correct them.

At the end of the project, save the takeoffs, summaries, and recapitulation sheets. They can be used as reference materials for future projects. Contractors can also use them to find out how profitable a company has been up to that time. This allows contractors to plan on future expenses because they know how much they have made.

Effective Writing

Cost estimates are the heart of a bid, but numbers alone do not sell a contractor. A bid, like anything that a contractor provides to a client or potential client, must be clearly written. A poorly written bid, no matter how right the price, is likely to make the client choose another contractor.

Listed By: _____	**RECAPITULATION SHEET**				Page _____ of _____
Checked By: _____	PROJECT _____ DATE _____				
	ARCHITECT_____ S/SF _____				

PAGE REF.	ITEM	MATERIAL	LABOR	SUB	EQUIPMENT	TOTAL

02401-14_F15.EPS

Figure 15 Sample recapitulation sheet.

Additional Resources

Building Contractor: Start and Run a Money-Making Business. 1993. R. Dodge Woodson. New York: TAB Books.

Plumbing Contractor: Start and Run a Money-Making Business. 1994. R. Dodge Woodson. Blue Ridge Summit, PA: TAB Books.

Plumbing Estimating Methods, Third Edition. Joseph J. Galeno and Sheldon T. Greene. Kingston, MA: R.S. Means Company.

2.0.0 Section Review

1. The guarantee that a contractor will pay sub-contractors brought in to do specific tasks is called a _____.

   a. premium
   b. cradle-to-grave responsibility
   c. contract
   d. payment bond

2. When preparing an estimate, the first step is to prepare a _____.

   a. work flow
   b. takeoff
   c. summary sheet
   d. recapitulation sheet

3. During the course of a project, if the contractor finds that expenses are not staying in line with the job's progress, the contractor may need to

   _____.

   a. adjust the cost structure and budget
   b. add more workers to the crew
   c. lay off staff
   d. confer with the architect or designer to find ways to scale back the project

SUMMARY

This module presented the techniques that plumbing businesses use to do the highest-quality work possible at a reasonable cost to the client. Employees should understand that they play a vital part in ensuring that a business is successful. The most basic measure of the success of a business is its profitability. Profit is the money that a business earns from work after paying off all expenses. Business owners use profit to buy new equipment, attract investors, and stay in business between projects.

Plumbing contractors bid for projects. To bid, a contractor must first develop an estimate. An estimate is a calculation of how much time and money will be needed to finish a project. An estimate takes into account the direct and indirect costs that must be covered. The estimate also includes the contractor's profit. When the client picks a winning bid, the contractor and client sign a legal agreement called a contract. Then the work begins.

Once a project is under way, cost control is an important part of a plumber's job. Ensure that materials are ordered on time and correctly so that the team does not waste time waiting for something to arrive. Store materials correctly, securely, and close to the work area. Use tools and equipment properly, and repair or replace damaged tools at once. Maintain company vehicles properly and operate them safely. When additional work is necessary, the client issues a change order. Complete the tasks in a change order just like any other task.

Task planning and scheduling are important on-the-job activities. Be sure you understand what you are expected to do. Find out beforehand what is required to accomplish a task so that you will be prepared. Read daily work reports to ensure that the project is on schedule, and plan tasks to allow you to work around other trades. If there is a schedule conflict, do not sit by idly. Do another task. Good task planning is one of many important skills that a plumber must develop to ensure that the business is successful.

1. The most important goal of a plumbing business is to _____.

   a. keep costs under control
   b. provide the highest-quality work possible at a reasonable cost to the client
   c. make clients happy so they will recommend you to friends
   d. keep prices competitive to obtain work

2. The two basic tools of accounting are the balance sheet and the _____.

   a. profit and loss statement
   b. budget document
   c. cash-flow report
   d. accounts-receivable ledger

3. The value of a business if all its debts are paid is its _____.

   a. profit margin
   b. capitalization
   c. book value
   d. equity

4. Amounts of money that are owed the company but have not yet been received are considered _____.

   a. accounts receivable
   b. pending credits
   c. due bills
   d. accounts payable

5. Gross income is the money that a company has made _____.

   a. after taxes are paid
   b. in a specified period
   c. before deducting expenses
   d. on a particular project

6. When ordering materials, choose suppliers based on the price of the materials and _____.

   a. delivery schedules
   b. warehouse location
   c. billing cycle
   d. location of the project

7. An itemized list of the prices of materials furnished and the terms of agreement is a(n) _____.

   a. invoice
   b. bill of lading
   c. account payable
   d. packing list

8. When additional work outside the scope of the original contract is needed, the client issues a(n) _____.

   a. amendment
   b. change order
   c. contract extension
   d. rework ticket

9. The person responsible for coordinating the work at the job site is the _____.

   a. job superintendent
   b. general contractor
   c. plumbing foreman
   d. project manager

10. Undertaking and completing jobs in the proper order and on time is the purpose of _____.

    a. budgeting
    b. priority assignment
    c. task planning
    d. time study

11. To help keep track of project progress, the job superintendent provides the contractor with a _____.

    a. weekly progress summary
    b. task-specific checklist
    c. labor-and-materials utilization form
    d. daily work report

12. On a CPM diagram, the main schedule line is referred to as the _____.

    a. backbone
    b. primary track
    c. critical path
    d. control vector

13. A client request for a bid from only one contractor is called _____.

   a. directed bidding
   b. sole-source bidding
   c. assigned bidding
   d. preferred-contractor bidding

14. Office rents, insurance, and permit costs are examples of _____.

   a. indirect costs
   b. out-of-pocket costs
   c. controllable costs
   d. plumbing contractor

15. The recapitulation sheet used in an estimate is a _____.

   a. compilation of anticipated labor hours
   b. detailed materials list
   c. task-by-task summary of the entire project
   d. listing of direct and indirect costs

Trade Terms Quiz

Fill in the blank with the correct term that you learned from your study of this module.

1. The process of organizing activities and materials so that tasks are completed in the right order and on time is called _____.

2. A(n) _____ is a formal offer to do work according to an estimate.

3. A report written by a job superintendent describing activities at a job site is called a(n) _____.

4. _____ is the process of compiling and analyzing financial records.

5. An expense for an activity or an item that supports the administrative and overhead activities of a business is called a(n) _____.

6. _____ is money owed by a business for payment of direct costs and indirect costs.

7. Money owed by a business, or a legal responsibility, is called a(n) _____.

8. A(n) _____ is an estimate of the time required for a crew to perform a task.

9. Money that is owed to a business but has not been paid is called _____.

10. _____ is all money earned before expenses are subtracted.

11. The costs of maintaining a business that are not applicable to a specific job, such as office and administrative expenses, are called _____.

12. _____ is a task-planning process in which diagrams show the relationships between tasks and estimated completion times.

13. A contractor's legal responsibility for the management of hazardous wastes from creation to disposal is called _____.

14. _____ is money left over after expenses are subtracted from income.

15. Any type of property owned by a business is called a(n) _____.

16. _____ is a calculation used when estimating materials for a job; it computes the total material cost divided by the number of units used.

17. Tasks in the critical path management process that could cause significant delay if not completed on time are called _____.

18. A(n) _____ is an agreement to perform work in addition to what was originally agreed to in a contract.

19. Calculating the costs to complete a specific project, which are submitted as part of a bid, is called _____.

20. _____ is the condition of an account when assets and liabilities are equal.

21. The type of bid in which a project is offered to a contractor of the client's choice is called _____.

22. _____ is the value of the business if all debts are paid.

23. Another term for profit is _____.

24. A(n) _____ is a legal agreement between a contractor and a client for work to be done.

25. An expense for an activity or an item related directly to a project is called a(n) _____.

26. A(n) _____ is a financial guarantee that work will be completed according to the terms of a contract.

27. An itemized list of materials for purchase or rent is called a(n) _____.

28. To _____ is to redo a previously completed task.

29. What the contractor pays the insurance company to ensure that the project is covered is called a(n) _____.

30. A(n) _____ is a summary of material, labor, and equipment costs for each task in a project.

31. An order form submitted to a supplier is called a(n) _____.

32. A(n) _____ is a form used to track assets and liabilities.

33. A list of all income and expenses over a given period of time is called a(n) _____.

34. _____ is the phase immediately following the awarding of a contract but before the start of construction, during which subcontractors are selected, purchase orders are issued, construction schedules are developed, and estimates are reviewed in depth.

35. The final stage of preplanning, during which all items required to start construction are moved to the job site, arrangements are made with the other trades to coordinate job-site activities, and all job-site details are organized in preparation for construction, is called _____.

Trade Terms

Accounting
Accounts receivable
Asset
Balance
Balance sheet
Bid
Bond
Change order

Contract
Cradle-to-grave responsibility
Critical activities
Critical path management
Daily work report
Debt

Direct cost
Equity
Estimating
Gross income
Indirect cost
Invoice
Liability
Mobilization

Net income
Overhead
Premium
Preplanning
Production figure
Profit
Profit and loss statement

Purchase order
Recapitulation sheet or recap sheet
Rework
Sole source
Task planning
Unit cost

Trade Terms Introduced in This Module

Accounting: The process of compiling and analyzing financial records.

Accounts receivable: Money that is owed to a business but has not been paid.

Asset: Any type of property owned by a business.

Balance: The condition of an account when assets and liabilities are equal.

Balance sheet: A form used to track assets and liabilities.

Bid: A formal offer to do work according to an estimate.

Bond: A financial guarantee that work will be completed according to the terms of a contract.

Change order: An agreement to perform work in addition to what was originally agreed to in a contract.

Contract: A legal agreement between a contractor and a client for work to be done.

Cradle-to-grave responsibility: A contractor's legal responsibility for the management of hazardous wastes from creation to disposal.

Critical activities: Tasks in the critical path management process that could cause significant delay if not completed on time.

Critical path management (CPM): A task-planning process in which diagrams show the relationships between tasks and estimated completion times.

Daily work report: A report written by a job superintendent describing activities at a job site.

Debt: Money owed by a business for payment of direct costs and indirect costs.

Direct cost: An expense for an activity or an item related directly to a project.

Equity: The value of the business if all debts are paid.

Estimating: Calculating the costs to complete a specific project, which are submitted as part of a bid.

Gross income: All money earned before expenses are subtracted.

Indirect cost: An expense for an activity or an item that supports the administrative and overhead activities of a business.

Invoice: An itemized list of materials for purchase or rent.

Liability: In accounting, money owed by a business. In insurance, a legal responsibility.

Mobilization: The final stage of preplanning, during which all items required to start construction are moved to the job site, arrangements are made with the other trades to coordinate job-site activities, and all job-site details are organized in preparation for construction.

Net income: Another term for profit.

Overhead: The costs of maintaining a business that are not applicable to a specific job. Office and administrative expenses are common examples of overhead.

Premium: What the contractor pays the insurance company to ensure that the project is covered.

Preplanning: The phase immediately following the awarding of a contract but before the start of construction, during which subcontractors are selected, purchase orders are issued, construction schedules are developed, and estimates are reviewed in depth.

Production figure: An estimate of the time required for a crew to perform a task.

Profit: Money left over after expenses are subtracted from income.

Profit and loss statement: A list of all income and expenses over a given period of time.

Purchase order: An order form submitted to a supplier.

Recapitulation sheet: A summary of material, labor, and equipment costs for each task in a project. Also called a recap sheet.

Rework: To redo a previously completed task.

Sole source: A type of bid in which a project is offered to a contractor of the client's choice.

Task planning: The process of organizing activities and materials so that tasks are completed in the right order and on time.

Unit cost: In estimating materials used in a job, the total material cost divided by the number of units used.

Additional Resources

This module presents thorough resources for task training. The following resource material is suggested for further study.

Building Contractor: Start and Run a Money-Making Business. 1993. R. Dodge Woodson. New York: TAB Books.

Construction Accounting & Financial Management, Third Edition. 2012. Steven Peterson. New York: Prentice Hall.

Financial Management and Accounting for the Construction Industry. 1988. New York: Construction Financial Management Association.

Financial Management in Construction Contracting. 2013. Andrew Ross and Peter Williams. Ames, IA: Wiley-Blackwell.

Plumbing Contractor: Start and Run a Money-Making Business. 1994. R. Dodge Woodson. Blue Ridge Summit, PA: TAB Books.

Plumbing Estimating Methods, Third Edition. Joseph J. Galeno and Sheldon T. Greene. Kingston, MA: R.S. Means Company.

Small Business Accounting Simplified, Fifth Edition. 2010. Daniel Sitarz. Carbondale, IL: Nova Publishing Company.

Figure Credits

Section Review Answer Key

Answer	Section Reference	Objective
Section One		
1. b	1.1.0	1a
2. d	1.2.4	1b
3. b	1.3.0	1c
Section Two		
1. d	2.1.0	2a
2. b	2.2.0	2b
3. a	2.3.0	2c

NCCER CURRICULA — USER UPDATE

NCCER makes every effort to keep its textbooks up-to-date and free of technical errors. We appreciate your help in this process. If you find an error, a typographical mistake, or an inaccuracy in NCCER's curricula, please fill out this form (or a photocopy), or complete the online form at **www.nccer.org/olf**. Be sure to include the exact module ID number, page number, a detailed description, and your recommended correction. Your input will be brought to the attention of the Authoring Team. Thank you for your assistance.

Instructors – If you have an idea for improving this textbook, or have found that additional materials were necessary to teach this module effectively, please let us know so that we may present your suggestions to the Authoring Team.

NCCER Product Development and Revision

13614 Progress Blvd., Alachua, FL 32615

Email: curriculum@nccer.org
Online: www.nccer.org/olf

❏ Trainee Guide   ❏ Lesson Plans   ❏ Exam   ❏ PowerPoints   Other _____

Craft / Level: _____   Copyright Date: _____

Module ID Number / Title: _____

Section Number(s): _____

Description: _____

_____

_____

_____

_____

Recommended Correction: _____

_____

_____

_____

_____

Your Name: _____

Address: _____

_____

Email: _____   Phone: _____

NCCER makes every effort to keep its textbooks up-to-date and free of technical errors. We appreciate your help in this process. If you find an error, a typographical mistake, or an inaccuracy in NCCER's curricula, please fill out this form (or a photocopy), or complete the online form at www.nccer.org/olf. Be sure to include the exact module ID number, page number, a detailed description, and your recommended correction. Your input will be brought to the attention of the Authoring Team. Thank you for your assistance.

Instructors – If you have an idea for improving this textbook, or if you have found that additional materials were necessary to teach this module effectively, please let us know so that we may present your suggestions to the Authoring Team.

NCCER Product Development and Revision
13614 Progress Blvd., Alachua, FL 32615

Email: curriculum@nccer.org
Online: www.nccer.org/olf

☐ Trainee Guide   ☐ Lesson Plans   ☐ Exam   ☐ PowerPoint   Other _____

Craft / Level: _____   Copyright Date: _____

Module ID Number / Title: _____

Section Number: _____

Description:

_____

_____

_____

_____

Recommended Correction:

_____

_____

_____

_____

Your Name: _____

Address: _____

Email: _____   Phone: _____

Fundamentals of Crew Leadership

46101-11

NCCER

President: Don Whyte
Director of Product Development: Daniele Dixon
Fundamentals of Crew Leadership Project Manager: Patty Bird
Senior Manager of Production: Tim Davis
Quality Assurance Coordinator: Debie Hicks
Editor: Chris Wilson
Desktop Publishing Coordinator: James McKay
Production Specialist: Megan Casey

Editorial and production services provided by Topaz Publications, Liverpool, NY
Lead Writer/Project Manager: Tom Burke
Desktop Publisher: Joanne Hart
Art Director: Megan Paye
Permissions Editors: Andrea LaBarge, Alison Richmond

FOREWORD

Work gets done most efficiently if workers are divided into crews with a common purpose. When a crew is formed to tackle a particular job, one person is appointed the leader. This person is usually an experienced craftworker who has demonstrated leadership qualities. To become an effective leader, it helps if you have natural leadership qualities, but there are specific job skills that you must learn in order to do the job well.

This module will teach you the skills you need to be an effective leader, including the ability to communicate effectively; provide direction to your crew; and effectively plan and schedule the work of your crew.

As a crew member, you weren't required to think much about project cost. However, as a crew leader, you need to understand how to manage materials, equipment, and labor in order to work in a cost-effective manner. You will also begin to view safety from a different perspective. The crew leader takes on the responsibility for the safety of the crew, making sure that workers follow company safety polices and have the latest information on job safety issues.

As a crew leader, you become part of the chain of command in your company, the link between your crew and those who supervise and manage projects. As such, you need to know how the company is organized and how you fit into the organization. You will also focus more on company policies than a crew member, because it is up to you to enforce them within your crew. You will represent your team at daily project briefings and then communicate relevant information to your crew. This means learning how to be an effective listener and an effective communicator.

Whether you are currently a crew leader wanting to learn more about the requirements, or a crew member preparing to move up the ladder, this module will help you reach your goal.

This program consists of a Participant's Manual and Lesson Plans for the Instructor. The Participant's Manual contains the material that the participant will study, along with self-check exercises and activities, to help in evaluating whether the participant has mastered the knowledge needed to become an effective crew leader. The Lesson Plans include instructional outlines, suggested classroom activities, and homework based on the material in the Participant's Manual.

For the participant to gain the most from this program, it is recommended that the material be presented in a formal classroom setting, using a trained and experienced instructor. If the student is so motivated, he or she can study the material on a self-learning basis by using the material in both the Participant's Manual and the Lesson Plans. Recognition through the National Registry is available for the participants provided the program is delivered through an Accredited Sponsor by a Master Trainer or ICTP instructor. More details on this program can be received by contacting NCCER at **www.nccer.org**.

Participants in this program should note that some examples provided to reinforce the material may not apply to the participant's exact work, although the process will. Every company has its own mode of operation. Therefore, some topics may not apply to every participant's company. Such topics have been included because they are important considerations for prospective crew leaders throughout the industries supported by NCCER.

Industry-Recognized Credentials

If you are training through an NCCER-accredited sponsor, you may be eligible for credentials from NCCER's Registry. The ID number for this module is 46101-11. Note that this module may have been used in other NCCER curricula and may apply to other level completions. Contact NCCER's Registry at 888.622.3720 or go to **www.nccer.org** for more information.

Contents

Topics to be presented in this module include:

Contents (continued)

Contents (continued)

Contents (continued)

Figures and Tables

Acknowledgments

This curriculum was revised as a result of the farsightedness
and leadership of the following sponsors:

ABC South Texas Chapter
HB Training & Consulting
Turner Industries Group, LLC

University of Georgia
Vision Quest Academy
Willmar Electric Service

This curriculum would not exist were it not for the dedication and unselfish energy of
those volunteers who served on the Authoring Team. A sincere thanks is extended to the following:

John Ambrosia
Harold (Hal) Heintz
Mark Hornbuckle
Jonathan Liston

Jay Tornquist
Wayne Tyson
Antonio "Tony" Vazquez

NCCER Partners

American Fire Sprinkler Association
Associated Builders and Contractors, Inc.
Associated General Contractors of America
Association for Career and Technical Education
Association for Skilled and Technical Sciences
Carolinas AGC, Inc.
Carolinas Electrical Contractors Association
Center for the Improvement of Construction
Management and Processes
Construction Industry Institute
Construction Users Roundtable
Construction Workforce Development Center
Design Build Institute of America
GSSC – Gulf States Shipbuilders Consortium
Manufacturing Institute
Mason Contractors Association of America
Merit Contractors Association of Canada
NACE International
National Association of Minority Contractors
National Association of Women in Construction
National Insulation Association
National Ready Mixed Concrete Association
National Technical Honor Society
National Utility Contractors Association
NAWIC Education Foundation
North American Technician Excellence

Painting & Decorating Contractors of America
Portland Cement Association
SkillsUSA®
Steel Erectors Association of America
U.S. Army Corps of Engineers
University of Florida, M. E. Rinker School of
Building Construction
Women Construction Owners & Executives, USA

Objectives

Upon completion of this section, you should be able to:

1. Describe the opportunities in the construction and power industries.
2. Describe how workers' values change over time.
3. Explain the importance of training and safety for the leaders in the construction and power industries.
4. Describe how new technologies are beneficial to the construction and power industries.
5. Identify the gender and minority issues associated with a changing workforce.
6. Describe what employers can do to prevent workplace discrimination.
7. Differentiate between formal and informal organizations.
8. Describe the difference between authority, responsibility, and accountability.
9. Explain the purpose of job descriptions and what they should include.
10. Distinguish between company policies and procedures.

1.0.0 INDUSTRY TODAY

Today's managers, supervisors, and crew leaders face challenges different from those of previous generations of leaders. To be a leader in industry today, it is essential to be well prepared. Today's crew leaders must understand how to use various types of new technology. In addition, they must have the knowledge and skills needed to manage, train, and communicate with a culturally diverse workforce whose attitudes toward work may differ from those of earlier generations and cultures. These needs are driven by changes in the workforce itself and in the work environment, and include the following:

• A shrinking workforce
• The growth of technology
• Changes in employee attitudes and values
• The emphasis on bringing women and minorities into the workforce
• The growing number of foreign-born workers
• Increased emphasis on workplace health and safety
• Greater focus on education and training

1.1.0 The Need for Training

Effective craft training programs are necessary if the industry is to meet the forecasted worker demands. Many of the skilled, knowledgeable craftworkers, crew leaders, and managers—the so-called baby boomers—have reached retirement age. In 2010, these workers who were born between 1946 and 1964, represented 38 percent of the workforce. Their departure leaves a huge vacuum across the industry spectrum. The Department of Labor (DOL) concludes that the best way for industry to reduce shortages of skilled workers is to create more education and training opportunities. The DOL suggests that companies and community groups form partnerships and create apprenticeship programs. Such programs could provide younger workers, including women and minorities, with the opportunity to develop job skills by giving them hands-on experience.

When training workers, it is important to understand that people learn in different ways. Some people learn by doing, some people learn by watching or reading, and others need step-by-step instructions as they are shown the process. Most people learn best by a combination of styles. It is important to understand what kind of a learner you are teaching, because if you learn one way, you tend to teach the way you learn. Have you ever tried to teach somebody and failed, and then another person successfully teaches the same thing in a different way? A person who acts as a mentor or trainer needs to be able to determine what kind of learner they are addressing and teach according to those needs.

The need for training is not limited to craftworkers. There must be supervisory training to ensure there are qualified leaders in the industry to supervise the craftworkers.

1.1.1 Motivation

As a supervisor or crew leader, it is important to understand what motivates your crew. Money is often thought to be a good motivator. Although that may be true to some extent, it has been proven to be a temporary solution. Once a person has reached a level of financial security, other factors come into play. Studies show that many people tend to be motivated by environment and conditions. For those people, a great workplace may provide more satisfaction than pay. If you give someone a raise, they tend to work harder for a period of time. Then the satisfaction dissipates and they may want another raise. People are often motivated by feeling a sense of accomplishment. That is why setting and working toward recognizable goals tends to make employees more productive. A person with a feeling of involvement or a sense of achievement is likely to be better motivated and help to motivate others.

1.1.2 Understanding Workers

Many older workers grew up in an environment in which they were taught to work hard and stay with the job until retirement. They expected to stay with a company for a long time, and companies were structured to create a family-type environment.

Times have changed. Younger workers have grown up in a highly mobile society and are used to rapid rewards. This generation of workers can sometimes be perceived as lazy and unmotivated, but in reality, they simply have a different perspective. For such workers, it may be better to give them small projects or break up large projects into smaller pieces so that they feel repetitively rewarded, thus enhancing their perception of success.

- *Goal setting* – Set short-term and long-term goals, including tasks to be done and expected time frames. Help the trainees understand that things can happen to offset the short-term goals. This is one reason to set long-term goals as well. Don't set them up for failure, as this leads to frustration, and frustration can lead to reduced productivity.
- *Feedback* – Timely feedback is important. For example, telling someone they did a good job last year, or criticizing them for a job they did a month ago, is meaningless. Simple recognition isn't always enough. Some type of reward should accompany positive feedback, even if it is simply recognizing the employee in a public way. Constructive feedback should be given in private and be accompanied by some positive action, such as providing one-on-one training to correct a problem.

1.1.3 Craft Training

Craft training is often informal, taking place on the job site, outside of a traditional training classroom. According to the American Society for Training and Development (ASTD), craft training is generally handled through on-the-job instruction by a qualified co-worker or conducted by a supervisor.

The Society of Human Resources Management (SHRM) offers the following tips to supervisors in charge of training their employees:

- *Help crew members establish career goals.* Once the goals are established, the training required to meet the goals can be readily identified.
- *Determine what kind of training to give.* Training can be on the job under the supervision of a co-worker. It can be one-on-one with the supervisor. It can involve cross-training to teach a new trade or skill, or it can involve delegating new or additional responsibilities.
- *Determine the trainee's preferred method of learning.* Some people learn best by watching, others from verbal instructions, and others by doing. When training more than one person at a time, try to use all three methods.

Communication is a critical component of training employees. The SHRM advises that supervisors do the following when training their employees:

- *Explain the task, why it needs to be done, and how it should be done.* Confirm that the trainees understand these three areas by asking questions. Allow them to ask questions as well.
- *Demonstrate the task.* Break the task down into manageable parts and cover one part at a time.
- *Ask trainees to do the task while you observe them.* Try not to interrupt them while they are doing the task unless they are doing something that is unsafe and potentially harmful.
- *Give the trainees feedback.* Be specific about what they did and mention any areas where they need to improve.

1.1.4 Supervisory Training

Given the need for skilled craftworkers and qualified supervisory personnel, it seems logical that companies would offer training to their employees through in-house classes, or by subsidizing outside training programs. While some contractors have their own in-house training programs or participate in training offered by associations and other organizations, many contractors do not offer training at all.

There are a number of reasons that companies do not develop or provide training programs, including the following:

- Lack of money to train
- Lack of time to train
- Lack of knowledge about the benefits of training programs
- High rate of employee turnover
- Workforce too small

- Past training involvement was ineffective
- The company hires only trained workers
- Lack of interest from workers
- Lack of company interest in training

For craftworkers to move up into supervisory and managerial positions, it will be necessary for them to continue their education and training. Those who are willing to acquire and develop new skills have the best chance of finding stable employment. It is therefore critical that employees take advantage of training opportunities, and that companies employ training as part of their business culture.

Your company has recognized the need for training. Your participation in a leadership training program such as this will begin to fill the gap between craft and supervisory training.

1.2.0 Impact of Technology

Many industries, including the construction industry, have made the move to technology as a means of remaining competitive. Benefits include increased productivity and speed, improved quality of documents, greater access to common data, and better financial controls and communication. As technology becomes a greater part of supervision, crew leaders need to be able to use it properly. One important concern with electronic communication is to keep it brief, factual, and legal. Because the receiver has no visual or auditory clues as to the sender's intent, the sender can be easily misunderstood. In other words, it is more difficult to tell if someone is just joking via e-mail because you can't see their face or hear the tone of their voice.

Cellular telephones, voicemail, and handheld communication devices have made it easy to keep in touch. They are particularly useful communication sources for contractors or crew leaders who are on a job site, away from their offices, or constantly on the go.

Cellular telephones allow the users to receive incoming calls as well as make outgoing calls. Unless the owner is out of the cellular provider's service area, the cell phone may be used any time to answer calls, make calls, and send and receive voicemail or email. Always check the company's policy with regard to the use of cell phones on the job.

Handheld communication devices known as smart phones allow supervisors to plan their calendars, schedule meetings, manage projects, and access their email from remote locations. These computers are small enough to fit in the palm of the hand, yet powerful enough to hold years of information from various projects. Information can be transmitted electronically to others on the project team or transferred to a computer.

2.0.0 GENDER AND CULTURAL ISSUES

During the past several years, the construction industry in the United States has experienced a shift in worker expectations and diversity. These two issues are converging at a rapid pace. At no time has there been such a generational merge in the workforce, ranging from The Silent Generation (1925–1945), Baby Boomers (1946–1964), Gen X (1965–1979), and the Millennials, also known as Generation Y (1980–2000).

This trend, combined with industry diversity initiatives, has created a climate in which companies recognize the need to embrace a diverse workforce that crosses generational, gender, and ethnic boundaries. To do this effectively, they are using their own resources, as well as relying on associations with the government and trade organizations. All current research indicates that industry will be more dependent on the critical skills of a diverse workforce—a workforce that is both culturally and ethnically fused. Across the United States, construction and other industries are aggressively seeking to bring new workers into their ranks, including women and racial and ethnic minorities. Diversity is no longer solely driven by social and political issues, but by consumers who need hospitals, malls, bridges, power plants, refineries, and many other commercial and residential structures.

Some issues relating to a diverse workforce will need to be addressed on the job site. These issues include different communication styles of men and women, language barriers associated with cultural differences, sexual harassment, and gender or racial discrimination.

2.1.0 Communication Styles of Men and Women

As more and more women move into construction, it becomes increasingly important that communication barriers between men and women are broken down and that differences in behaviors are understood so that men and women can work together more effectively. The Jamestown, New York Area Labor Management Committee (JALMC) offers the following explanations and tips:

- *Women tend to ask more questions than men do.* Men are more likely to proceed with a job and figure it out as they go along, while women are more likely to ask questions first.
- *Men tend to offer solutions before empathy; women tend to do the opposite.* Both men and women should say what they want up front, whether it's the solution to a problem, or simply a sympathetic ear. That way, both genders will feel understood and supported.
- *Women are more likely to ask for help when they need it.* Women are generally more pragmatic when it comes to completing a task. If they need help, they will ask for it. Men are more likely to attempt to complete a task by themselves, even when assistance is needed.
- *Men tend to communicate more competitively, and women tend to communicate more cooperatively.* Both parties need to hear one another out without interruption.

This does not mean that one method is more effective than the other. It simply means that men and women use different approaches to achieve the same result.

2.2.0 Language Barriers

Language barriers are a real workplace challenge for crew leaders. Millions of workers speak languages other than English. Spanish is commonly spoken in the United States. As the makeup of the immigrant population continues to change, the number of non-English speakers will rise dramatically, and the languages being spoken will also change. Bilingual job sites are increasingly common.

Companies have the following options to overcome this challenge:

- Offer English classes either at the work site or through school districts and community colleges.
- Offer incentives for workers to learn English.

As the workforce becomes more diverse, communicating with people for whom English is a second language will be even more critical. The following tips will help when communicating across language barriers:

- Be patient. Give workers time to process the information in a way that they can comprehend.
- Avoid humor. Humor is easily misunderstood and may be misinterpreted as a joke at the worker's expense.
- Don't assume that people are unintelligent simply because they don't understand what you are saying.
- Speak slowly and clearly, and avoid the tendency to raise your voice.
- Use face-to-face communication whenever possible. Over-the-phone communication is often more difficult when a language barrier is involved.
- Use pictures or drawings to get your point across.
- If a worker speaks English poorly but understands reasonably well, ask the worker to demonstrate his or her understanding through other means.

2.3.0 Cultural Differences

As workers from a multitude of backgrounds and cultures are brought together, there are bound to be differences and conflicts in the workplace.

To overcome cultural conflicts, the SHRM suggests the following approach to resolving cultural conflicts between individuals:

- *Define the problem from both points of view.* How does each person involved view the conflict? What does each person think is wrong? This involves moving beyond traditional thought processes to consider alternate ways of thinking.
- *Uncover cultural interpretations.* What assumptions are being made based on cultural programming? By doing this, the supervisor may realize what motivated an employee to act in a particular manner.

- *Create cultural synergy.* Devise a solution that works for both parties involved. The purpose is to recognize and respect other's cultural values, and work out mutually acceptable alternatives.

2.4.0 Sexual Harassment

In today's business world, men and women are working side-by-side in careers of all kinds. As women make the transition into traditionally male industries, such as construction, the likelihood of sexual harassment increases. Sexual harassment is defined as unwelcome behavior of a sexual nature that makes someone feel uncomfortable in the workplace by focusing attention on their gender instead of on their professional qualifications. Sexual harassment can range from telling an offensive joke or hanging a poster of a swimsuit-clad man or woman, to making sexual comments or physical advances.

Historically, sexual harassment was thought to be an act performed by men of power within an organization against women in subordinate positions. However, the number of sexual harassment cases over the years, have shown that this is no longer the case.

Sexual harassment can occur in a variety of circumstances, including but not limited to the following:

- The victim as well as the harasser may be a woman or a man. The victim does not have to be of the opposite sex.
- The harasser can be the victim's supervisor, an agent of the employer, a supervisor in another area, a co-worker, or a non-employee.
- The victim does not have to be the person harassed, but could be anyone affected by the offensive conduct.
- Unlawful sexual harassment may occur without economic injury to or discharge of the victim.
- The harasser's conduct must be unwelcome.

The Equal Employment Opportunity Commission (EEOC) enforces sexual harassment laws within industries. When investigating allegations of sexual harassment, the EEOC looks at the whole record, including the circumstances and the context in which the alleged incidents occurred. A decision on the allegations is made from the facts on a case-by-case basis. A crew leader who is aware of sexual harassment and does nothing to stop it can be held responsible. The crew leader therefore should not only take action to stop sexual harassment, but should serve as a good example for the rest of the crew.

Prevention is the best tool to eliminate sexual harassment in the workplace. The EEOC encourages employers to take steps to prevent sexual harassment from occurring. Employers should clearly communicate to employees that sexual harassment will not be tolerated. They do so by developing a policy on sexual harassment, establishing an effective complaint or grievance process, and taking immediate and appropriate action when an employee complains.

Both swearing and off-color remarks and jokes are not only offensive to co-workers, but also tarnish a worker's image. Crew leaders need to emphasize that abrasive or crude behavior may affect opportunities for advancement. If disciplinary action becomes necessary, it should be covered by company policy. A typical approach is a three-step process in which the perpetrator is first given a verbal reprimand. In the event of further violations, a written reprimand and warning are given. Dismissal typically accompanies subsequent violations.

2.5.0 Gender and Minority Discrimination

More attention is being placed on fair recruitment, equal pay for equal work, and promotions for women and minorities in the workplace. Consequently, many business practices, including the way employees are treated, the organization's hiring and promotional practices, and the way people are compensated, are being analyzed for equity.

Once a male-dominated industry, construction companies are moving away from this image and are actively recruiting and training women, younger workers, people from other cultures, and workers with disabilities. This means that organizations hire the best person for the job, without regard for race, sex, religion, age, etc.

To prevent discrimination cases, employers must have valid job-related criteria for hiring, compensation, and promotion. These measures must be used consistently for every applicant interview, employee performance appraisal, and hiring or promotion decision. Therefore, all workers responsible for recruitment, selection, and supervision of employees, and evaluating job performance, must be trained on how to use the job-related criteria legally and effectively.

3.0.0 BUSINESS ORGANIZATIONS

An organization is the relationship among the people within the company or project. The crew leader needs to be aware of two types of organizations. These are formal organizations and informal organizations.

A formal organization exists when the activities of the people within the work group are directed toward achieving a common goal. An example of a formal organization is a work crew consisting of four carpenters and two laborers led by a crew leader, all working together toward a common goal.

A formal organization is typically documented on an organizational chart, which outlines all the positions that make up an organization and shows how those positions are related. Some organizational charts even depict the people within each position and the person to whom they report, as well as the people that the person supervises. *Figures 1* and 2 show examples of organization charts for fictitious companies. Note that each of these positions represents an opportunity for advancement in the construction industry that a crew leader can eventually achieve.

An informal organization allows for communication among its members so they can perform as a group. It also establishes patterns of behavior that help them to work as a group, such as agreeing to use a specific training program.

An example of an informal organization is a trade association such as Associated Builders and Contractors (ABC), Associated General Contractors (AGC), and the National Association of Women in Construction (NAWIC). Those, along with the thousands of other trade associations in the U.S., provide a forum in which members with common concerns can share information, work on issues, and develop standards for their industry.

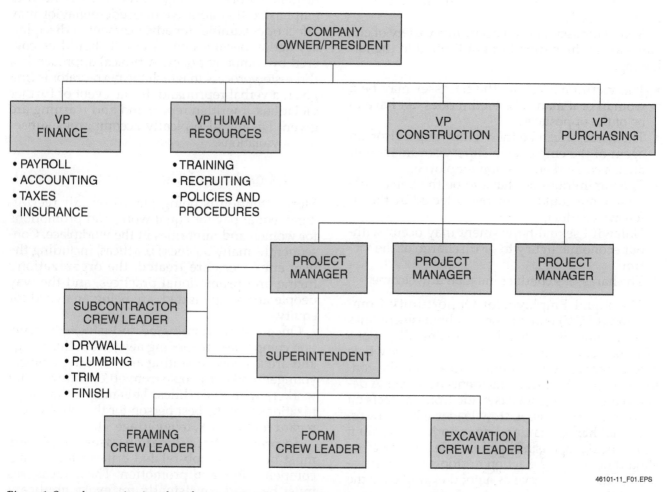

46101-11_F01.EPS

Figure 1 Sample organization chart for a construction company.

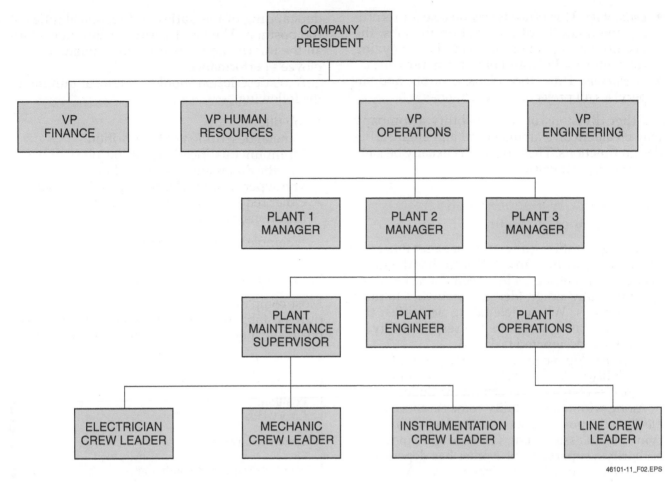

Figure 2 Sample organization chart for an industrial company.

Both types of organizations establish the foundation for how communication flows. The formal structure is the means used to delegate authority and responsibility and to exchange information. The informal structure is used to exchange information.

Members in an organization perform best when each member:

- Knows the job and how it will be done
- Communicates effectively with others in the group
- Understands his or her role in the group
- Recognizes who has the authority and responsibility

3.1.0 Division of Responsibility

The conduct of a business involves certain functions. In a small organization, responsibilities may be divided between one or two people. However, in a larger organization with many different and complex activities, responsibilities may be grouped into similar activity groups, and the responsibility for each group assigned to department managers. In either case, the following major departments exist in most companies:

- *Executive* – This office represents top management. It is responsible for the success of the company through short-range and long-range planning.
- *Human Resources* – This office is responsible for recruiting and screening prospective employees; managing employee benefits programs; advising management on pay and benefits; and developing and enforcing procedures related to hiring practices.
- *Accounting* – This office is responsible for all record keeping and financial transactions, including payroll, taxes, insurance, and audits.
- *Contract Administration* – This office prepares and executes contractual documents with owners, subcontractors, and suppliers.
- *Purchasing* – This office obtains material prices and then issues purchase orders. The purchasing office also obtains rental and leasing rates on equipment and tools.

- *Estimating*: This office is responsible for recording the quantity of material on the jobs, the takeoff, pricing labor and material, analyzing subcontractor bids, and bidding on projects.
- *Operations*: This office plans, controls, and supervises all project-related activities.

Other divisions of responsibility a company may create involve architectural and engineering design functions. These divisions usually become separate departments.

3.2.0 Authority, Responsibility, and Accountability

As an organization grows, the manager must ask others to perform many duties so that the manager can concentrate on management tasks. Managers typically assign (delegate) activities to their subordinates. When delegating activities, the crew leader assigns others the responsibility to perform the designated tasks.

Responsibility means obligation, so once the responsibility is delegated, the person to whom it is assigned is obligated to perform the duties.

Along with responsibility comes authority. *Authority* is the power to act or make decisions in carrying out an assignment. The type and amount of authority a supervisor or worker has depends on the company for which he or she works. Authority and responsibility must be balanced so employees can carry out their tasks. In addition, delegation of sufficient authority is needed to make an employee accountable to the crew leader for the results.

Accountability is the act of holding an employee responsible for completing the assigned activities. Even though authority and responsibility may be delegated to crew members, the final responsibility always rests with the crew leader.

3.3.0 Job Descriptions

Many companies furnish each employee with a written job description that explains the job in detail. Job descriptions set a standard for the employee. They make judging performance easier, clarify the tasks each person should handle, and simplify the training of new employees.

Each new employee should understand all the duties and responsibilities of the job after reviewing the job description. Thus, the time it takes for the employee to make the transition from being a new and uninformed employee to a more experienced member of a crew is shortened.

A job description need not be long, but it should be detailed enough to ensure there is no misun-

derstanding of the duties and responsibilities of the position. The job description should contain all the information necessary to evaluate the employee's performance.

A job description should contain, at minimum, the following:

- Job title
- General description of the position
- Minimum qualifications for the job
- Specific duties and responsibilities
- The supervisor to whom the position reports
- Other requirements, such as qualifications, certifications, and licenses

A sample job description is shown in *Figure 3*.

3.4.0 Policies and Procedures

Most companies have formal policies and procedures established to help the crew leader carry out his or her duties. A *policy* is a general state-

---

Position:
Crew Leader

General Summary:
First line of supervision on a construction crew installing concrete formwork.

Reports To:
Job Superintendent

Physical and Mental Responsibilities:
- Ability to stand for long periods
- Ability to solve basic math and geometry problems

Duties and Responsibilities:
- Oversee crew
- Provide instruction and training in construction tasks as needed
- Make sure proper materials and tools are on the site to accomplish tasks
- Keep project on schedule
- Enforce safety policies and procedures

Knowledge, Skills, and Experience Required:
- Extensive travel throughout the Eastern United States, home base in Atlanta
- Ability to operate a backhoe and trencher
- Valid commercial driver's license with no DUI violations
- Ability to work under deadlines with the knowledge and ability to foresee problem areas and develop a plan of action to solve the situation

46101-11_F03.EPS

Figure 3 Example of a job description.

ment establishing guidelines for a specific activity. Examples include policies on vacations, breaks, workplace safety, and checking out tools.

Procedures are the ways that policies are carried out. For example, a procedure written to implement a policy on workplace safety would include guidelines for reporting accidents and general safety procedures that all employees are expected to follow.

A crew leader must be familiar with the company policies and procedures, especially with regard to safety practices. When OSHA inspectors visit a job site, they often question employees and crew leaders about the company policies related to safety. If they are investigating an accident, they will want to verify that the responsible crew leader knew the applicable company policy and followed it.

Review Questions

1. The construction industry should provide training for craftworkers and supervisors _____.

   a. to ensure that there are enough future workers
   b. to avoid discrimination lawsuits
   c. in order to update the skills of older workers who are retiring at a later age than they previously did
   d. even though younger workers are now less likely to seek jobs in other areas than they were 10 years ago

2. Companies traditionally offer craftworker training _____.

   a. that a supervisor leads in a classroom setting
   b. that a craftworker leads in a classroom setting
   c. in a hands-on setting, where craftworkers learn from a co-worker or supervisor
   d. on a self-study basis to allow craftworkers to proceed at their own pace

3. One way to provide effective training is to _____.

   a. avoid giving negative feedback until trainees are more experienced in doing the task
   b. tailor the training to the career goals and needs of trainees
   c. choose one training method and use it for all trainees
   d. encourage trainees to listen, saving their questions for the end of the session

4. One way to prevent sexual harassment in the workplace is to _____.

   a. require employee training in which the potentially offensive subject of stereotypes is carefully avoided
   b. develop a consistent policy with appropriate consequences for engaging in sexual harassment
   c. communicate to workers that the victim of sexual harassment is the one who is being directly harassed, not those affected in a more indirect way
   d. educate workers to recognize sexual harassment for what it is—unwelcome conduct by the opposite sex

5. Employers can minimize all types of workplace discrimination by hiring based on a consistent list of job-related requirements.

    a. True
    b. False

6. Members tend to function best within an organization when they _____.

    a. are allowed to select their own style of clothing for each project
    b. understand their role
    c. do not disagree with the statements of other workers or supervisors
    d. are able to work without supervision

7. A formal organization is defined as a group of individuals who work independently, but share the same goal

    a. True
    b. False

8. A formal organization uses an organizational chart to _____.

    a. depict all companies with which it conducts business
    b. show all customers with which it conducts business
    c. track projects between departments
    d. show the relationships among the existing positions in the company

9. Which of the following is a function typically performed by the operations department of a company?

    a. Purchase materials
    b. Plan projects
    c. Prepare payrolls
    d. Recruiting and screening new hires

10. The company department that manages employee benefits and personnel recruiting is _____.

    a. Engineering
    b. Human Resources
    c. Purchasing
    d. Contract Administration

11. The power to make decisions and act on them in carrying out an assignment is _____.

    a. delegating
    b. responsibility
    c. decisiveness
    d. authority

12. Accountability is defined as _____.

    a. the power to act or make decisions in carrying out assignments
    b. giving an employee a particular task to perform
    c. the act of an employee responsible for the completion and results of a particular duty
    d. having the power to promote someone

13. A good job description should include _____.

    a. a complete organization chart
    b. any information needed to judge job performance
    c. the company dress code
    d. the company's sexual harassment policy

14. The purpose of a policy is to _____.

    a. establish company guidelines regarding a particular activity
    b. specify what tools and equipment are required for a job
    c. list all information necessary to judge an employee's performance
    d. inform employees about the future plans of the company

15. One example of a procedure would be the rules for taking time off.

    a. True
    b. False

SECTION TWO
LEADERSHIP SKILLS

Objectives

Upon completion of this section, you should be able to:

1. Describe the role of a crew leader.
2. List the characteristics of effective leaders.
3. Be able to discuss the importance of ethics in a supervisor's role.
4. Identify the three styles of leadership.
5. Describe the forms of communication.
6. Describe the four parts of verbal communication.
7. Describe the importance of active listening.
8. Explain how to overcome the barriers to communication.
9. List ways that leaders can motivate their employees.
10. Explain the importance of delegating and implementing policies and procedures.
11. Distinguish between problem solving and decision making.

1.0.0 INTRODUCTION TO LEADERSHIP

For the purpose of this program, it is important to define some of the positions that will be discussed. The term *craftworker* refers to a person who performs the work of his or her trade(s). The crew leader is a person who supervises one or more craftworkers on a crew. A superintendent is essentially an on-site supervisor who is responsible for one or more crew leaders or front-line supervisors. Finally, a project manager or general superintendent may be responsible for managing one or more projects. This training will concentrate primarily on the role of the crew leader.

Craftworkers and crew leaders differ in that the crew leader manages the activities that the skilled craftworkers on the crews actually perform. In order to manage a crew of craftworkers, a crew leader must have first-hand knowledge and experience in the activities being performed. In addition, he or she must be able to act directly in organizing and directing the activities of the various crew members.

This section explains the importance of developing effective leadership skills as a new crew leader. Effective ways to communicate with all levels of employees and co-workers, build teams, motivate crew members, make decisions, and resolve problems are covered in depth.

2.0.0 THE SHIFT IN WORK ACTIVITIES

The crew leader is generally selected and promoted from a work crew. The selection will often be based on that person's ability to accomplish tasks, to get along with others, to meet schedules, and to stay within the budget. The crew leader must lead the team to work safely and provide a quality product.

Making the transition from a craftworker to a crew leader can be difficult, especially when the new crew leader is in charge of supervising a group of peers. Crew leaders are no longer responsible for their work alone; rather, they are accountable for the work of an entire crew of people with varying skill levels and abilities, a multitude of personalities and work styles, and different cultural and educational backgrounds. Crew leaders must learn to put their personal relationships aside and work for the common goals of the team.

New crew leaders are often placed in charge of workers who were formerly their friends and peers on a crew. This situation can create some conflicts. For example, some of the crew may try to take advantage of the friendship by seeking special favors. They may also want to be privy to information that should be held closely. These problems can be overcome by working with the crew to set mutual performance goals and by freely communicating with them within permitted limits. Use their knowledge and strengths along with your own so that they feel like they are key players on the team.

As an employee moves from a craftworker position to the role of a crew leader, he or she will find that more hours will be spent supervising the work of others than actually performing the technical skill for which he or she has been trained. *Figure 4* represents the percentage of time craftworkers, crew leaders, superintendents, and project managers spend on technical and supervisory work as their management responsibilities increase.

The success of the new crew leader is directly related to the ability to make the transition from crew member into a leadership role.

3.0.0 BECOMING A LEADER

A crew leader must have leadership skills to be successful. Therefore, one of the primary goals of a person who wants to become a crew leader should be to develop strong leadership skills and learn to use them effectively.

There are many ways to define a leader. One straightforward definition is a person who influences other people in the achievement of a goal.

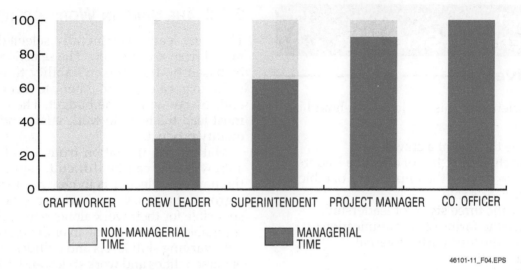

Figure 4 Percentage of time spent on technical and supervisory work.

3.1.0 Characteristics of Leaders

Leadership traits are similar to the skills that a crew leader needs in order to be effective. Although the characteristics of leadership are many, there are some definite commonalities among effective leaders.

First and foremost, effective leaders lead by example. In other words, they work and live by the standards that they establish for their crew members or followers, making sure they set a positive example.

Effective leaders also tend to have a high level of drive and determination, as well as a stick-to-it attitude. When faced with obstacles, effective leaders don't get discouraged; instead, they identify the potential problems, make plans to overcome them, and work toward achieving the intended goal. In the event of failure, effective leaders learn from their mistakes and apply that knowledge to future situations. They also learn from their successes.

Effective leaders are typically effective communicators who clearly express the goals of a project to their crew members. Accomplishing this may require that the leader overcome issues such as language barriers, gender bias, or differences in personalities to ensure that each member of the crew understands the established goals of the project.

Effective leaders have the ability to motivate their crew members to work to their full potential and become effective members of the team. Crew leaders try to develop crew member skills and encourage them to improve and learn as a means to contribute more to the team effort. Effective leaders strive for excellence from themselves and their team, so they work hard to provide the skills and leadership necessary to do so.

In addition, effective leaders must possess organizational skills. They know what needs to be accomplished, and they use their resources to make it happen. Because they cannot do it alone, leaders enlist the help of their team members to share in the workload. Effective leaders delegate work to their crew members, and they implement company policies and procedures to ensure that the work is completed safely, effectively, and efficiently.

Finally, effective leaders have the authority and self-confidence that allows them to make decisions and solve problems. In order to accomplish their goals, leaders must be able to calculate risks, absorb and interpret information, assess courses of action, make decisions, and assume the responsibility for those decisions.

3.1.1 Leadership Traits

There are many other traits of effective leaders. Some other major characteristics of leaders include the following:

- Ability to plan and organize
- Loyalty to their company and crew
- Ability to motivate
- Fairness

- Enthusiasm
- Willingness to learn from others
- Ability to teach others
- Initiative
- Ability to advocate an idea
- Good communication skills

3.1.2 Expected Leadership Behavior

Followers have expectations of their leaders. They look to their leaders to:

- Lead by example
- Suggest and direct
- Plan and organize the work
- Communicate effectively
- Make decisions and assume responsibility
- Have the necessary technical knowledge
- Be a loyal member of the group
- Abide by company policies and procedures

3.2.0 Functions of a Leader

The functions of a leader will vary with the environment, the group being led, and the tasks to be performed. However, there are certain functions common to all situations that the leader will be called upon to perform. Some of the major functions are:

- Organize, plan, staff, direct, and control work
- Empower group members to make decisions and take responsibility for their work
- Maintain a cohesive group by resolving tensions and differences among its members and between the group and those outside the group
- Ensure that all group members understand and abide by company policies and procedures
- Accept responsibility for the successes and failures of the group's performance
- Represent the group
- Be sensitive to the differences of a diverse workforce

3.3.0 Leadership Styles

There are three main styles of leadership. At one extreme is the autocratic or commander style of leadership, where the crew leader makes all of the decisions independently, without seeking the opinions of crew members. At the other extreme is the hands-off or facilitator style, where the crew leader empowers the employees to make decisions. In the middle is the democratic or collaborative style, where the crew leader seeks crew member opinions and makes the appropriate decisions based on their input.

The following are some characteristics of each of the three leadership styles:

Commander types:

- Expect crew members to work without questioning procedures
- Seldom seek advice from crew members
- Insist on solving problems alone
- Seldom permit crew members to assist each other
- Praise and criticize on a personal basis
- Have no sincere interest in creatively improving methods of operation or production

Partner types:

- Discuss problems with their crew members
- Listen to suggestions from crew members
- Explain and instruct
- Give crew members a feeling of accomplishment by commending them when they do a job well
- Are friendly and available to discuss personal and job-related problems

Facilitator types:

- Believe no supervision is best
- Rarely give orders
- Worry about whether they are liked by their crew members

Effective leadership takes many forms. The correct style for a particular situation or operation depends on the nature of the crew as well as the work it has to accomplish. For example, if the crew does not have enough experience for the job ahead, then a commander style may be appropriate. The autocratic style of leadership is also effective when jobs involve repetitive operations that require little decision-making.

However, if a worker's attitude is an issue, a partner style may be appropriate. In this case, providing the missing motivational factors may increase performance and result in the improvement of the worker's attitude. The democratic style of leadership is also used when the work is of a creative nature, because brainstorming and exchanging ideas with such crew members can be beneficial.

The facilitator style is effective with an experienced crew on a well-defined project.

The company must give a crew leader sufficient authority to do the job. This authority must be commensurate with responsibility, and it must be made known to crew members when they are hired so that they understand who is in charge.

A crew leader must have an expert knowledge of the activities to be supervised in order to be effective. This is important because the crew members need to know that they have someone to turn to when they have a question or a problem, when they need some guidance, or when modifications or changes are warranted by the job.

Respect is probably the most useful element of authority. Respect usually derives from being fair to employees, by listening to their complaints and suggestions, and by using incentives and rewards appropriately to motivate crew members. In addition, crew leaders who have a positive attitude and a favorable personality tend to gain the respect of their crew members as well as their peers. Along with respect comes a positive attitude from the crew members.

3.4.0 Ethics in Leadership

The crew leader should practice the highest standards of ethical conduct. Every day the crew leader has to make decisions that may have ethical implications. When an unethical decision is made, it not only hurts the crew leader, but also other workers, peers, and the company for which he or she works.

There are three basic types of ethics:

1. Business or legal
2. Professional or balanced
3. Situational

Business, or legal, ethics concerns adhering to all laws and regulations related to the issue.

Professional, or balanced, ethics relates to carrying out all activities in such a manner as to be honest and fair to everyone concerned.

Situational ethics pertains to specific activities or events that may initially appear to be a gray area. For example, you may ask yourself, "How will I feel about myself if my actions are published in the newspaper or if I have to justify my actions to my family, friends, and colleagues?"

The crew leader will often be put into a situation where he or she will need to assess the ethical consequences of an impending decision. For instance, should a crew leader continue to keep one of his or her crew working who has broken into a cold sweat due to overheated working conditions just because the superintendent says the activity is behind schedule? Or should a crew leader, who is the only one aware that the reinforcing steel placed by his or her crew was done incorrectly, correct the situation before the concrete is placed in the form? If a crew leader is ever asked to carry

through on an unethical decision, it is up to him or her to inform the superintendent of the unethical nature of the issue, and if still requested to follow through, refuse to act.

4.0.0 COMMUNICATION

Successful crew leaders learn to communicate effectively with people at all levels of the organization. In doing so, they develop an understanding of human behavior and acquire communication skills that enable them to understand and influence others.

There are many definitions of communication. Communication is the act of accurately and effectively conveying or transmitting facts, feelings, and opinions to another person. Simply stated, communication is the method of exchanging information and ideas.

Just as there are many definitions of communication, it also comes in many forms, including verbal, nonverbal, and written. Each of these forms of communication are discussed in this section.

4.1.0 Verbal Communication

Verbal communication refers to the spoken words exchanged between two or more people. Verbal communication consists of four distinct parts:

1. Sender
2. Message
3. Receiver
4. Feedback

Figure 5 depicts the relationship of these four parts within the communication process. In verbal communication, the focus is on feedback, which is used to verify that the sender's message was received as intended.

> **Did you know?**
>
> Research shows that the typical supervisor spends about 80 percent of his or her day communicating through writing, speaking, listening, or using body language. Of that time, studies suggest that approximately 20 percent of communication is written, and 80 percent involves speaking or listening.

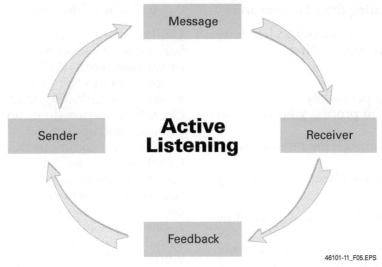

Figure 5 Communication process.

4.1.1 The Sender

The sender is the person who creates the message to be communicated. In verbal communication, the sender actually says the message aloud to the person(s) for whom it is intended.

The sender must be sure to speak in a clear and concise manner that can be easily understood by others. This is not an easy task; it takes practice. Some basic speaking tips are:

- Avoid talking with anything in your mouth (food, gum, etc.).
- Avoid swearing and acronyms.
- Don't speak too quickly or too slowly. In extreme cases, people tend to focus on the rate of speech rather than what is being communicated.
- Pronounce words carefully to prevent misunderstandings.
- Speak with enthusiasm. Avoid speaking in a harsh voice or in a monotone.

4.1.2 The Message

The message is what the sender is attempting to communicate to the audience. A message can be a set of directions, an opinion, or a feeling. Whatever its function, a message is an idea or fact that the sender wants the audience to know.

Before speaking, determine what must be communicated, then take the time to organize what to say, ensuring that the message is logical and complete. Taking the time to clarify your thoughts prevents rambling, not getting the message across effectively, or confusing the audience. It also permits the sender to get to the point quickly.

In delivering the message, the sender should assess the audience. It is important not to talk down to them. Remember that everyone, whether in a senior or junior position, deserves respect and courtesy. Therefore, the sender should use words and phrases that the audience can understand and avoid technical language or slang. In addition, the sender should use short sentences, which gives the audience time to understand and digest one point or fact at a time.

4.1.3 The Receiver

The receiver is the person to whom the message is communicated. For the communication process to be successful, it is important that the receiver understands the message as the sender intended. Therefore, the receiver must listen to what is being said.

There are many barriers to effective listening, particularly on a busy construction job site. Some of these obstacles include the following:

- Noise, visitors, cell phones, or other distractions
- Preoccupation, being under pressure, or daydreaming
- Reacting emotionally to what is being communicated
- Thinking about how to respond instead of listening
- Giving an answer before the message is complete
- Personal biases to the sender's communication style
- Finishing the sender's sentence

Some tips for overcoming these barriers are:

- Take steps to minimize or remove distractions; learn to tune out your surroundings
- Listen for key points
- Take notes
- Try not to take things personally
- Allow yourself time to process your thoughts before responding
- Let the sender communicate the message without interruption
- Be aware of your personal biases, and try to stay open-minded

There are many ways for a receiver to show that he or she is actively listening to what is being said. This can even be accomplished without saying a word. Examples include maintaining eye contact, nodding your head, and taking notes. It may also be accomplished through feedback.

4.1.4 Feedback

Feedback refers to the communication that occurs after the message has been sent by the sender and received by the receiver. It involves the receiver responding to the message.

Feedback is a very important part of the communication process because it allows the receiver to communicate how he or she interpreted the message. It also allows the sender to ensure that the message was understood as intended. In other words, feedback is a checkpoint to make sure the receiver and sender are on the same page.

The receiver can use the opportunity of providing feedback to paraphrase back what was heard. When paraphrasing what you heard, it is best to use your own words. That way, you can show the sender that you interpreted the message correctly and could explain it to others if needed.

In addition, the receiver can clarify the meaning of the message and request additional information when providing feedback. This is generally accomplished by asking questions.

One opportunity to provide feedback is in the performance of crew evaluations. Many companies have formal evaluation forms that are used on a yearly basis to evaluate workers for pay increases. These evaluations should not come as a once-a-year surprise. An effective crew leader provides constant performance feedback, which is ultimately reflected in the annual performance evaluation. It is also important to stress the importance of self-evaluation with your crew.

4.2.0 Nonverbal Communication

Unlike verbal or written communication, nonverbal communication does not involve the spoken or written word. Rather, non-verbal communication refers to things that you can actually see when communicating with others. Examples include facial expressions, body movements, hand gestures, and eye contact.

Nonverbal communication can provide an external signal of an individual's inner emotions. It occurs simultaneously with verbal communication; often, the sender of the nonverbal communication is not even aware of it.

Because it can be physically observed, nonverbal communication is just as important as the words used in conveying the message. Often, people are influenced more by nonverbal signals than by spoken words. Therefore, it is important to be conscious of nonverbal cues because you don't want the receiver to interpret your message incorrectly based on your posture or an expression on your face. After all, these things may have nothing to do with the communication exchange; instead, they may be carrying over from something else going on in your day.

4.3.0 Written or Visual Communication

Some communication will have to be written or visual. Written or visual communication refers to communication that is documented on paper or transmitted electronically using words or visuals.

Many messages on a job have to be communicated in text form. Examples include weekly reports, requests for changes, purchase orders, and correspondence on a specific subject. These items are written because they must be recorded for business and historical purposes. In addition, some communication on the job will have to be visual. Items that are difficult to explain verbally or by the written word can best be explained through diagrams or graphics. Examples include the plans or drawings used on a job.

When writing or creating a visual message, it is best to assess the reader or the audience before beginning. The reader must be able to read the message and understand the content; otherwise, the communication process will be unsuccessful. Therefore, the writer should consider the actual meaning of words or diagrams and how others might interpret them. In addition, the writer should make sure that all handwriting is legible if the message is being handwritten.

Here are some basic tips for writing:

- Avoid emotion-packed words or phrases.
- Be positive whenever possible.
- Avoid using technical language or jargon.
- Stick to the facts.
- Provide an adequate level of detail.
- Present the information in a logical manner.
- Avoid making judgments unless asked to do so.
- Proofread your work; check for spelling and grammatical errors.
- Make sure that the document is legible.
- Avoid using acronyms.
- Make sure the purpose of the message is clearly stated.
- Be prepared to provide a verbal or visual explanation, if needed.

Here are some basic tips for creating visuals:

- Provide an adequate level of detail.
- Ensure that the diagram is large enough to be seen.
- Avoid creating complex visuals; simplicity is better.
- Present the information in a logical order.
- Be prepared to provide a written or verbal explanation of the visual, if needed.

4.4.0 Communication Issues

It is important to note that each person communicates a little differently; that is what makes us unique as individuals. As the diversity of the workforce changes, communication will become even more challenging because the audience may include individuals from different ethnic groups, cultural backgrounds, educational levels, and economic status groups. Therefore, it is necessary to assess the audience in order to determine how to communicate effectively with each individual.

The key to effective communication is to acknowledge that people are different and to be able to adjust the communication style to meet the needs of the audience or the person on the receiving end of your message. This involves relaying the message in the simplest way possible, avoiding the use of words that people may find confusing. Be aware of how you use technical language, slang, jargon, and words that have multiple meanings. Present the information in a clear, concise manner. Avoid rambling and always speak clearly, using good grammar.

In addition, be prepared to communicate the message in multiple ways or adjust your level of detail or terminology to ensure that everyone understands the meaning as intended. For instance, a visual person who cannot comprehend directions in a verbal or written form may need a map. It may be necessary to overcome language barriers on the job site by using graphics or visual aids to relay the message.

Figure 6 shows how to tailor the message to the audience.

VERBAL INSTRUCTIONS Experienced Crew	VERBAL INSTRUCTIONS Newer Crew	WRITTEN INSTRUCTIONS	DIAGRAM/MAP
"Please drive to the supply shop to pick up our order."	"Please drive to the supply shop. Turn right here and left at Route 1. It's at 75th Street and Route 1. Tell them the company name and that you're there to pick up our order."	1. Turn right at exit. 2. Drive 2 miles to Route 1. Turn LEFT. 3. Drive 1 mile (pass the tire shop) to 75th Street. 4. Look for supply store on right. . . .	

Different people learn in different ways. Be sure to communicate so you can be understood.

46101-11_F06.EPS

Figure 6 Tailor your message.

Read the following verbal conversations, and identify any problems:

Conversation I:

Judy: Hey, Roger...

Roger: What's up?

Judy: Has the site been prepared for the job trailer yet?

Roger: Job trailer?

Judy: The job trailer—it's coming in today. What time will the job site be prepared?

Roger: The trailer will be here about 1:00 PM.

Judy: The job site! What time will the job site be prepared?

_____

_____

_____

_____

_____

_____

Conversation II:

John: Hey, Mike, I need your help.

Mike: What is it?

John: You and Joey go over and help Al's crew finish laying out the site.

Mike: Why me? I can't work with Joey. He can't understand a word I say.

John: Al's crew needs some help, and you and Joey are the most qualified to do the job.

Mike: I told you, I can't work with Joey.

_____

_____

_____

_____

_____

_____

Conversation III:

Ed: Hey, Jill.

Jill: Sir?

Ed: Have you received the latest DOL, EEO requirement to be sure the OFCP administrator finds our records up to date when he reviews them in August?

Jill: DOL, EEO, and OFCP?

Ed: Oh, and don't forget the MSHA, OSHA, and EPA reports are due this afternoon.

Jill: MSHA, OSHA, and EPA?

_____

_____

_____

_____

_____

_____

_____

Conversation IV:

Susan: Hey, Bob, would you do me a favor?

Bob: Okay, Sue. What is it?

Susan: I was reading the concrete inspection report and found the concrete in Bays 4A, 3B, 6C, and 5D didn't meet the 3,000 psi strength requirements. Also, the concrete inspector on the job told me the two batches that came in today had to be refused because they didn't meet the slump requirements as noted on page 16 of the spec. I need to know if any placement problems happened on those bays, how long the ready mix trucks were waiting today, and what we plan to do to stop these problems in the future.

_____

_____

_____

_____

_____

_____

Read the following written memos, and identify any problems:

Memo I:

Let's start with the transformer vault $285.00 due. For what you ask? Answer: practically nothing I admit, but here is the story. Paul the superintendent decided it was not the way good ole Comm Ed wanted it, we took out the ladder and part of the grading (as Paul instructed us to do) we brought it back here to change it. When Comm Ed the architect or Doe found out that everything would still work the way it was, Paul instructed us to reinstall the work. That is the whole story there is please add the $285.00 to my next payout.

_____
_____
_____
_____
_____
_____
_____
_____

Memo II:

Let's take rooms C 307-C-312 and C-313 we made the light track supports and took them to the job to erect them when we tried to put them in we found direct work in the way, my men spent all day trying to find out what to do so ask your Superintendent (Frank) he will verify seven hours pay for these men as he went back and forth while my men waited. Now the Architect has changed the system of hanging and has the gall to say that he has made my work easier, I can't see how. Anyway, we want an extra two (2) men for seven (7) hours for April 21 at $55.00 per hour or $385.00 on April 28th Doe Reference 197 finally resolved this problem. We will have no additional charges on Doe Reference 197, please note.

_____
_____
_____
_____
_____
_____

5.0.0 MOTIVATION

The ability to motivate others is a key skill that leaders must develop. Motivation is the ability to influence. It also describes the amount of effort that a person is willing to put forth to accomplish something. For example, a crew member who skips breaks and lunch in an effort to complete a job on time is thought to be highly motivated, but a crew member who does the bare minimum or just enough to keep his or her job is considered unmotivated.

Employee motivation has dimension because it can be measured. Examples of how motivation can be measured include determining the level of absenteeism, the percentage of employee turnover, and the number of complaints, as well as the quality and quantity of work produced.

5.1.0 Employee Motivators

Different things motivate different people in different ways. Consequently, there is no one-size-fits-all approach to motivating crew members. It is im-

portant to recognize that what motivates one crew member may not motivate another. In addition, what works to motivate a crew member once may not motivate that same person again in the future.

Frequently, the needs that motivate individuals are the same as those that create job satisfaction. They include the following:

- Recognition and praise
- Accomplishment
- Opportunity for advancement
- Job importance
- Change
- Personal growth
- Rewards

A crew leader's ability to satisfy these needs increases the likelihood of high morale within a crew. Morale refers to an individual's attitude toward the tasks he or she is expected to perform. High morale, in turn, means that employees will be motivated to work hard, and they will have a positive attitude about coming to work and doing their jobs.

5.1.1 Recognition and Praise

Recognition and praise refer to the need to have good work appreciated, applauded, and acknowledged by others. This can be accomplished by simply thanking employees for helping out on a project, or it can entail more formal praise, such as an award for Employee of the Month.

Some tips for giving recognition and praise include the following:

- Be available on the job site so that you have the opportunity to witness good work.
- Know good work and praise it when you see it.
- Look for good work and look for ways to praise it.
- Give recognition and praise only when truly deserved; otherwise, it will lose its meaning.
- Acknowledge satisfactory performance, and encourage improvement by showing confidence in the ability of the crew members to do above-average work.

5.1.2 Accomplishment

Accomplishment refers to a worker's need to set challenging goals and achieve them. There is nothing quite like the feeling of achieving a goal, particularly a goal one never expected to accomplish in the first place.

Crew leaders can help their crew members attain a sense of accomplishment by encouraging them to develop performance plans, such as goals for the year that will be used in performance evaluations. In addition, crew leaders can provide the support and tools (such as training and coaching) necessary to help their crew members achieve these goals.

5.1.3 Opportunity for Advancement

Opportunity for advancement refers to an employee's need to gain additional responsibility and develop new skills and abilities. It is important that employees know that they are not limited to their current jobs. Let them know that they have a chance to grow with the company and to be promoted as recognition for excelling in their work.

Effective leaders encourage their crew members to work to their full potentials. In addition, they share information and skills with their employees in an effort to help them to advance within the organization.

5.1.4 Job Importance

Job importance refers to an employee's need to feel that his or her skills and abilities are valued and make a difference. Employees who do not feel valued tend to have performance and attendance issues. Crew leaders should attempt to make every crew member feel like an important part of the team, as if the job wouldn't be possible without their help.

5.1.5 Change

Change refers to an employee's need to have variety in work assignments. Change is what keeps things interesting or challenging. It prevents the boredom that results from doing the same task day after day with no variety.

5.1.6 Personal Growth

Personal growth refers to an employee's need to learn new skills, enhance abilities, and grow as a person. It can be very rewarding to master a new competency on the job. Similar to change, personal growth prevents the boredom associated with doing the same thing day after day without developing any new skills.

Crew leaders should encourage the personal growth of their employees as well as themselves. Learning should be a two-way street on the job site; crew leaders should teach their crew members and learn from them as well. In addition, crew members should be encouraged to learn from each other.

5.1.7 Rewards

Rewards are compensation for hard work. Rewards can include a crew member's base salary or go beyond that to include bonuses or other incentives. They can be monetary in nature (salary raises, holiday bonuses, etc.), or they can be nonmonetary, such as free merchandise (shirts, coffee mugs, jackets, etc.) or other prizes. Attendance at training courses can be another form of reward.

5.2.0 Motivating Employees

To increase motivation in the workplace, crew leaders must individualize how they motivate different crew members. It is important that crew leaders get to know their crew members and determine what motivates them as individuals. Once again, as diversity increases in the workforce, this becomes even more challenging; therefore, effective communication skills are essential.

Here is a list of some tips for motivating employees:

- Keep jobs challenging and interesting. Boredom is a guaranteed de-motivator.
- Communicate your expectations. People need clear goals in order to feel a sense of accomplishment when the goals are achieved.
- Involve the employees. Feeling that their opinions are valued leads to pride in ownership and active participation.
- Provide sufficient training. Give employees the skills and abilities they need to be motivated to perform.
- Mentor the employees. Coaching and supporting employees boosts their self-esteem, their self-confidence, and ultimately their motivation.
- Lead by example. Become the kind of leader employees admire and respect, and they will be motivated to work for you.
- Treat employees well. Be considerate, kind, caring, and respectful; treat employees the way that you want to be treated.
- Avoid using scare tactics. Threatening employees with negative consequences can backfire, resulting in employee turnover instead of motivation.
- Reward your crew for doing their best by giving them easier tasks from time to time. It is tempting to give your best employees the hardest or dirtiest jobs because you know they will do the jobs correctly.
- Reward employees for a job well done.

Participant Exercise B

You are the crew leader of a masonry crew. Sam Williams is the person whom the company holds responsible for ensuring that equipment is operable and distributed to the jobs in a timely manner.

Occasionally, disagreements with Sam have resulted in tools and equipment arriving late. Sam, who has been with the company 15 years, resents having been placed in the job and feels that he outranks all the crew leaders.

Sam figured it was about time he talked with someone about the abuse certain tools and other items of equipment were receiving on some of the jobs. Saws were coming back with guards broken and blades chewed up, bits were being sheared in half, motor housings were bent or cracked, and a large number of tools were being returned covered with mud. Sam was out on your job when he observed a mason carrying a portable saw by the cord. As he watched, he saw the mason bump the swinging saw into a steel column. When the man arrived at his workstation, he dropped the saw into the mud.

You are the worker's crew leader. Sam approached as you were coming out of the work trailer. He described the incident. He insisted, as crew leader, you are responsible for both the work of the crew and how its members use company property. Sam concluded, "You'd better take care of this issue as soon as possible! The company is sick and tired of having your people mess up all the tools!"

You are aware that some members of your crew have been mistreating the company equipment.

1. How would you respond to Sam's accusations?

_____

_____

_____

_____

2. What action would you take regarding the misuse of the tools?

_____

_____

_____

_____

3. How can you motivate the crew to take better care of their tools? Explain.

_____

_____

_____

_____

 Fundamentals of Crew Leadership

6.0.0 TEAM BUILDING

Organizations are making the shift from the traditional boss-worker mentality to one that promotes teamwork. The manager becomes the team leader, and the workers become team members. They all work together to achieve the common goals of the team.

There are a number of benefits associated with teamwork. They include the ability to complete complex projects more quickly and effectively, higher employee satisfaction, and a reduction in turnover.

6.1.0 Successful Teams

Successful teams are made up of individuals who are willing to share their time and talents in an effort to reach a common goal—the goal of the team. Members of successful teams possess an *Us* or *We* attitude rather than an *I* or *You* attitude; they consider what's best for the team and put their egos aside.

Some characteristics of successful teams include the following:

- Everyone participates and every team member counts.
- There is a sense of mutual trust and interdependence.
- Team members are empowered.
- They communicate.
- They are creative and willing to take risks.
- The team leader develops strong people skills and is committed to the team.

6.2.0 Building Successful Teams

To be successful in the team leadership role, the crew leader should contribute to a positive attitude within the team.

There are several ways in which the team leader can accomplish this. First, he or she can work with the team members to create a vision or purpose of what the team is to achieve. It is important that every team member is committed to the purpose of the team, and the team leader is instrumental in making this happen.

Team leaders within the construction industry are typically assigned a crew. However, it can be beneficial for the team leader to be involved in selecting the team members. Selection should be based on a willingness of people to work on the team and the resources that they are able to bring to the team.

When forming a new team, team leaders should do the following:

- Explain the purpose of the team. Team members need to know what they will be doing, how long they will be doing it (if they are temporary or permanent), and why they are needed.
- Help the team establish goals or targets. Teams need a purpose, and they need to know what it is they are responsible for accomplishing.
- Define team member roles and expectations. Team members need to know how they fit into the team and what is expected of them as members of the team.
- Plan to transfer responsibility to the team as appropriate. Teams should be responsible for the tasks to be accomplished.

7.0.0 GETTING THE JOB DONE

Crew leaders must implement policies and procedures to make sure that the work is done correctly. Construction jobs have crews of people with various experiences and skill levels available to perform the work. The crew leader's job is to draw from this expertise to get the job done well and in a timely manner.

7.1.0 Delegating

Once the various activities that make up the job have been determined, the crew leader must identify the person or persons who will be responsible for completing each activity. This requires that the crew leader be aware of the skills and abilities of the people on the crew. Then, the crew leader must put this knowledge to work in matching the crew's skills and abilities to specific tasks that must be accomplished to complete the job.

After matching crew members to specific activities, the crew leader must then delegate the assignments to the responsible person(s). Delegation is generally communicated verbally by the crew leader talking directly to the person who has been assigned the activity. However, there may be times when work is assigned indirectly through written instructions or verbally through someone other than the crew leader.

When delegating work, remember to:

- Delegate work to a crew member who can do the job properly. If it becomes evident that he or she does not perform to the standard desired, either teach the crew member to do the work correctly or turn it over to someone else who can.

- Make sure the crew member understands what to do and the level of responsibility. Make sure desired results are clear, specify the boundaries and deadlines for accomplishing the results, and note the available resources.
- Identify the standards and methods of measurement for progress and accomplishment, along with the consequences of not achieving the desired results. Discuss the task with the crew member and check for understanding by asking questions. Allow the crew member to contribute feedback or make suggestions about how the task should be performed in a safe and quality manner.
- Give the crew member the time and freedom to get started without feeling the pressure of too much supervision. When making the work assignment, be sure to tell the crew member how much time there is to complete it, and confirm that this time is consistent with the job schedule.
- Examine and evaluate the result once a task is complete. Then, give the crew member some feedback as to how well it has been done. Get the crew member's comments. The information obtained from this is valuable and will enable the crew leader to know what kind of work to assign that crew member in the future. It will also give the crew leader a means of measuring his or her own effectiveness in delegating work.

7.2.0 Implementing Policies and Procedures

Every company establishes policies and procedures that employees are expected to follow and the crew leaders are expected to implement. Company policies and procedures are essentially guides for how the organization does business. They can also reflect organizational philosophies such as putting safety first or making the customer the top priority. Examples of policies and procedures include safety guidelines, credit standards, and billing processes.

Here are some tips for implementing policies and procedures:

- Learn the purpose of each policy. That way, you can follow it and apply it properly and fairly.
- If you're not sure how to apply a company policy or procedure, check the company manual or ask your supervisor.
- Apply company policies and procedures. Remember that they combine what's best for the customer and the company. In addition, they provide direction on how to handle specific situations and answer questions.

- If you are uncertain how to apply a policy, check with your supervisor.

Crew leaders may need to issue orders to their crew members. Basically, an order initiates, changes, or stops an activity. Orders may be general or specific, written or oral, and formal or informal. The decision of how an order will be issued is up to the crew leader, but it is governed by the policies and procedures established by the company.

When issuing orders:

- Make them as specific as possible.
- Avoid being general or vague unless it is impossible to foresee all of the circumstances that could occur in carrying out the order.
- Recognize that it is not necessary to write orders for simple tasks unless the company requires that all orders be written.
- Write orders for more complex tasks that will take considerable time to complete or orders that are permanent.
- Consider what is being said, the audience to whom it applies, and the situation under which it will be implemented to determine the appropriate level of formality for the order.

8.0.0 PROBLEM SOLVING AND DECISION MAKING

Problem solving and decision making are a large part of every crew leader's daily work. There will always be problems to be resolved and decisions to be made, especially in fast-paced, deadline-oriented industries.

8.1.0 Decision Making vs. Problem Solving

Sometimes, the difference between decision making and problem solving is not clear. Decision making refers to the process of choosing an alternative course of action in a manner appropriate for the situation. Problem solving involves determining the difference between the way things are and the way things should be, and finding out how to bring the two together. The two activities are interrelated because in order to make a decision, you may also have to use problem-solving techniques.

8.2.0 Types of Decisions

Some decisions are routine or simple. Such decisions can be made based on past experiences. An example would be deciding how to get to and from work. If you've worked at the same place for a long time, you are already aware of the options for traveling to and from work (take the bus, drive a car, carpool with a co-worker, take a taxi, etc.). Based on past experiences with the options identified, you can make a decision about how best to get to and from work.

On the other hand, some decisions are more difficult. These decisions require more careful thought about how to carry out an activity by using a formal problem-solving technique. An example is planning a trip to a new vacation spot. If you are not sure how to get there, where to stay, what to see, etc., one option is to research the area to determine the possible routes, hotel accommodations, and attractions. Then, you will have to make a decision about which route to take, what hotel to choose, and what sites to visit, without the benefit of direct past experience.

8.3.0 Problem Solving

The ability to solve problems is an important skill in any workplace. It's especially important for craftworkers, whose workday is often not predictable or routine. In this section, you will learn a five-step process for solving problems, which you can apply to both workplace and personal issues. Review the following steps and then see how they can be applied to a job-related problem. Keep in mind that a problem will not be solved until everyone involved admits that there is a problem.

Step 1 **Define the problem.** This isn't as easy as it sounds. Thinking through the problem often uncovers additional problems.

Step 2 **Think about different ways to solve the problem.** There is often more than one solution to a problem, so you must think through each possible solution and pick the best one. The best solution might be taking parts of two different solutions and combining them to create a new solution.

Step 3 **Pick the solution that seems best and figure out an action plan.** It is best to receive input both from those most affected by the problem and from those who will be most affected by any potential solution.

Step 4 **Test the solution to determine whether it actually works.** Many solutions sound great in theory but in practice don't turn out to be effective. On the other hand, you might discover from trying to apply

a solution that it is acceptable with a little modification. If a solution does not work, think about how you could improve it, and then implement your new plan.

Step 5 **Evaluate the process.** Review the steps you took to discover and implement the solution. Could you have done anything better? If the solution turns out to be satisfactory, you can add the solution to your knowledge base.

Next, you will see how to apply the problem-solving process to a workplace problem. Read the following situation and apply the five-step problem-solving process to come up with a solution to the issues posed by the situation.

Situation:

You are part of a team of workers assigned to a new shopping mall project. The project will take about 18 months to complete. The only available parking is half a mile from the job site. The crew has to carry heavy toolboxes and safety equipment from their cars and trucks to the work area at the start of the day, and then carry them back at the end of their shifts.

Step 1 **Define the problem.** Workers are wasting time and energy hauling all their equipment to and from the work site.

Step 2 **Think about different ways to solve the problem.** Several solutions have been proposed:
- Install lockers for tools and equipment closer to the work site.
- Have workers drive up to the work site to drop off their tools and equipment before parking.
- Bring in another construction trailer where workers can store their tools and equipment for the duration of the project.
- Provide a round-trip shuttle service to ferry workers and their tools.

> **NOTE**
>
> Each solution will have pros and cons, so it is important to receive input from the workers affected by the problem. For example, workers will probably object to any plan (like the drop-off plan) that leaves their tools vulnerable to theft.

Step 3 **Pick the solution that seems best and figure out an action plan.** The workers decide that the shuttle service makes the most sense. It should solve the time and energy problem, and workers can keep their tools with them. To put the plan into effect, the project supervisor arranges for a large van and driver to provide the shuttle service.

Step 4 **Test the solution to determine whether it actually works.** The solution works, but there is a problem. All the workers are scheduled to start and leave at the same time, so there is not enough room in the van for all the workers and their equipment. To solve this problem, the supervisor schedules trips spaced 15 minutes apart. The supervisor also adjusts worker schedules to correspond with the trips. That way, all the workers will not try to get on the shuttle at the same time.

Step 5 **Evaluate the process.** This process gave both management and workers a chance to express an opinion and discuss the various solutions. Everyone feels pleased with the process and the solution.

8.4.0 Special Leadership Problems

Because they are responsible for leading others, it is inevitable that crew leaders will encounter problems and be forced to make decisions about how to respond to the problem. Some problems will be relatively simple to resolve, like covering for a sick crew member who has taken a day off from work. Other problems will be complex and much more difficult to handle.

Some complex problems are relatively common. A few of the major employee problems include:

- Inability to work with others
- Absenteeism and turnover
- Failure to comply with company policies and procedures

8.4.1 Inability to Work with Others

Crew leaders will sometimes encounter situations where an employee has a difficult time working with others on the crew. This could be a result of personality differences, an inability to communicate, or some other cause. Whatever the reason, the crew leader must address the issue and get the crew working as a team.

The best way to determine the reason for why individuals don't get along or work well together is to talk to the parties involved. The crew leader should speak openly with the employee, as well as the other individual(s) to uncover the source of the problem and discuss its resolution.

Once the reason for the conflict is found, the crew leader can determine how to respond. There may be a way to resolve the problem and get the workers communicating and working as a team again. On the other hand, there may be nothing that can be done that will lead to a harmonious solution. In this case, the crew leader would either have to transfer the employee to another crew or have the problem crew member terminated. This latter option should be used as a last measure and should be discussed with one's superiors or Human Resources Department.

8.4.2 Absenteeism and Turnover

Absenteeism and turnover are big problems. Without workers available to do the work, jobs are delayed, and money is lost.

Absenteeism refers to workers missing their scheduled work time on a job. Absenteeism has many causes, some of which are inevitable. For instance, people get sick, they have to take time off for family emergencies, and they have to attend family events such as funerals. However, there are some causes of absenteeism that can be prevented by the crew leader.

The most effective way to control absenteeism is to make the company's policy clear to all employees. Companies that do this find that chronic absenteeism is reduced. New employees should have the policy explained to them. This explanation should include the number of absences allowed and the reasons for which sick or personal days can be taken. In addition, all workers should know how to inform their crew leaders when they miss work and understand the consequences of exceeding the number of sick or personal days allowed.

Once the policy on absenteeism is explained to employees, crew leaders must be sure to implement it consistently and fairly. If the policy is administered equally, employees will likely follow it. However, if the policy is not administered equally and some employees are given exceptions, then it will not be effective. Consequently, the rate of absenteeism is likely to increase.

Despite having a policy on absenteeism, there will always be employees who are chronically late or miss work. In cases where an employee abuses the absenteeism policy, the crew leader should discuss the situation directly with the employee. The crew leader should confirm that the employee understands the company's policy and insist that the employee comply with it. If the employee's behavior continues, disciplinary action may be in order.

Turnover refers to the loss of an employee that is initiated by that employee. In other words, the employee quits and leaves the company to work elsewhere or is fired for cause.

Like absenteeism, there are some causes of turnover that cannot be prevented and others that can. For instance, it is unlikely that a crew leader could keep an employee who finds a job elsewhere earning twice as much money. However, crew leaders can prevent some employee turnover situations. They can work to ensure safe working conditions for their crew, treat their workers fairly and consistently, and help promote good working conditions. The key is communication. Crew leaders need to know the problems if they are going to be able to successfully resolve them.

Some of the major causes of turnover include the following:

- Unfair/inconsistent treatment by the immediate supervisor
- Unsafe project sites
- Lack of job security

For the most part, the actions described for absenteeism are also effective for reducing turnover. Past studies have shown that maintaining harmonious relationships on the job site goes a long way in reducing both turnover and absenteeism. This requires effective leadership on the part of the crew leader.

8.4.3 Failure to Comply With Company Policies and Procedures

Policies are the rules that define the relationship between the company, its employees, its clients, and its subcontractors. Procedures are the instructions for carrying out the policies. Some companies have dress codes that are reflected in their policies. The dress code may be partly to ensure safety, and partly to define the image a company wants to project to the outside world.

Companies develop procedures to ensure that everyone who performs a task does it safely and efficiently. Many procedures deal with safety. A lockout/tagout procedure is an example. In this procedure, the company defines who may perform a lockout, how it is done, and who has the authority to remove or override it. Workers who fail to follow the procedure endanger themselves, as well as their co-workers.

Among a typical company's policies is the policy on disciplinary action. This policy defines steps to be taken in the event that an employee violates the company's policies or procedures. The steps range from counseling by a supervisor for the first offense, to a written warning, to dismissal for repeat offenses. This will vary from one company to another. For example, some companies will fire an employee for any violation of safety procedures.

The crew leader has the first-line responsibility for enforcing company policies and procedures. The crew leader should take the time with a new crew member to discuss the policies and procedures and show the crew member how to access them. If a crew member shows a tendency to neglect a policy or procedure, it is up to the crew leader to counsel that individual. If the crew member continues to violate a policy or procedure, the crew leader has no choice but to refer that individual to the appropriate authority within the company for disciplinary action.

Case I:

On the way over to the job trailer, you look up and see a piece of falling scrap heading for one of the laborers. Before you can say anything, the scrap material hits the ground about five feet in front of the worker. You notice the scrap is a piece of conduit. You quickly pick it up, assuring the worker you will take care of this matter.

Looking up, you see your crew on the third floor in the area from which the material fell. You decide to have a talk with them. Once on the deck, you ask the crew if any of them dropped the scrap. The men look over at Bob, one of the electricians in your crew. Bob replies, "I guess it was mine. It slipped out of my hand."

It is a known fact that the Occupational Safety and Health Administration (OSHA) regulations state that an enclosed chute of wood shall be used for material waste transportation from heights of 20 feet or more. It is also known that Bob and the laborer who was almost hit have been seen arguing lately.

1. Assuming Bob's action was deliberate, what action would you take?

_____

_____

_____

2. Assuming the conduit accidentally slipped from Bob's hand, how can you motivate him to be more careful?

_____

_____

_____

3. What follow-up actions, if any, should be taken relative to the laborer who was almost hit?

_____

_____

_____

4. Should you discuss the apparent OSHA violation with the crew? Why or why not?

_____

_____

_____

5. What acts of leadership would be effective in this case? To what leadership traits are they related?

_____

_____

_____

Case II:

Mike has just been appointed crew leader of a tile-setting crew. Before his promotion into management, he had been a tile setter for five years. His work had been consistently of superior quality.

Except for a little good-natured kidding, Mike's co-workers had wished him well in his new job. During the first two weeks, most of them had been cooperative while Mike was adjusting to his supervisory role.

At the end of the second week, a disturbing incident took place. Having just completed some of his duties, Mike stopped by the job-site wash station. There he saw Steve and Ron, two of his old friends who were also in his crew, washing.

"Hey, Ron, Steve, you should not be cleaning up this soon. It's at least another thirty minutes until quitting time," said Mike. "Get back to your work station, and I'll forget I saw you here."

"Come off it, Mike," said Steve. "You used to slip up here early on Fridays. Just because you have a little rank now, don't think you can get tough with us." To this Mike replied, "Things are different now. Both of you get back to work, or I'll make trouble." Steve and Ron said nothing more, and they both returned to their work stations.

From that time on, Mike began to have trouble as a crew leader. Steve and Ron gave him the silent treatment. Mike's crew seemed to forget how to do the most basic activities. The amount of rework for the crew seemed to be increasing. By the end of the month, Mike's crew was behind schedule.

1. How do you think Mike should have handled the confrontation with Ron and Steve?

_____

_____

2. What do you suggest Mike could do about the silent treatment he got from Steve and Ron?

_____

_____

3. If you were Mike, what would you do to get your crew back on schedule?

_____

_____

4. What acts of leadership could be used to get the crew's willing cooperation?

_____

_____

5. To which leadership traits do they correspond?

_____

_____

1. A crew leader differs from a craftworker in that a _____.

   a. crew leader need not have direct experience in those job duties that a craftworker typically performs
   b. crew leader can expect to oversee one or more craftworkers in addition to performing some of the typical duties of the craftworker
   c. crew leader is exclusively in charge of overseeing, since performing technical work is not part of this role
   d. crew leader's responsibilities do not include being present on the job site

2. Among the many traits effective leaders should have is _____.

   a. the ability to communicate the goals of a project
   b. the drive necessary to carry the workload by themselves in order to achieve a goal
   c. a perfectionist nature, which ensures that they will not make useless mistakes
   d. the ability to make decisions without needing to listen to the opinions of others

3. Of the three styles of leadership, the _____ style would be effective in dealing with a craftworker's negative attitude.

   a. facilitator
   b. commander
   c. partner
   d. dictator

4. One way to overcome barriers to effective communication is to _____.

   a. avoid taking notes on the content of the message, since this can be distracting
   b. avoid reacting emotionally to the message
   c. anticipate the content of the message and interrupt if necessary in order to show interest
   d. think about how to respond to the message while listening

5. Feedback is important in verbal communication because it requires the _____.

   a. sender to repeat the message
   b. receiver to restate the message
   c. sender to avoid technical jargon
   d. sender to concentrate on the message

6. A good way to motivate employees is to use a one-size-fits-all approach, since employees are members of a team with a common goal.

   a. True
   b. False

7. A crew leader can effectively delegate responsibilities by _____.

   a. refraining from evaluating the employee's performance once the task is completed, since it is a new task for the employee
   b. doing the job for the employee to make sure the task is done correctly
   c. allowing the employee to give feedback and suggestions about the task
   d. communicating information to the employee, generally in written form

8. Problem solving differs from decision making in that _____.

   a. problem solving involves identifying discrepancies between the way a situation is and the way it should be
   b. decision making involves separating facts from non-facts
   c. decision making involves eliminating differences
   d. problem solving involves determining an alternative course of action for a given situation

SAFETY

Objectives

Upon completion of this section, you will be able to:

1. Explain the importance of safety.
2. Give examples of direct and indirect costs of workplace accidents.
3. Identify safety hazards of the construction industry.
4. Explain the purpose of OSHA.
5. Discuss OSHA inspection procedures.
6. Identify the key points of a safety program.
7. List steps to train employees on how to perform new tasks safely.
8. Identify a crew leader's safety responsibilities.
9. Explain the importance of having employees trained in first aid and cardiopulmonary resuscitation (CPR).
10. Describe the indications of substance abuse.
11. List the essential parts of an accident investigation.
12. Describe ways to maintain employee interest in safety. Distinguish between company policies and procedures.

1.0.0 SAFETY OVERVIEW

Businesses lose millions of dollars every year because of on-the-job accidents. Work-related injuries, sickness, and deaths have caused untold suffering for workers and their families. Project delays and budget overruns from injuries and fatalities result in huge losses for employers, and work-site accidents erode the overall morale of the crew.

Craftworkers are exposed to hazards as part of the job. Examples of these hazards include falls from heights, working on scaffolds, using cranes in the presence of power lines, operating heavy machinery, and working on electrically-charged or pressurized equipment. Despite these hazards, experts believe that applying preventive safety measures could drastically reduce the number of accidents.

As a crew leader, one of your most important tasks is to enforce the company's safety program and make sure that all workers are performing their tasks safely. To be successful, the crew leader should:

- Be aware of the costs of accidents.
- Understand all federal, state, and local governmental safety regulations.
- Be involved in training workers in safe work methods.
- Conduct training sessions.
- Get involved in safety inspections, accident investigations, and fire protection and prevention.

Crew leaders are in the best position to ensure that all jobs are performed safely by their crew members. Providing employees with a safe working environment by preventing accidents and enforcing safety standards will go a long way towards maintaining the job schedule and enabling a job's completion on time and within budget.

1.1.0 Accident Statistics

Each day, workers in construction and industrial occupations face the risk of falls, machinery accidents, electrocutions, and other potentially fatal occupational hazards.

The National Institute of Occupational Safety and Health (NIOSH) statistics show that about 1,000 construction workers are killed on the job each year, more fatalities than in any other industry. Falls are the leading cause of deaths in the construction industry, accounting for over 40 percent of the fatalities. Nearly half of the fatal falls occurred from roofs, scaffolds, or ladders. Roofers, structural metal workers, and painters experienced the greatest number of fall fatalities.

In addition to the number of fatalities that occur each year, there are a staggering number of work-related injuries. In 2007, for example, more than 135,000 job-related injuries occurred in the construction industry. NIOSH reports that approximately 15 percent of all worker's compensation costs are spent on injured construction workers. The causes of injuries on construction sites include falls, coming into contact with electric current, fires, and mishandling of machinery or equipment. According to NIOSH, back injuries are the leading safety problem in workplaces.

Did you know?

When OSHA inspects a job site, they focus on the types of safety hazards that are most likely to cause fatal injuries. These hazards fall into the following classifications:

- Falls from elevations
- Struck-by hazards
- Caught in/between hazards
- Electrical shock hazards

2.0.0 COSTS OF ACCIDENTS

Occupational accidents are estimated to cost more than $100 billion every year. These costs affect the employee, the company, and the construction industry as a whole.

Organizations encounter both direct and indirect costs associated with workplace accidents. Examples of direct costs include workers' compensation claims and sick pay; indirect costs include increased absenteeism, loss of productivity, loss of job opportunities due to poor safety records, and negative employee morale attributed to workplace injuries. There are many other related costs involved with workplace accidents. A company can be insured against some of them, but not others. To compete and survive, companies must control these as well as all other employment-related costs.

2.1.0 Insured Costs

Insured costs are those costs either paid directly or reimbursed by insurance carriers. Insured costs related to injuries or deaths include the following:

- Compensation for lost earnings (known as worker's comp)
- Medical and hospital costs
- Monetary awards for permanent disabilities
- Rehabilitation costs
- Funeral charges
- Pensions for dependents

Insurance premiums or charges related to property damages include:

- Fire
- Loss and damage
- Use and occupancy
- Public liability
- Replacement cost of equipment, material, and structures

2.2.0 Uninsured Costs

The costs related to accidents can be compared to an iceberg, as shown in *Figure 7*. The tip of the iceberg represents direct costs, which are the visible costs. The more numerous indirect costs are not readily measureable, but they can represent a greater burden than the direct costs.

Uninsured costs related to injuries or deaths include the following:

- First aid expenses
- Transportation costs
- Costs of investigations
- Costs of processing reports
- Down time on the job site
- Costs to train replacement workers

Uninsured costs related to wage losses include:

- Idle time of workers whose work is interrupted
- Time spent cleaning the accident area
- Time spent repairing damaged equipment
- Time lost by workers receiving first aid
- Costs of training injured workers in a new career

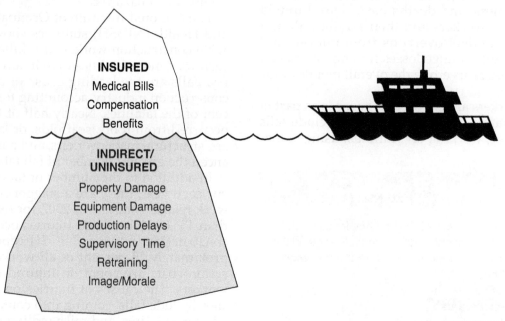

INSURED
Medical Bills
Compensation
Benefits

INDIRECT/ UNINSURED
Property Damage
Equipment Damage
Production Delays
Supervisory Time
Retraining
Image/Morale

46101-11_F07.EPS

Figure 7 Costs associated with accidents.

NCCER – *Plumbing Level Four* 46101-11

Uninsured costs related to production losses include:

- Product spoiled by accident
- Loss of skill and experience
- Lowered production or worker replacement
- Idle machine time

Associated costs may include the following:

- Difference between actual losses and amount recovered
- Costs of rental equipment used to replace damaged equipment
- Costs of new workers used to replace injured workers
- Wages or other benefits paid to disabled workers
- Overhead costs while production is stopped
- Impact on schedule
- Loss of bonus or payment of forfeiture for delays

Uninsured costs related to off-the-job activities include:

- Time spent on injured workers' welfare
- Loss of skill and experience of injured workers
- Costs of training replacement workers

Uninsured costs related to intangibles include:

- Lowered employee morale
- Increased labor conflict
- Unfavorable public relations
- Loss of bid opportunities because of poor safety records
- Loss of client goodwill

3.0.0 SAFETY REGULATIONS

To reduce safety and health risks and the number of injuries and fatalities on the job, the federal government has enacted laws and regulations, including the *Occupational Safety and Health Act of 1970*. The purpose of OSHA is "to assure so far as possible every working man and woman in the Nation safe and healthful working conditions and to preserve our human resources."

To promote a safe and healthy work environment, OSHA issues standards and rules for working conditions, facilities, equipment, tools, and work processes. It does extensive research into occupational accidents, illnesses, injuries, and deaths in an effort to reduce the number of occurrences and adverse effects. In addition, OSHA regulatory agencies conduct workplace inspections to ensure that companies follow the standards and rules.

3.1.0 Workplace Inspections

To enforce OSHA regulations, the government has granted regulatory agencies the right to enter public and private properties to conduct workplace safety investigations. The agencies also have the right to take legal action if companies are not in compliance with the Act. These regulatory agencies employ OSHA Compliance Safety and Health Officers (CSHOs), who are chosen for their knowledge in the occupational safety and health field. The CSHOs are thoroughly trained in OSHA standards and in recognizing safety and health hazards.

States with their own occupational safety and health programs conduct inspections. To do so, they enlist the services of qualified state CSHOs.

Companies are inspected for a multitude of reasons. They may be randomly selected, or they may be chosen because of employee complaints, due to an imminent danger, or as a result of major accidents or fatalities.

OSHA can assess significant financial penalties for safety violations. In some cases, a superintendent or crew leader can be held criminally liable for repeat violations.

3.2.0 Penalties for Violations

OSHA has established monetary fines for the violation of their regulations. The penalties as of 2010 are shown in *Table 1*.

In addition to the fines, there are possible criminal charges for willful violations resulting in death or serious injury. There can also be personal liability for failure to comply with OSHA regulations. The attitude of the employer and their safety history can have a significant effect on the outcome of a case.

Table 1 OSHA Penalties for Violations

Violation	Penalty
Willful Violations	Maximum $70,000
Repeated Violations	Minimum $70,000
Serious, Other-than-Serious, Other Specific Violations	Minimum $7,000
OSHA Notice Violation	$1,000
Failure to Post *OSHA 300A Summary of Work-Related Injuries and Illnesses*	$1,000
Failure to Properly Maintain *OSHA 300 Log of Work-Related Injuries and Illnesses*	$1,000
Failure to Promptly and Properly Report Fatality/Catastrophe	$5,000
Failure to Permit Access to Records Under *OSHA 1904* Regulations	$1,000
Failure to Follow Advance Notification Requirements Under *OSHA 1903.6* Regulations	$2,000
Failure to Abate – for Each Calendar Day Beyond Abatement Date	$7,000
Retaliation Against Individual for Filing OSHA Complaint	$10,000

Did you know?

Nearly half the states in the U.S. have state-run OSHA programs. These programs are set up under federal OSHA guidelines and must establish job health and safety standards that are at least as effective as the federal standards. The states have the option of adopting more stringent standards or setting standards for hazards not addressed in the federal program. Of the 22 states with state-run OSHA programs, eight of them limit their coverage to public employees.

4.0.0 EMPLOYER SAFETY RESPONSIBILITIES

Each employer must set up a safety and health program to manage workplace safety and health and to reduce work-related injuries, illnesses, and fatalities. The program must be appropriate for the conditions of the workplace. It should consider the number of workers employed and the hazards to which they are exposed while at work.

To be successful, the safety and health program must have management leadership and employee participation. In addition, training and informational meetings play an important part in effective programs. Being consistent with safety policies is the key. Regardless of the employer's responsibility, however, the individual worker is ultimately responsible for his or her own safety.

4.1.0 Safety Program

The crew leader plays a key role in the successful implementation of the safety program. The crew leader's attitudes toward the program set the standard for how crew members view safety. Therefore, the crew leader should follow all program guidelines and require crew members to do the same.

Safety programs should consist of the following:

- Safety policies and procedures
- Hazard identification and assessment
- Safety information and training
- Safety record system
- Accident investigation procedures
- Appropriate discipline for not following safety procedures
- Posting of safety notices

4.1.1 Safety Policies and Procedures

Employers are responsible for following OSHA and state safety standards. Usually, they incorporate OSHA and state regulations into a safety policies and procedures manual. Such a manual is presented to employees when they are hired.

Basic safety requirements should be presented to new employees during their orientation to the company. If the company has a safety manual, the new employee should be required to read it and sign a statement indicating that it is understood. If the employee cannot read, the employer should have someone read it to the employee and answer

---

any questions that arise. The employee should then sign a form stating that he or she understands the information.

It is not enough to tell employees about safety policies and procedures on the day they are hired and never mention them again. Rather, crew leaders should constantly emphasize and reinforce the importance of following all safety policies and procedures. In addition, employees should play an active role in determining job safety hazards and find ways that the hazards can be prevented and controlled.

4.1.2 Hazard Identification and Assessment

Safety policies and procedures should be specific to the company. They should clearly present the hazards of the job. Therefore, crew leaders should also identify and assess hazards to which employees are exposed. They must also assess compliance with OSHA and state standards.

To identify and assess hazards, OSHA recommends that employers conduct inspections of the workplace, monitor safety and health information logs, and evaluate new equipment, materials, and processes for potential hazards before they are used.

Crew leaders and employees play important roles in identifying hazards. It is the crew leader's responsibility to determine what working conditions are unsafe and to inform employees of hazards and their locations. In addition, they should encourage their crew members to tell them about hazardous conditions. To accomplish this, crew leaders must be present and available on the job site.

The crew leader also needs to help the employee be aware of and avoid the built-in hazards to which craftworkers are exposed. Examples include working at elevations, working in confined spaces such as tunnels and underground vaults, on caissons, in excavations with earthen walls, and other naturally dangerous projects.

In addition, the crew leader can take safety measures, such as installing protective railings to prevent workers from falling from buildings, as well as scaffolds, platforms, and shoring.

4.1.3 Safety Information and Training

The employer must provide periodic information and training to new and long-term employees. This must be done as often as necessary so that all employees are adequately trained. Special training and informational sessions must be provided when safety and health information changes or workplace conditions create new hazards. It is important to note that safety-related information must be presented in a manner that the employee will understand.

Whenever a crew leader assigns an experienced employee a new task, the crew leader must ensure that the employee is capable of doing the work in a safe manner. The crew leader can accomplish this by providing safety information or training for groups or individuals.

The crew leader should do the following:

- Define the task.
- Explain how to do the task safely.
- Explain what tools and equipment to use and how to use them safely.
- Identify the necessary personal protective equipment.
- Explain the nature of the hazards in the work and how to recognize them.
- Stress the importance of personal safety and the safety of others.
- Hold regular safety training sessions with the crew's input.
- Review material safety data sheets (MSDSs) that may be applicable.

4.1.4 Safety Record Systems

OSHA regulations (CFR 29, Part 1904) require that employers keep records of hazards identified and document the severity of the hazard. The information should include the likelihood of employee exposure to the hazard, the seriousness of the harm associated with the hazard, and the number of exposed employees.

In addition, the employer must document the actions taken or plans for action to control the hazards. While it is best to take corrective action immediately, it is sometimes necessary to develop a plan for the purpose of setting priorities and deadlines and tracking progress in controlling hazards.

Employers who are subject to the recordkeeping requirements of the Occupational Safety and Health Act of 1970 must maintain a log of all recordable occupational injuries and illnesses. This is known as the OSHA 300/300A form.

An MSDS is designed to provide both workers and emergency personnel with the proper procedures for handling or working with a substance that may be dangerous. An MSDS will include information such as physical data (melting point, boiling point, flash point, etc.), toxicity, health effects, first aid, reactivity, storage, disposal, protective equipment, and spill/leak procedures. These sheets are of particular use if a spill or other accident occurs.

Any company with 11 or more employees must post an *OSHA 300A* form, *Log of Work-Related Injuries and Illnesses,* between February 1 and April 30 of each year. Employees have the right to review this form. Check your company's policies with regard to these reports.

OSHA's Form 300A (Rev. 01/2004)

Summary of Work-Related Injuries and Illnesses

Year 20____

U.S. Department of Labor
Occupational Safety and Health Administration

Form approved OMB no. 1218-0176

All establishments covered by Part 1904 must complete this Summary page, even if no work-related injuries or illnesses occurred during the year. Remember to review the Log to verify that the entries are complete and accurate before completing this summary.

Using the Log, count the individual entries you made for each category. Then write the totals below, making sure you've added the entries from every page of the Log. If you had no cases, write "0."

Employees, former employees, and their representatives have the right to review the OSHA Form 300 in its entirety. They also have limited access to the OSHA Form 301 or its equivalent. See 29 CFR Part 1904.35, in OSHA's recordkeeping rule, for further details on the access provisions for these forms.

Number of Cases

Total number of deaths	Total number of cases with days away from work	Total number of cases with job transfer or restriction	Total number of other recordable cases
(G)	(H)	(I)	(J)

Number of Days

Total number of days away from work	Total number of days of job transfer or restriction
(K)	(L)

Injury and Illness Types

Total number of...
(M)

(1) Injuries

(2) Skin disorders

(3) Respiratory conditions

(4) Poisonings

(5) Hearing loss

(6) All other illnesses

Establishment information

Your establishment name ____

Street ____

City ____ State ____ ZIP ____

Industry description (e.g., Manufacture of motor truck trailers) ____

Standard Industrial Classification (SIC), if known (e.g., 3715) ____

OR

North American Industrial Classification (NAICS), if known (e.g., 336212) ____

Employment information (If you don't have these figures, see the Worksheet on the back of this page to estimate.)

Annual average number of employees ____

Total hours worked by all employees last year ____

Sign here

Knowingly falsifying this document may result in a fine.

I certify that I have examined this document and that to the best of my knowledge the entries are true, accurate, and complete.

Company executive ____ Title ____

Phone ____ Date ____

Post this Summary page from February 1 to April 30 of the year following the year covered by the form.

Public reporting burden for this collection of information is estimated to average 58 minutes per response, including time to review the instructions, search and gather the data needed, and complete and review the collection of information. Persons are not required to respond to the collection of information unless it displays a currently valid OMB control number. If you have any comments about these estimates or any other aspects of this data collection, contact: US Department of Labor, OSHA Office of Statistical Analysis, Room N-3644, 200 Constitution Avenue, NW, Washington, DC 20210. Do not send the completed forms to this office.

46101-11_SA01.EPS

Logs must be maintained and retained for five years following the end of the calendar year to which they relate. Logs must be available (normally at the establishment) for inspection and copying by representatives of the Department of Labor, the Department of Health and Human Services, or states accorded jurisdiction under the Act. Employees, former employees, and their representatives may also have access to these logs.

4.1.5 Accident Investigation

In the event of an accident, the employer is required to investigate the cause of the accident and determine how to avoid it in the future. According to OSHA, the employer must investigate each work-related death, serious injury or illness, or incident having the potential to cause death or serious physical harm. The employer should document any findings from the investigation, as well as the action plan to prevent future occurrences. This should be done immediately, with photos or video if possible. It is important that the investigation uncover the root cause of the accident so that it can be avoided in the future. In many cases, the root cause can be traced to a flaw in the system that failed to recognize the unsafe condition or the potential for an unsafe act (*Figure 8*).

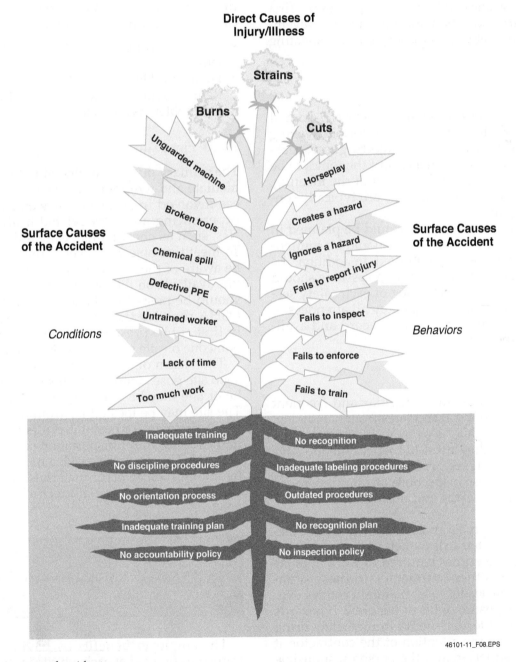

46101-11_F08.EPS

Figure 8 Root causes of accidents.

 Fundamentals of Crew Leadership

5.0.0 CREW LEADER INVOLVEMENT IN SAFETY

To be effective leaders, crew leaders must be actively involved in the safety program. Crew leader involvement includes conducting frequent safety training sessions and inspections; promoting first aid and fire protection and prevention; preventing substance abuse on the job; and investigating accidents.

5.1.0 Safety Training Sessions

A safety training session may be a brief, informal gathering of a few employees or a formal meeting with instructional films and talks by guest speakers. The size of the audience and the topics to be addressed determine the format of the meeting. Small, informal safety sessions are typically conducted weekly.

Safety training sessions should be planned in advance, and the information should be communicated to all employees affected. In addition, the topics covered in these training sessions should be timely and practical. A log of each safety session must be kept and signed by all attendees. It must then be maintained as a record and available for inspection.

5.2.0 Inspections

Crew leaders must make regular and frequent inspections to prevent accidents from happening. They must also take steps to avoid accidents. For that purpose, they need to inspect the job sites where their workers perform tasks. It is recommended that these inspections be done before the start of work each day and during the day at different times.

Crew leaders must protect workers from existing or potential hazards in their work areas. Crew leaders are sometimes required to work in areas controlled by other contractors. In these situations, the crew leader must maintain control over the safety exposure of his or her crew. If hazards exist, the crew leader should immediately bring the hazards to the attention of the contractor at fault, their superior, and the person responsible for the job site.

Crew leader inspections are only valuable if action is taken to correct potential hazards. Therefore, crew leaders must be alert for unsafe acts on their work sites. When an employee performs an unsafe action, the crew leader must explain to the employee why the act was unsafe, ask that the employee not do it again, and request cooperation in promoting a safe working environment. The crew leader must document what happened and what the employee was asked to do to correct the situation. It is then very important that crew leaders follow up to make certain the employee is complying with the safety procedures. Never allow a safety violation to go uncorrected. There are three courses of action that you, as a crew leader, can take in an unsafe situation:

• Get the appropriate party to correct the problem.
• Fix the problem yourself.
• Refuse to have the crew work in the area until the problem is corrected.

5.3.0 First Aid

The primary purpose of first aid is to provide immediate and temporary medical care to employees involved in accidents, as well as employees experiencing non-work-related health emergencies, such as chest pains or breathing difficulty. To meet this objective, every crew leader should be aware of the location and contents of first aid kits available on the job site. Emergency numbers should be posted in the job trailer. In addition, OSHA requires that at least one person trained in first aid be present at the job site at all times. Someone on site should also be trained in CPR.

The victim of an accident or sudden illness at a job site poses more problems than normal since he or she may be working in a remote location. The site may be located far from a rescue squad, fire department, or hospital, presenting a problem in the rescue and transportation of the victim to a hospital. The worker may also have been injured by falling rock or other materials, so special rescue equipment or first-aid techniques are often needed.

The employer benefits by having personnel trained in first aid at each job site in the following ways:

- The immediate and proper treatment of minor injuries may prevent them from developing into more serious conditions. As a result, medical expenses, lost work time, and sick pay may be eliminated or reduced.
- It may be possible to determine if professional medical attention is needed.
- Valuable time can be saved when a trained individual prepares the patient for treatment when professional medical care arrives. As a result, lives can be saved.

The American Red Cross, Medic First Aid, and the United States Bureau of Mines provide basic and advanced first aid courses at nominal costs. These courses include both first aid and CPR. The local area offices of these organizations can provide further details regarding the training available.

5.4.0 Fire Protection and Prevention

Fires and explosions kill and injure many workers each year, so it is important that crew leaders understand and practice fire-prevention techniques as required by company policy.

The need for protection and prevention is increasing as new building materials are introduced. Some building materials are highly flammable. They produce great amounts of smoke and gases, which cause difficulties for fire fighters, and can quickly overcome anyone present. Other materials melt when ignited and may spread over floors, preventing fire-fighting personnel from entering areas where this occurs.

OSHA has specific standards for fire safety. They require that employers provide proper exits, fire-fighting equipment, and employee training on fire prevention and safety. For more information, consult OSHA guidelines.

5.5.0 Substance Abuse

Unfortunately, drug and alcohol abuse is a continuing problem in the workplace. Drug abuse means inappropriately using drugs, whether they are legal or illegal. Some people use illegal street drugs, such as cocaine or marijuana. Others use legal prescription drugs incorrectly by taking too many pills, using other people's medications, or self-medicating. Others consume alcohol to the point of intoxication.

It is essential that crew leaders enforce company policies and procedures regarding substance abuse. Crew leaders must work with management to deal with suspected drug and alcohol abuse and should not handle these situations themselves. These cases are normally handled by the Human Resources Department or designated manager. There are legal consequences of drug and alcohol abuse and the associated safety implications. If you suspect that an employee is suffering from drug or alcohol abuse, immediately contact your supervisor and/or Human Resources Department for assistance. That way, the business and the employee's safety are protected.

It is the crew leader's responsibility to make sure that safety is maintained at all times. This may include removing workers from a work site where they may be endangering themselves or others.

For example, suppose several crew members go out and smoke marijuana or have a few beers during lunch. Then, they return to work to erect scaffolding for a concrete pour in the afternoon. If you can smell marijuana on the crew member's clothing or alcohol on their breath, you must step in and take action. Otherwise, they might cause an accident that could delay the project or cause serious injury or death to themselves or others.

It is often difficult to detect drug and alcohol abuse because the effects can be subtle. The best way is to look for identifiable effects, such as those mentioned above or sudden changes in behavior that are not typical of the employee. Some examples of such behaviors include the following:

- Unscheduled absences; failure to report to work on time
- Significant changes in the quality of work
- Unusual activity or lethargy
- Sudden and irrational temper flare-ups
- Significant changes in personal appearance, cleanliness, or health

There are other more specific signs that should arouse suspicion, especially if more than one is exhibited. Among them are:

- Slurring of speech or an inability to communicate effectively
- Shiftiness or sneaky behavior, such as an employee disappearing to wooded areas, storage areas, or other private locations
- Wearing sunglasses indoors or on overcast days to hide dilated or constricted pupils
- Wearing long-sleeved garments, particularly on hot days, to cover marks from needles used to inject drugs
- Attempting to borrow money from co-workers
- The loss of an employee's tools and company equipment

5.6.0 Job-Related Accident Investigations

There are two times when a crew leader may be involved with an accident investigation. The first time is when an accident, injury, or report of work-connected illness takes place. If present on site, the crew leader should proceed immediately to the accident location to ensure that proper first aid is being provided. He or she will also want to make sure that other safety and operational measures are taken to prevent another incident.

If mandated by company policy, the crew leader will need to make a formal investigation and submit a report after an incident. An investigation looks for the causes of the accident by examining the situation under which it occurred and talking to the people involved. Investigations are perhaps the most useful tool in the prevention of future accidents.

There are four major parts to an accident investigation. The crew leader will be concerned with each one. They are:

- Describing the accident
- Determining the cause of the accident
- Determining the parties involved and the part played by each
- Determining how to prevent re-occurrences

Case Study

For years, a prominent safety engineer was confused as to why sheet metal workers fractured their toes frequently. The crew leader had not performed thorough accident investigations, and the injured workers were embarrassed to admit how the accidents really occurred. It was later discovered they used the metal reinforced cap on their safety shoes as a "third hand" to hold the sheet metal vertically in place when they fastened it. The sheet metal was inclined to slip and fall behind the safety cap onto the toes, causing fractures. Several injuries could have been prevented by performing a proper investigation after the first accident.

6.0.0 PROMOTING SAFETY

The best way for crew leaders to encourage safety is through example. Crew leaders should be aware that their behavior sets standards for their crew members. If a crew leader cuts corners on safety, then the crew members may think that it is okay to do so as well.

The key to effectively promote safety is good communication. It is important to plan and coordinate activities and to follow through with safety programs. The most successful safety promotions occur when employees actively participate in planning and carrying out activities.

Some activities used by organizations to help motivate employees on safety and help promote safety awareness include:

- Safety training sessions
- Contests
- Recognition and awards
- Publicity

6.1.0 Safety Training Sessions

Safety training sessions can help keep workers focused on safety and give them the opportunity to discuss safety concerns with the crew. This topic was addressed in a previous section.

6.2.0 Safety Contests

Contests are a great way to promote safety in the workplace. Examples of safety-related contests include the following:

- Sponsoring housekeeping contests for the cleanest job site or work area
- Challenging employees to come up with a safety slogan for the company or department
- Having a poster contest that involves employees or their children creating safety-related posters
- Recording the number of accident-free workdays or person-hours
- Giving safety awards (hats, T-shirts, other promotional items or prizes)

One of the positive aspects of contests is their ability to encourage employee participation. It is important, however, to ensure that the contest has a valid purpose. For example, the posters or slogans created in a poster contest can be displayed throughout the organization as safety reminders.

6.3.0 Incentives and Awards

Incentives and awards serve several purposes. Among them are acknowledging and encouraging good performance, building goodwill, reminding employees of safety issues, and publicizing the importance of practicing safety standards. There are countless ways to recognize and award safety. Examples include the following:

- Supplying food at the job site when a certain goal is achieved
- Providing a reserved parking space to acknowledge someone for a special achievement
- Giving gift items such as T-shirts or gift certificates to reward employees
- Giving plaques to a department or an individual (*Figure 9*)
- Sending a letter of appreciation
- Publicly honoring a department or an individual for a job well done

Creativity can be used to determine how to recognize and award good safety on the work site. The only precautionary measure is that the award should be meaningful and not perceived as a bribe. It should be representative of the accomplishment.

6.4.0 Publicity

Publicizing safety is the best way to get the message out to employees. An important aspect of publicity is to keep the message accurate and current. Safety posters that are hung for years on end tend to lose effectiveness. It is important to keep ideas fresh.

Examples of promotional activities include posters or banners, advertisements or information on bulletin boards, payroll mailing stuffers, and employee newsletters. In addition, merchandise can be purchased that promotes safety, including buttons, hats, T-shirts, and mugs.

46101-11_F09.EPS

Figure 9 Examples of safety plaques.

Described here are three scenarios that reflect unsafe practices by craft workers. For each of these scenarios write down how you would deal with the situation, first as the crew leader of the craft worker, and then as the leader of another crew.

1. You observe a worker wearing his hard hat backwards and his safety glasses hanging around his neck. He is using a concrete saw.

   _____

   _____

   _____

   _____

   _____

   _____

2. As you are supervising your crew on the roof deck of a building under construction, you notice that a section of guard rail has been removed. Another contractor was responsible for installing the guard rail.

   _____

   _____

   _____

   _____

   _____

   _____

   _____

   _____

3. Your crew is part of plant shutdown at a power station. You observe that a worker is welding without a welding screen in an area where there are other workers.

   _____

   _____

   _____

   _____

   _____

   _____

   _____

1. One of a crew leader's responsibilities is to enforce company safety policies.

   a. True
   b. False

2. Which of the following is an indirect cost of an accident?

   a. Medical bills
   b. Production delays
   c. Compensation
   d. Employee benefits

3. A crew leader can be held criminally liable for repeat safety violations.

   a. True
   b. False

4. OSHA inspection of a business or job site _____.

   a. can be done only by invitation
   b. is done only after an accident
   c. can be conducted at random
   d. can be conducted only if a safety violation occurs

5. The *OSHA 300* form deals with _____.

   a. penalties for safety violations
   b. workplace illnesses and injuries
   c. hazardous material spills
   d. safety training sessions

6. A crew leader's responsibilities include all of the following, *except* _____.

   a. conducting safety training sessions
   b. developing a company safety program
   c. performing safety inspections
   d. participating in accident investigations

7. In order to ensure workplace safety, the crew leader should _____.

   a. hold formal safety training sessions
   b. have crew members conduct on-site safety inspections
   c. notify contractors and their supervisor of hazards in a situation where a job is being performed in an unsafe area controlled by other contractors
   d. hold crew members responsible for making a formal report and investigation following an accident

8. Prohibitions on the abuse of drugs deals only with illegal drugs such as cocaine and marijuana.

   a. True
   b. False

PROJECT CONTROL

Objectives

Upon completion of this section, you will be able to:

1. Describe the three phases of a construction project.
2. Define the three types of project delivery systems.
3. Define planning and describe what it involves.
4. Explain why it is important to plan.
5. Describe the two major stages of planning.
6. Explain the importance of documenting job site work.
7. Describe the estimating process.
8. Explain how schedules are developed and used.
9. Identify the two most common schedules.
10. Explain how the critical path method (CPM) of scheduling is used.
11. Describe the different costs associated with building a job.
12. Explain the crew leader's role in controlling costs.
13. Illustrate how to control the main resources of a job: materials, tools, equipment, and labor.
14. Explain the differences between production and productivity and the importance of each.

Performance Tasks

1. Develop and present a look-ahead schedule.
2. Develop an estimate for a given work activity

1.0.0 PROJECT CONTROL OVERVIEW

The contractor, project manager, superintendent, and crew leader each have management responsibilities for their assigned jobs. For example, the contractor's responsibility begins with obtaining the contract, and it does not end until the client takes ownership of the project. The project manager is generally the person with overall responsibility for coordinating the project. Finally, the superintendent and crew leader are responsible for coordinating the work of one or more workers, one or more crews of workers within the company and, on occasion, one or more crews of subcontractors. The crew leader directs a crew in the performance of work tasks.

This section describes methods of effective and efficient project control. It examines estimating, planning and scheduling, and resource and cost control. All the workers who participate in the job are responsible at some level for controlling cost and schedule performance and for ensuring that the project is completed according to plans and specifications.

> **NOTE**
>
> The material in this section is based largely on building-construction projects. However, the project control principles described here apply generally to all types of projects.

Construction projects are made up of three phases: the development phase, the planning phase, and the construction phase.

1.1.0 Development Phase

A new building project begins when an owner has decided to build a new facility or add to an existing facility. The development process is the first stage of planning for a new building project. This process involves land research and feasibility studies to ensure that the project has merit. Architects or engineers develop the conceptual drawings that define the project graphically. They then provide the owner with sketches of room layouts and elevations and make suggestions about what construction materials should be used.

During the development phase, an estimate for the proposed project is developed in order to establish a preliminary budget. Once that budget has been established, the financing of the project is discussed with lending institutions. The architects/engineers and/or the owner begins preliminary reviews with government agencies. These reviews include zoning, building restrictions, landscape requirements, and environmental impact studies.

Also during the development phase, the owner must analyze the project's cost and potential retun on investment (ROI) to ensure that its costs will not exceed its market value and that the project provides a good return on investment. If the project passes this test, the architect/engineer will proceed to the planning phase.

1.2.0 Planning Phase

When the architect/engineer begins to develop the project drawings and specifications, other design professionals such as structural, mechanical, and electrical engineers are brought in. They perform the calculations, make a detailed technical analysis, and check details of the project for accuracy.

Fundamentals of Crew Leadership

The design professionals create drawings and specifications. These drawings and specifications are used to communicate the necessary information to the contractors, subcontractors, suppliers, and workers that contribute to a project.

During the planning phase, the owners hold many meetings to refine estimates, adjust plans to conform to regulations, and secure a construction loan. If the project is a condominium, an office building, or a shopping center, then a marketing program must be developed. In such cases, the selling of the project is often started before actual construction begins.

Next, a complete set of drawings, specifications, and bid documents is produced. Then the owner will select the method to obtain contractors. The owner may choose to negotiate with several contractors or select one through competitive bidding. Note that safety must also be considered as part of the planning process. A safety crew leader may walk through the site as part of the pre-bid process.

Contracts can take many forms. The three basic types from which all other types are derived are firm fixed price, cost reimbursable, and guaranteed maximum price.

- *Firm fixed price* – In this type of contract, the buyer generally provides detailed drawings and specifications, which the contractor uses to calculate the cost of materials and labor. To these costs, the contractor adds a percentage representing company overhead expenses such as office rent, insurance, and accounting/payroll costs. On top of all this, the contractor adds a profit factor. When submitting the bid, the contractor will state very specifically the conditions and assumptions on which the bid is based. These conditions and assumptions form the basis from which changes can be priced. Because the price is established in advance, any changes in the job requirements once the job is started will impact the contractor's profit margin. This is where the crew leader can play an important role by identifying problems that increase the amount of labor or material that was planned. By passing this information up the chain of command, the crew leader allows the company to determine if the change is outside the scope of the bid. If so, they can submit a change order request to cover the added cost.
- *Cost reimbursable* – In this type of contract, the buyer reimburses the contractor for labor, materials, and other costs encountered in the performance of the contract. Typically, the contractor and buyer agree in advance on hourly or daily labor rates for different categories of worker.

These rates include an amount representing the contractor's overhead expense. The buyer also reimburses the contractor for the cost of materials and equipment used on the job. The buyer and contractor also negotiate a profit margin. On this type of contract, the profit margin is likely to be lower than that of a fixed-price contract because the contractor's cost risk is significantly reduced. The profit margin is often subject to incentive or penalty clauses that make the amount of profit awarded subject to performance by the contractor. Performance is usually tied to project schedule milestones.

- *Guaranteed maximum price (GMP)* – This form of contract, also called a not-to-exceed contract, is most often used on projects that have been negotiated with the owner. Involvement in the process usually includes preconstruction, and the entire team develops the parameters that define the basis for the work. In some instances, the owner will require a competitively-bid GMP. In such cases, the scope of work has not been fully defined, but bids are taken for general conditions (direct costs) and fee based on an assumed volume of work. The advantages of the GMP contract vehicle are:
  - Reduced design time
  - Allows for phased construction
  - Uses a team approach to a project
  - Reduction in changes related to incomplete drawings

1.3.0 Construction Phase

The designated contractor enlists the help of mechanical, electrical, elevator, and other specialty subcontractors to complete the construction phase. The contractor may perform one or more parts of the construction, and rely on subcontractors for the remainder of the work. However, the general contractor is responsible for managing all the trades necessary to complete the project.

As construction nears completion, the architect/engineer, owner, and government agencies start their final inspections and acceptance of the project. If the project has been managed by the general contractor, the subcontractors have performed their work, and the architect/ engineers have regularly inspected the project to ensure that local codes have been followed, then the inspection procedure can be completed in a timely manner. This results in a satisfied client and a profitable project for all.

On the other hand, if the inspection reveals faulty workmanship, poor design or use of materials, or violation of codes, then the inspection and

---

acceptance will become a lengthy battle and may result in a dissatisfied client and an unprofitable project.

Figure 10 shows the flow of a typical project.

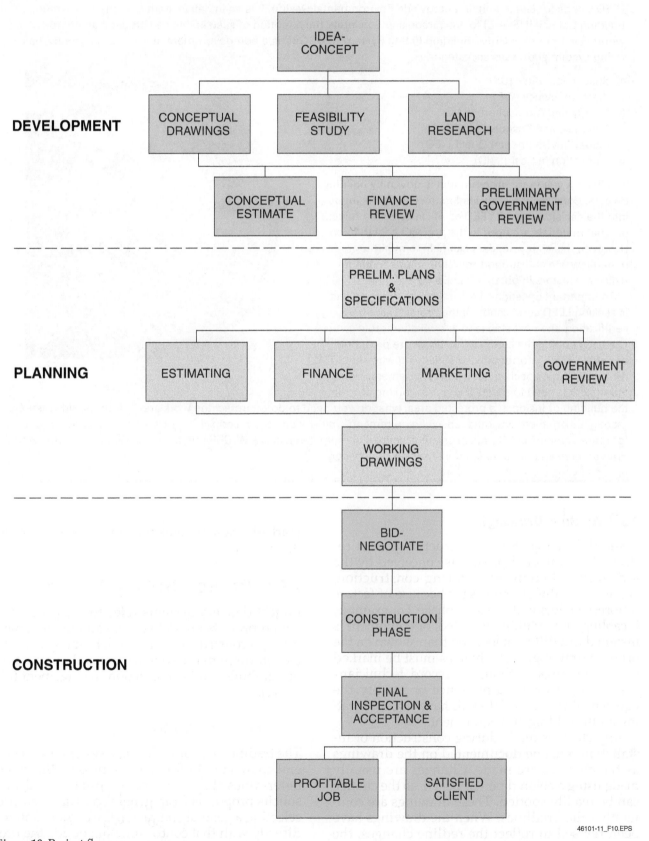

Figure 10 Project flow.

46101-11_F10.EPS

LEED stands for Leadership in Energy and Environmental Design. It is an initiative started by the U.S. Green Building Council (USGBC) to encourage and accelerate the adoption of sustainable construction standards worldwide through a Green Building Rating System™. USGBC is a non-government, not-for-profit group. Their rating system addresses six categories:

1. Sustainable Sites (SS)
2. Water Efficiency (WE)
3. Energy and Atmosphere (EA)
4. Materials and Resources (MR)
5. Indoor Environmental Quality (EQ)
6. Innovation in Design (ID)

   LEED is a voluntary program that is driven by building owners. Construction crew leaders may not have input into the decision to seek LEED certification for a project, or what materials are used in the project's construction. However, these crew leaders can help to minimize material waste and support recycling efforts, both of which are factors in obtaining LEED certification.

   An important question to ask is whether your project is seeking LEED certification. If the project is seeking certification, the next step is to ask what your role will be in getting the certification. If you are procuring materials, what information is needed and who should receive it? What specifications and requirements do the materials need to meet? If you are working outside the building or inside in a protected area, what do you need to do to protect the work area? How should waste be managed? Are there any other special requirements that will be your responsibility? Do you see any opportunities for improvement? LEED principles are described in more detail in the NCCER publications, *Your Role in the Green Environment* and *Sustainable Construction Supervisor.*

46101-11_SA02.EPS

1.3.1 As-Built Drawings

A set of drawings for a construction project reflects the completed project as conceived by the architect and engineers. During construction, changes usually are necessary because of factors unforeseen during the design phase. For example, if cabling or conduit is re-routed, or equipment is installed in a different location than shown on the original drawing, such changes must be marked on the drawings. Without this record, technicians called to perform maintenance or modify the equipment at a later date will have trouble locating all the cabling and equipment.

   Any changes made during construction or installation must be documented on the drawings as the changes are made. Changes are usually made using a colored pen or pencil, so the change can be readily spotted. These drawings are commonly called redlines. When the drawings have been revised to reflect the redline changes, the final drawings are called as-builts, and are so marked. These become the drawings of record for the project.

2.0.0 PROJECT DELIVERY SYSTEMS

Project delivery systems refer to the process by which projects are delivered, from development through construction. Project delivery systems focus on three main systems: general contracting, design-build, and construction management (*Figure 11*).

2.1.0 General Contracting

The traditional project delivery system uses a general contractor. In this type of project, the owner determines the design of the project, and then solicits proposals from general contractors. After selecting a general contractor, the owner contracts directly with that contractor, who builds the project as the prime, or controlling, contractor.

	GENERAL CONTRACTING	DESIGN-BUILD	CONSTRUCTION MANAGEMENT
OWNER	Designs project (or hires architect)	Hires general contractor	Hires construction management company
GENERAL CONTRACTOR	Builds project (with owner's design)	Involved in project design, builds project	Builds, may design (hired by construction management company)
CONSTRUCTION MANAGEMENT COMPANY			Hires and manages general contractor and architect

46101-11_F11.EPS

Figure 11 Project delivery systems.

2.2.0 Design-Build

The design-build project delivery system is different from the general contracting delivery system. In the design-build system, both the design and construction of a project are managed by a single entity. GMP contracts are commonly used in these situations.

2.3.0 Construction Management

The construction management project delivery system uses a construction manager to facilitate the design and construction of a project. Construction managers are very involved in project control; their main concerns are controlling time, cost, and the quality of the project.

3.0.0 COST ESTIMATING AND BUDGETING

Before a project is built, an estimate must be prepared. An estimate is the process of calculating the cost of a project. There are two types of costs to consider, including direct and indirect costs. Direct costs, also known as general conditions, are those that can clearly be assigned to a budget. Indirect costs are overhead costs that are shared by all projects. These costs are generally applied as an overhead percentage to labor and material costs.

Direct costs include the following:

- Materials
- Labor
- Tools
- Equipment

Indirect costs refer to overhead items such as:

- Office rent
- Utilities
- Telecommunications
- Accounting
- Office supplies, signs

The bid price includes the estimated cost of the project as well as the profit. Profit refers to the amount of money that the contractor will make after all of the direct and indirect costs have been paid. If the direct and indirect costs exceed those estimated to perform the job, the difference between the actual and estimated costs must come out of the company's profit. This reduces what the contractor makes on the job.

Profit is the fuel that powers a business. It allows the business to invest in new equipment and facilities, provide training, and to maintain a reserve fund for times when business is slow. In the case of large companies, profitability attracts investors who provide the capital necessary for the business to grow. For these reasons, contractors cannot afford to lose money on a consistent basis. Those who cannot operate profitably are forced out of business. Crew leaders can help their companies remain profitable by managing budget, schedule, quality, and safety adhering to the drawings, specifications, and project schedule.

3.1.0 The Estimating Process

The cost estimate must consider a number of factors. Many companies employ professional cost estimators to do this work. They also maintain performance data for previous projects. This data

---

can be used as a guide in estimating new projects. A complete estimate is developed as follows:

Step 1 Using the drawings and specifications, an estimator records the quantity of the materials needed to construct the job. This is called a quantity takeoff. The information is placed on a takeoff sheet like the one shown in *Figure 12*.

Step 2 Productivity rates are used to estimate the amount of labor required to complete the project. Most companies keep records of these rates for the type and size of the jobs that they perform. The company's estimating department keeps these records.

Step 3 The amount of work to be done is divided by the productivity rate to determine labor hours. For example, if the productivity rate for concrete finishing is 40 square feet per hour, and there are 10,000 square feet of concrete to be finished, then 250 hours of concrete finishing labor is required. This number would be multiplied by the hourly rate for concrete finishing to determine the cost of that labor category. If this work is subcontracted, then the subcontractor's cost estimate, raised by an overhead factor, would be used in place of direct labor cost.

Step 4 The total material quantities are taken from the quantity takeoff sheet and placed on a summary or pricing sheet, an example of which is shown in *Figure 13*. Material prices are obtained from local suppliers, and the total cost of materials is calculated.

Step 5 The cost of equipment needed for the project is determined. This number could reflect rental cost or a factor used by the company when their own equipment is to be used.

Step 6 The total cost of all resources—materials, equipment, tools, and labor—is then totaled on the summary sheet. The unit cost—the total cost divided by the total number of units of material to be put into place—can also be calculated.

Step 7 The cost of taxes, bonds, insurance, subcontractor work, and other indirect costs are added to the direct costs of the materials, equipment, tools, and labor.

Step 8 Direct and indirect costs are summed to obtain the total project cost. The contractor's expected profit is then added to that total.

> **NOTE**
>
> There are software programs available to simplify the cost estimating process. Many of them are tailored to specific industries such as construction and manufacturing, and to specific trades within the industries. For example, there are programs available for electrical and HVAC contractors. Estimating programs are typically set up to include a takeoff form and a form for estimating labor by category. Most of these programs include a data base that contains current prices for labor and materials, so they automatically price the job and produce a bid. Once the job is awarded, the programs can generate purchase orders for materials.

3.1.1 Estimating Material Quantities

The crew leader may be required to estimate quantities of materials.

A set of construction drawings and specifications is needed in order to estimate the amount of a certain type of material required to perform a job. The appropriate section of the technical specifications and page(s) of drawings should be carefully reviewed to determine the types and quantities of materials required. The quantities are then placed on the worksheet. For example, the specification section on finished carpentry should be reviewed along with the appropriate pages of drawings before taking off the linear feet of door and window trim.

If an estimate is required because not enough materials were ordered to complete the job, the estimator must also determine how much more work is necessary. Once this is known, the crew leader can then determine the materials needed. The construction drawings will also be used in this process.

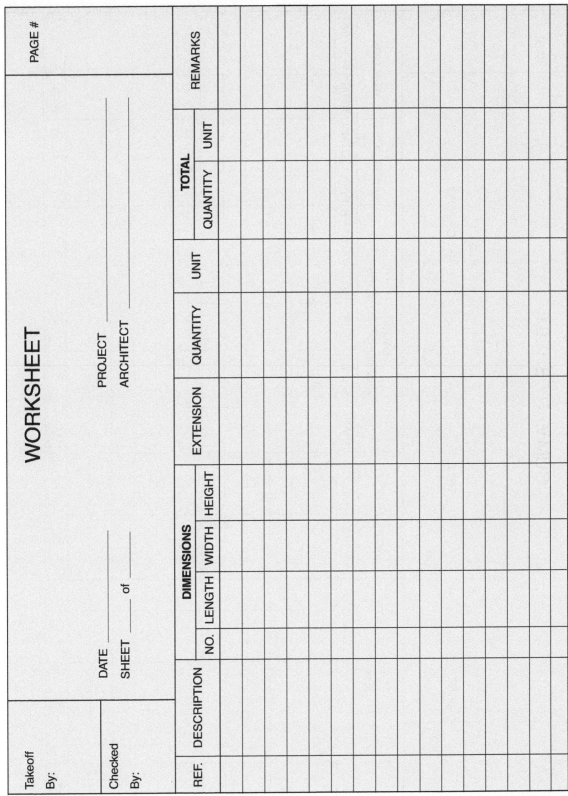

Figure 12 Quantity takeoff sheet.

46101-11_F12.EPS

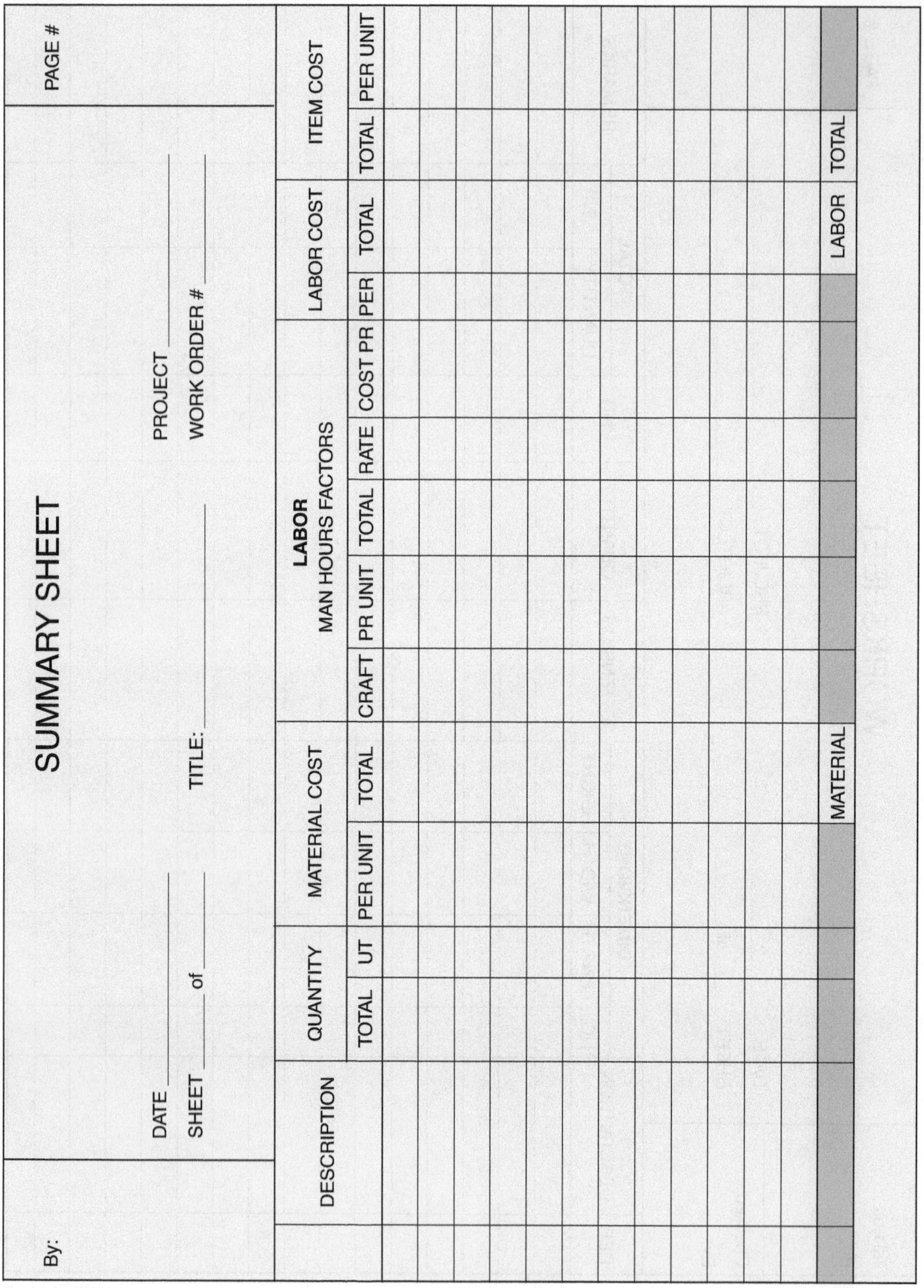

SUMMARY SHEET

By:

DATE _____

SHEET _____ of _____

PROJECT _____

WORK ORDER # _____

TITLE: _____

PAGE #

DESCRIPTION	QUANTITY		MATERIAL COST		LABOR MAN HOURS FACTORS						LABOR COST		ITEM COST	
	TOTAL	UT	PER UNIT	TOTAL	CRAFT	PR UNIT	TOTAL	RATE	COST PR	PER	TOTAL		TOTAL	PER UNIT
	MATERIAL										LABOR		TOTAL	

46101-11_F13.EPS

Figure 13 Summary sheet.

NCCER – *Plumbing Level Four* 46101-11

Assume you are the leader of a crew building footing formwork for the construction shown in *Figure 14.* You have used all of the materials provided for the job, yet you have not completed it. You study the drawings and see that the formwork consists of two side forms, each 12" high. The total length of footing for the entire project is 115'-0". You have completed 88'-0" to date; therefore, you have 27'-0" remaining (115'-0" – 88'-0" = 27'-0"). Your job is to prepare an estimate of materials that you will need to complete the job. In this case, only the side forms will be estimated (the miscellaneous materials will not be considered here).

- Footing length to complete: 27'-0"
- Footing height: 1'-0"

Refer to the worksheet in *Figure 15* for a final tabulation of the side forms needed to complete the job.

1. Using the same footing as described in the example above, calculate the quantity (square feet) of formwork needed to finish 203 linear feet of the footing. Place this information directly on the worksheet.
2. You are the crew leader of a carpentry crew whose task is to side a warehouse with plywood sheathing. The wall height is 16 feet, and there is a total of 480 linear feet of wall to side. You have done 360 linear feet of wall and have run out of materials. Calculate how many more feet of plywood you will need to complete the job. If you are using 4' × 8' plywood panels, how many will you need to order, assuming no waste? Write your estimate on the worksheet.

Show your calculations to the instructor.

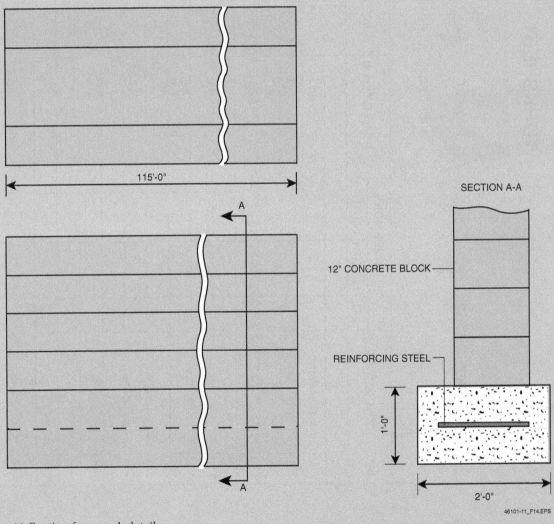

Figure 14 Footing formwork detail.

Participant Exercise E Worksheet

Takeoff By: RWH	WORKSHEET	PAGE #1
Checked By:	DATE 2/1/15 SHEET 01 of 01 PROJECT Sam's Diner ARCHITECT 654b	

REF.	DESCRIPTION	DIMENSIONS				EXTENSION	QUANTITY	UNIT	TOTAL		REMARKS
		NO	LENGTH	WIDTH	HEIGHT				QUANTITY	UNIT	
	Footing Side Forms	*2*	*27'0"*		*1'0"*	*2x27x1*	*54*	*SF*	*54*	*SF*	

46101-11_F15.EPS

Figure 15 Worksheet with entries.

4.0.0 PLANNING

Planning can be defined as determining the method used to carry out the different tasks to complete a project. It involves deciding what needs to be done and coming up with an organized sequence of events or plan for doing the work.

Planning involves the following:

- Determining the best method for performing the job
- Identifying the responsibilities of each person on the work crew
- Determining the duration and sequence of each activity
- Identifying what tools and equipment are needed to complete a job
- Ensuring that the required materials are at the work site when needed
- Making sure that heavy construction equipment is available when required

- Working with other contractors in such a way as to avoid interruptions and delays

4.1.0 Why Plan?

With a plan, a crew leader can direct work efforts efficiently and can use resources such as personnel, materials, tools, equipment, work area, and work methods to their full potential.

Some reasons for planning include the following:

- Controlling the job in a safe manner so that it is built on time and within cost
- Lowering job costs through improved productivity
- Preparing for bad weather or unexpected occurrences
- Promoting and maintaining favorable employee morale
- Determining the best and safest methods for performing the job

---

Participant Exercise F

1. In your own words, define planning, and describe how a job can be done better if it is planned. Give an example.

   _____

   _____

   _____

   _____

2. Consider a job that you recently worked on to answer the following:
   a. List the material(s) used.
   b. List each member of the crew with whom you worked and what each person did.
   c. List the kinds of equipment used.

   _____

   _____

   _____

   _____

   _____

3. List some suggestions for how the job could have been done better, and describe how you would plan for each of the suggestions.

   _____

   _____

   _____

   _____

 Fundamentals of Crew Leadership

4.2.0 Stages of Planning

There are various times when planning is done for a construction job. The two most important occur in the pre-construction phase and during the construction work.

4.2.1 Pre-Construction Planning

The pre-construction stage of planning occurs before the start of construction. Except in a fairly small company or for a relatively small job, the crew leader usually does not get directly involved in the pre-construction planning process, but it is important to understand what it involves.

There are two phases of pre-construction planning. The first is when the proposal, bid, or negotiated price for the job is being developed. This is when the estimator, the project manager, and the field superintendent develop a preliminary plan for how the work will be done. This is accomplished by applying experience and knowledge from previous projects. It involves determining what methods, personnel, tools, and equipment will be used and what level of productivity they can achieve.

The second phase occurs after the contract is awarded. This phase requires a thorough knowledge of all project drawings and specifications. During this stage, the actual work methods and resources needed to perform the work are selected. Here, crew leaders might get involved, but their planning must adhere to work methods, production rates, and resources that fit within the estimate prepared before the contract was awarded. If the project requires a method of construction different from what is normal, the crew leader will usually be informed of what method to use.

4.2.2 Construction Planning

During construction, the crew leader is directly involved in planning on a daily basis. This planning consists of selecting methods for completing tasks before beginning work. Effective planning exposes likely difficulties, and enables the crew leader to minimize the unproductive use of personnel and equipment. Effective planning also provides a gauge to measure job progress. Effective crew leaders develop what is known as look-ahead (short-term) schedules. These schedules consider actual circumstances as well as projections two to three weeks into the future. Developing a look-ahead schedule helps ensure that all resources are available on the project when needed.

> **Did you know?**
>
> A good crew leader will always have a backup plan in case circumstances prevent the original plan from working. There are many circumstances that can cause a plan to go awry, including adverse weather, equipment failure, absent crew members, and schedule slippage by other crafts.

One of the characteristics of an effective crew leader is the ability to reduce each job to its simpler parts and organize a plan for handling each task.

Project planners establish time and cost limits for the project; the crew leader's planning must fit within those constraints. Therefore, it is important to consider the following factors that may affect the outcome:

- Site and local conditions, such as soil types, accessibility, or available staging areas
- Climate conditions that should be anticipated during the project
- Timing of all phases of work
- Types of materials to be installed and their availability
- Equipment and tools required and their availability
- Personnel requirements and availability
- Relationships with the other contractors and their representatives on the job

On a simple job, these items can be handled almost automatically. However, larger or more complex jobs require the planner to give these factors more formal consideration and study.

5.0.0 THE PLANNING PROCESS

The planning process consists of the following five steps:

Step 1 Establish a goal.

Step 2 Identify the work activities that must be completed in order to achieve the goal.

Step 3 Identify the tasks that must be done to accomplish those activities.

Step 4 Communicate responsibilities.

Step 5 Follow up to see that the goal is achieved.

5.1.0 Establish a Goal

The term *goal* has different meanings for different people. In general, a goal is a specific outcome that one works toward. For example, the project superintendent of a home construction project could establish the goal to have a house dried-in by a certain date. (Dried-in means ready for the application of roofing and siding.) In order to meet that goal, the leader of the framing crew and the superintendent would need to agree to a goal to have the framing completed by a given date. The crew leader would then establish sub-goals (objectives) for the crew to complete each element of the framing (floors, walls, roof) by a set time. The superintendent would need to set similar goals with the crews that install sheathing, building wrap, windows, and exterior doors. However, if the framing crew does not meet its goal, the other crews will be delayed.

5.2.0 Identify the Work to be Done

The second step in planning is to identify the work to be done to achieve the goal. In other words, it is a series of activities that must be done in a certain sequence. The topic of breaking down a job into activities is covered later in this section. At this point, the crew leader should know that, for each activity, one or more objectives must be set.

An objective is a statement of what is desired at a specific time. An objective must:

- Mean the same thing to everyone involved
- Be measurable, so that everyone knows when it has been reached
- Be achievable
- Have everyone's full support

Examples of practical objectives include the following:

- By 4:30 P.M. today, the crew will have completed installation of the floor joists.
- By closing time Friday, the roof framing will be complete.

Notice that both examples meet the first three requirements of an objective. In addition, it is assumed that everyone involved in completing the task is committed to achieving the objective. The advantage in developing objectives for each work activity is that it allows the crew leader to evaluate whether or not the plan and schedules are being followed. In addition, objectives serve as sub-goals that are usually under the crew leader's control.

Some construction work activities, such as installing 12"-deep footing forms, are done so often that they require little planning. However, other jobs, such as placing a new type of mechanical equipment, require substantial planning. This type of job requires that the crew leader set specific objectives.

Whenever faced with a new or complex activity, take the time to establish objectives that will serve as guides for accomplishing the job. These guides can be used in the current situation, as well as in similar situations in the future.

5.3.0 Identify Tasks to be Performed

To plan effectively, the crew leader must be able to break a work activity assignment down into smaller tasks. Large jobs include a greater number of tasks than small ones, but all jobs can be broken down into manageable components.

When breaking down an assignment into tasks, make each task identifiable and definable. A task is identifiable when the types and amounts of resources it requires are known. A task is definable if it has a specific duration. For purposes of efficiency, the job breakdown should not be too detailed or complex, unless the job has never been done before or must be performed with strictest efficiency.

For example, a suitable breakdown for the work activity to install 12" × 12" vinyl floor tile in a cafeteria might be the following:

Step 1 Prepare the floor.

Step 2 Lay out the tile.

Step 3 Spread the adhesive.

Step 4 Lay the tile.

Step 5 Clean the tile.

Step 6 Wax the floor.

The crew leader could create even more detail by breaking down any one of the tasks, such as lay the tile, into subtasks. In this case, however, that much detail is unnecessary and wastes the crew leader's time and the project's money. However, breaking tasks down further might be necessary in a case where the job is very complex or the analysis of the job needs to be very detailed.

Every work activity can be divided into three general parts:

- Preparing
- Performing
- Cleaning up

One of the most frequent mistakes made in the planning process is forgetting to prepare and to clean up. The crew leader must be certain that preparation and cleanup are not overlooked.

After identifying the various tasks that make up the job and developing an objective for each task, the crew leader must determine what resources the job requires. Resources include labor, equipment, materials, and tools. In most jobs, these resources are identified in the job estimate. The crew leader must make sure that these resources are available on the site when needed.

5.4.0 Communicating Responsibilities

A crew leader is unable to complete all of the activities within a job independently. Other people must be relied upon to get everything done. Therefore, most jobs have a crew of people with various experiences and skill levels to assist in the work. The crew leader's job is to draw from this expertise to get the job done well and in a safe and timely manner.

Once the various activities that make up the job have been determined, the crew leader must identify the person or persons responsible for completing each activity. This requires that the crew leader be aware of the skills and abilities of the people on the crew. Then, the crew leader must put this knowledge to work in matching the crew's skills and abilities to specific tasks that must be performed to complete the job.

After matching crew members to specific activities, the crew leader must then communicate the assignments to the crew. Communication of responsibilities is generally handled verbally; the crew leader often talks directly to the person to which the activity has been assigned. There

may be times when work is assigned indirectly through written instructions or verbally through someone other than the crew leader. Either way, the crew members should know what it is they are responsible for accomplishing on the job.

5.5.0 Follow-Up Activities

Once the activities have been delegated to the appropriate crew members, the crew leader must follow up to make sure that they are completed effectively and efficiently. Follow-up work involves being present on the job site to make sure all the resources are available to complete the work; ensuring that the crew members are working on their assigned activities; answering any questions; and helping to resolve any problems that occur while the work is being done. In short, follow-up activity means that the crew leader is aware of what's going on at the job site and is doing whatever is necessary to make sure that the work is completed on schedule.

Figure 16 reviews the planning steps.

The crew leader should carry a small note pad or electronic device to be used for planning and note taking. That way, thoughts about the project can be recorded as they occur, and pertinent details will not be forgotten. The crew leader may also choose to use a planning form such as the one illustrated in *Figure 17*.

As the job is being built, refer back to these plans and notes to see that the tasks are being done in sequence and according to plan. This is referred to as analyzing the job. Experience shows that jobs that are not built according to work plans usually end up costing more and taking more time; therefore, it is important that crew leaders refer back to the plans periodically.

The crew leader is involved with many activities on a day-to-day basis. As a result, it is easy to forget important events if they are not documented. To help keep track of events such as job changes, interruptions, and visits, the crew leader should keep a job diary.

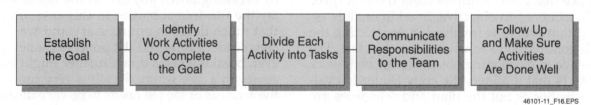

| Establish the Goal | Identify Work Activities to Complete the Goal | Divide Each Activity into Tasks | Communicate Responsibilities to the Team | Follow Up and Make Sure Activities Are Done Well |

Figure 16 Steps to effective planning.

DAILY WORK PLAN

"PLAN YOUR WORK AND WORK YOUR PLAN = EFFICIENCY"

Plan of _____   Date _____

PRIORITY	DESCRIPTION	✓ When Completed ✗ Carried Forward

46101-11_F17.EPS

Figure 17 Planning form.

A job diary is a notebook in which the crew leader records activities or events that take place on the job site that may be important later. When recording in a job diary, make sure that the information is accurate, factual, complete, consistent, organized, and up to date. Follow company policy in determining which events should be recorded. However, if there is a doubt about what to include, it is better to have too much information than too little.

Figure 18 shows a sample page from a job diary.

6.0.0 PLANNING RESOURCES

Once a job has been broken down into its tasks or activities, the next step is to assign the various resources needed to perform them.

6.1.0 Safety Planning

Using the company safety manual as a guide, the crew leader must assess the safety issues associated with the job and take necessary measures to minimize any risk to the crew. This may involve working with the company or site safety officer and may require a formal hazard analysis.

6.2.0 Materials Planning

The materials required for the job are identified during pre-construction planning and are listed on the job estimate. The materials are usually ordered from suppliers who have previously provided quality materials on schedule and within estimated cost.

July 8, 2015

Weather: Hot and Humid

Project: Company XYZ Building

- The paving contractor crew arrived late (10 am).

- The owner representative inspected the footing foundation at approximately 1 pm.

- The concrete slump test did not pass. Two trucks had to be ordered to return to the plant, causing a delay.

- John Smith had an accident on the second floor. I sent him to the doctor for medical treatment. The cause of the accident is being investigated.

46101-11_F18.EPS

Figure 18 Sample page from a job diary.

The crew leader is usually not involved in the planning and selection of materials, since this is done in the pre-construction phase. The crew leader does, however, have a major role to play in the receipt, storage, and control of the materials after they reach the job site.

The crew leader is also involved in planning materials for tasks such as job-built formwork and scaffolding. In addition, the crew leader may run out of a specific material, such as fasteners, and need to order more. In such cases, a higher authority should be consulted, since most companies have specific purchasing policies and procedures.

6.3.0 Site Planning

There are many planning elements involved in site work. The following are some of the key elements:

- Emergency procedures
- Access roads
- Parking
- Stormwater runoff
- Sedimentation control
- Material and equipment storage
- Material staging
- Site security

6.4.0 Equipment Planning

Much of the planning for use of construction equipment is done during the pre-construction phase. This planning includes the types of equipment needed, the use of the equipment, and the length of time it will be on the site. The crew leader must work with the home office to make certain that the equipment reaches the job site on time. The crew leader must also ensure that sure equipment operators are properly trained.

Coordinating the use of the equipment is also very important. Some equipment is used in combination with other equipment. For example, dump trucks are generally required when loaders and excavators are used. The crew leader should also coordinate equipment with other contractors on the job. Sharing equipment can save time and money and avoid duplication of effort.

Habitat for Humanity, a charitable organization that builds homes for disadvantaged families, accepts surplus building materials for use in their projects. In some cities, they have stores called ReStores, which serve as retail outlets for such materials. They may also accept materials salvaged during demolition of a structure. LEED credits can be obtained through practices such as salvaging building materials, segregating scrap materials for recycling, and taking steps to minimize waste.

46101-11_SA03.EPS

The crew leader must designate time for equipment maintenance in order to prevent equipment failure. In the event of an equipment failure, the crew leader must know who to contact to resolve the problem. An alternate plan must be ready in case one piece of equipment breaks down, so that the other equipment does not sit idle. This planning should be done in conjunction with the home office or the crew leader's immediate superior.

6.5.0 Tool Planning

A crew leader is responsible for planning what tools will be used on the job. This task includes:

- Determining the tools required
- Informing the workers who will provide the tools (company or worker)
- Making sure the workers are qualified to use the tools safely and effectively
- Determining what controls to establish for tools

6.6.0 Labor Planning

All jobs require some sort of labor because the crew leader cannot complete all the work alone. When planning for labor, the crew leader must:

- Identify the skills needed to perform the work.
- Determine how many people having those specific skills are needed.
- Decide who will actually be on the crew.

In many companies, the project manager or job superintendent determines the size and makeup of the crew. Then, the crew leader is expected to accomplish the goals and objectives with the crew provided. Even though the crew leader may not be involved in staffing the crew, the crew leader is responsible for training the crew members to ensure that they have the skills needed to do the job.

In addition, the crew leader is responsible for keeping the crew adequately staffed at all times so that jobs are not delayed. This involves dealing with absenteeism and turnover, two common problems that affect industry today.

7.0.0 SCHEDULING

Planning and scheduling are closely related and are both very important to a successful job. Planning involves determining the activities that must be completed and how they should be accomplished. Scheduling involves establishing start and finish times or dates for each activity.

A schedule for a project typically shows:

- Operations listed in sequential order
- Units of construction
- Duration of activities
- Estimated date to start and complete each activity
- Quantity of materials to be installed

There are different types of schedules used today. They include the bar chart; the network schedule, which is sometimes called the critical path method (CPM) or precedence diagram; and the short-term, or look-ahead schedule.

7.1.0 The Scheduling Process

The following is a brief summary of the steps a crewleader must complete to develop a schedule.

Step 1 Make a list of all of the activities that will be performed to build the job, including individual work activities and special tasks, such as inspections or the delivery of materials.

At this point, the crew leader should just be concerned with generating a list, not with determining how the activities will be accomplished, who will perform them, how long they will take, or in what sequence they will be completed.

Step 2 Use the list of activities created in Step 1 to reorganize the work activities into a logical sequence.

When doing this, keep in mind that certain steps cannot happen until others have been completed. For example, footings must be excavated before concrete can be placed.

Step 3 Assign a duration or length of time that it will take to complete each activity and determine the start time for each. Each activity will then be placed into a schedule format. This step is important because it helps the crew leader ensure that the activities are being completed on schedule.

The crew leader must be able to read and interpret the job schedule. On some jobs, the beginning and expected end date for each activity, along with the expected crew or worker's production rate, is provided on the form. The crew leader can use this

information to plan work more effectively, set realistic goals, and measure whether or not they were accomplished within the scheduled time.

Before starting a job, the crew leader must:

- Determine the materials, tools, equipment, and labor needed to complete the job.
- Determine when the various resources are needed.
- Follow up to ensure that the resources are available on the job site when needed.

Availability of needed resources should be verified three to four working days before the start of the job. It should be done even earlier for larger jobs. This advance preparation will help avoid situations that could potentially delay the job or cause it to fall behind schedule.

7.2.0 Bar Chart Schedule

Bar chart schedules, also known as Gantt charts, can be used for both short-term and long-term jobs. However, they are especially helpful for jobs of short duration.

Bar charts provide management with the following:

- A visual concept of the overall time required to complete the job through the use of a logical method rather than a calculated guess
- A means to review each part of the job
- Coordination requirements between crafts
- Alternative methods of performing the work

A bar chart can be used as a control device to see whether the job is on schedule. If the job is not on schedule, immediate action can be taken in the office and the field to correct the problem and ensure that the activity is completed on schedule.

A bar chart is illustrated in *Figure 19*.

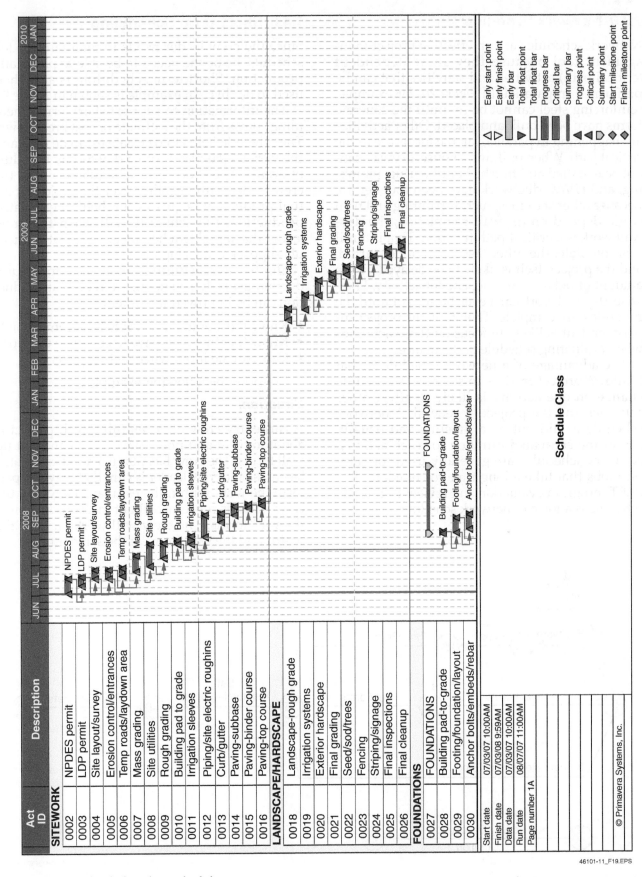

Figure 19 Example of a bar chart schedule.

46101-11_F19.EPS

7.3.0 Network Schedule

Network schedules are an effective project management tool because they show dependent (critical path) activities and activities that can be performed in parallel. In *Figure 20,* for example, reinforcing steel cannot be set until the concrete forms have been built and placed. Other activities are happening in parallel, but the forms are in the critical path. When building a house, drywall cannot be installed and finished until wiring, plumbing, and HVAC ductwork have been roughed-in. Because other activities, such as painting and trim work, depend on drywall completion, the drywall work is a critical-path function. That is, until it is complete, the other tasks cannot be started, and the project itself is likely to be delayed by the amount of delay in any dependent activity. Likewise, drywall work can't even be started until the rough-ins are complete. Therefore, the project superintendent is likely to focus on those activities when evaluating schedule performance.

The advantage of a network schedule is that it allows project leaders to see how a schedule change on one activity is likely to affect other activities and the project in general. A network schedule is laid out on a timeline and usually shows the estimated duration for each activity. Network schedules are generally used for complex jobs that take a long time to complete. The PERT (program evaluation and review technique) schedule is a form of network schedule.

7.4.0 Short-Term Scheduling

Since the crew leader needs to maintain the job schedule, he or she needs to be able to plan daily production. Short-term scheduling is a method used to do this. An example is shown in *Figure 21.*

The information to support short-term scheduling comes from the estimate or cost breakdown. The schedule helps to translate estimate data and the various job plans into a day-to-day schedule of events. The short-term schedule provides the crew leader with visibility over the project. If actual production begins to slip behind estimated production, the schedule will warn the crew leader that a problem lies ahead and that a schedule slippage is developing.

Short-term scheduling can be used to set production goals. It is generally agreed that production can be improved if workers:

- Know the amount of work to be accomplished
- Know the time they have to complete the work
- Can provide input when setting goals

Consider the following example:

Situation:

A carpentry crew on a retaining wall project is about to form and pour catch basins and put up wall forms. The crew has put in a number of catch basins, so the crew leader is sure that they can perform the work within the estimate. However, the crew leader is concerned about their production of the wall forms. The crew will work on both the basins and the wall forms at the same time.

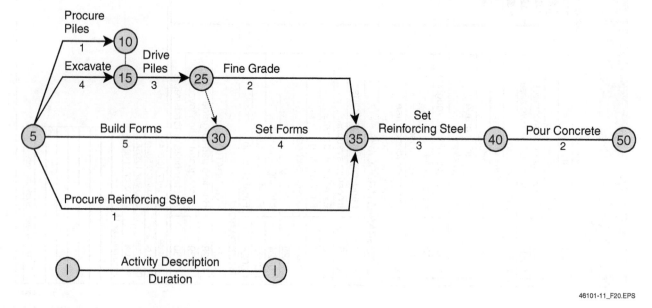

Figure 20 Example of a network schedule.

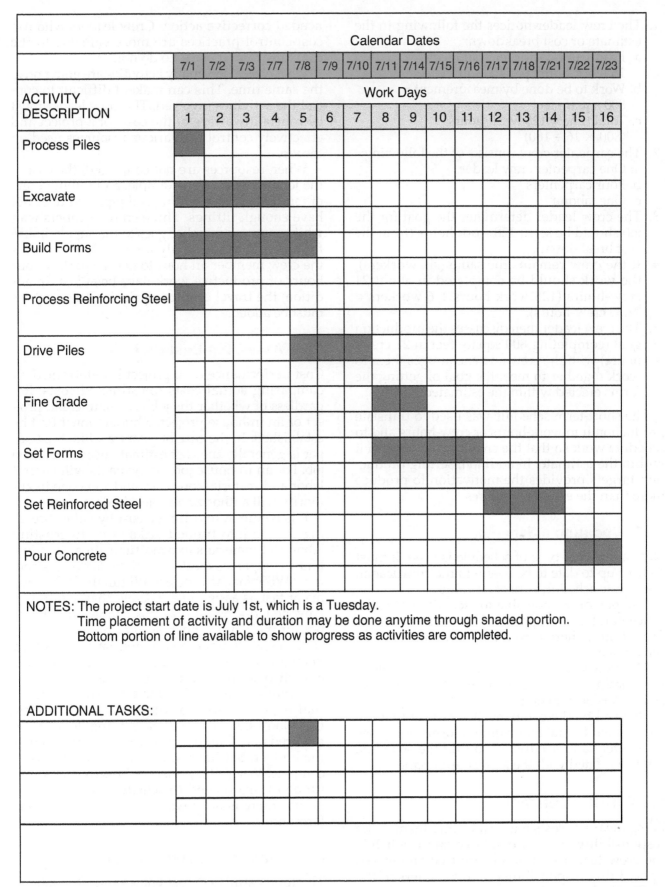

Calendar Dates

	7/1	7/2	7/3	7/7	7/8	7/9	7/10	7/11	7/14	7/15	7/16	7/17	7/18	7/21	7/22	7/23

ACTIVITY DESCRIPTION — Work Days

	1	2	3	4	5	6	7	8	9	10	11	12	13	14	15	16
Process Piles	■															
Excavate	■	■	■													
Build Forms	■	■	■	■												
Process Reinforcing Steel	■															
Drive Piles					■	■	■									
Fine Grade								■	■							
Set Forms								■	■	■						
Set Reinforced Steel												■	■	■		
Pour Concrete															■	■

NOTES: The project start date is July 1st, which is a Tuesday.
      Time placement of activity and duration may be done anytime through shaded portion.
      Bottom portion of line available to show progress as activities are completed.

ADDITIONAL TASKS:

46101-11_F21.EPS

Figure 21 Short-term schedule.

1. The crew leader notices the following in the estimate or cost breakdown:
   a. Production factor for wall forms = 16 work-hours per 100 square feet
   b. Work to be done by measurement = 800 square feet
   c. Total time = 128 work-hours (800 × 16 ÷ 100)
2. The carpenter crew consists of the following:
   a. One carpenter crew leader
   b. Four carpenters
   c. One laborer
3. The crew leader determines the goal for the job should be set at 128 work-hours (from the cost breakdown).
4. If the crew remains the same (six workers), the work should be completed in about 21 crew-hours (128 work hours ÷ 6 workers = 21.33 crew-hours).
5. The crew leader then discusses the production goal (completing 800 square feet in 21 crew-hours) with the crew and encourages them to work together to meet the goal of getting the forms erected within the estimated time.

The short-term schedule was used to translate production into work-hours or crew-hours and to schedule work so that the crew can accomplish it within the estimate. In addition, setting production targets provides the motivation to produce more than the estimate requires.

7.5.0 Updating a Schedule

No matter what type of schedule is used, it must be kept up to date to be useful to the crew leader. Inaccurate schedules are of no value.

The person responsible for scheduling in the office handles the updates. This person uses information gathered from job field reports to do the updates.

The crew leader is usually not directly involved in updating schedules. However, he or she may be responsible for completing field or progress reports used by the company. It is critical that the crew leader fill out any required forms or reports completely and accurately so that the schedule can be updated with the correct information.

8.0.0 Cost Control

Being aware of costs and controlling them is the responsibility of every employee on the job. It is the crew leader's job to ensure that employees uphold this responsibility. Control refers to the comparison of estimated performance against actual performance and following up with any needed corrective action. Crew leaders who use cost-control practices are more valuable to the company than those who do not.

On a typical job, many activities are going on at the same time. This can make it difficult to control the activities involved. The crew leader must be constantly aware of the costs of a project and effectively control the various resources used on the job.

When resources are not controlled, the cost of the job increases. For example, a plumbing crew of four people is installing soil pipe and does not have enough fittings. Three crew members wait while one crew member goes to the supply house for a part that costs only a few dollars. It takes the crew member an hour to get the part, so four hours of productive work have been lost. In addition, the travel cost for retrieving the supplies must be added.

8.1.0 Assessing Cost Performance

Cost performance on a project is determined by comparing actual costs to estimated costs. Regardless of whether the job is a contract bid project or an in-house project, a budget must first be established. In the case of a contract bid, the budget is generally the cost estimate used to bid the job. For an in-house job, participants will submit labor and material forecasts, and someone in authority will authorize a project budget.

It is common to estimate cost by either breaking the job into funded tasks or by forecasting labor and materials expenditures on a timeline. Many companies create a work breakdown structure (WBS) for each project. Within the WBS, each major task is assigned a discrete charge number. Anyone working on that task charges that number on their time sheet, so that project managers can readily track cost performance. However, knowing how much has been spent does not necessarily determine cost performance.

Although financial reports can show that actual expenses are tracking forecast expenses, they don't show if the work is being done at the required rate. Thus it is possible to have spent half the budget, but have less than half of the work compete. When the project is broken down into funded tasks related to schedule activities and events, there is far greater control over cost performance.

8.2.0 Field Reporting System

The total estimated cost comes from the job estimates, but the actual cost of doing the work is obtained from an effective field reporting system.

A field reporting system is made up of a series of forms, which are completed by the crew leader and others. Each company has its own forms and methods for obtaining information. The general information and the process of how they are used are described here.

First, the number of hours each person worked on each task must be known. This information comes from daily time cards. Once the accounting department knows how many hours each employee worked on an activity, it can calculate the total cost of the labor by multiplying the number of hours worked by the wage rate for each worker. The cost for the labor to do each task can be calculated as the job progresses. This cost will be compared with the estimated cost. This comparison will also be done at the end of the job.

When material is put in place, a designated person will measure the quantities from time to time and send this information to the home office and, possibly, the crew leader. This information, along with the actual cost of the material and the amount of hours it took the workers to install it, is compared to the estimated cost. If the cost is greater than the estimate, management and the crew leader will have to take action to reduce the cost.

A similar process is used to determine if the costs to operate equipment or the production rate are comparable to the estimated cost and production rate.

For this comparison process to be of use, the information obtained from field personnel must be correct. It is important that the crew leader be accurate in reporting. The crew leader is responsible for carrying out his or her role in the field reporting system. One of the best ways to do this is to maintain a daily diary, using a notebook or electronic device. In the event of a legal/contractual conflict with the client, such diaries are considered as evidence in court proceedings, and can be helpful in reaching a settlement.

Here is an example. You are running a crew of five concrete finishers for a subcontractor. When you and your crew show up to finish a slab, the GC says, "We're a day behind on setting the forms, so I need you and your crew to stand down until tomorrow." What do you do?

Of course, you would first call your office to let them know about the delay. Then, you would immediately record it in your job diary. A six-man crew for one day represents 48 labor hours. If your company charges $30 an hour, that's a potential loss of $1,440, which the company would want to recover from the GC. If there is a dispute, your entry in the job diary could result in a favorable decision for your employer.

8.3.0 Crew Leader's Role in Cost Control

The crew leader is often the company representative in the field, where the work takes place. Therefore, the crew leader has a great deal to do with determining job costs. When work is assigned to a crew, the crew leader should be given a budget and schedule for completing the job. It is then up to the crew leader to make sure the job is done on time and stays within budget. This is done by actively managing the use of labor, materials, tools, and equipment.

If the actual costs are at or below the estimated costs, the job is progressing as planned and scheduled, and the company will realize the expected profit. However, if the actual costs exceed the estimated costs, one or more problems may result in the company losing its expected profit, and maybe more. No company can remain in business if it continually loses money. One of the factors that can increase cost is client-related changes. The crew leaders must be able to assess the potential impact of such changes and, if necessary, confer with their employer to determine the course of action. If losses are occurring, the crew leader and superintendent will need to work together to get the costs back in line.

Noted below are some of the reasons why actual costs can exceed estimated costs and suggestions on what the crew leader can do to bring the costs back in line. Before starting any action, however, the crew leader should check with his or her superior to see that the action proposed is acceptable and within the company's policies and procedures.

- *Cause* – Late delivery of materials, tools, and/or equipment
  Corrective Action: Plan ahead to ensure that job resources will be available when needed
- *Cause* – Inclement weather
  Corrective Action: Work with the superintendent and have alternate plans ready
- *Cause* – Unmotivated workers
  Corrective Action: Counsel the workers
- *Cause* – Accidents
  Corrective Action: Enforce the existing safety program

There are many other methods to get the job done on time if it gets off schedule. Examples include working overtime, increasing the size of the crew, pre-fabricating assemblies, or working staggered shifts. However, these examples may increase the cost of the job, so they should not be done without the approval of the project manager.

9.0.0 Resource Control

The crew leader's job is to ensure that assigned tasks are completed safely according to the plans and specifications, on schedule, and within the scope of the estimate. To accomplish this, the crew leader must closely control how resources of materials, equipment, tools, and labor are used. Waste must be minimized whenever possible.

Control involves measuring performance and correcting deviations from plans and specifications to accomplish objectives. Control anticipates deviation from plans and specifications and takes measures to prevent it from occurring.

An effective control process can be broken down into the following steps:

Step 1 Establish standards and divide them into measurable units.

For example, a baseline can be created using experience gained on a typical job, where 2,000 LF of 1¼" copper water tube was installed in five days. Dividing 2,000 by 5 gives 400. Thus, the average installation rate in this case for 1¼" copper water tube was 400 LF/day.

Step 2 Measure performance against a standard.

On another job, 300 square feet of the same tube was placed during an average day. Thus, this average production of 300 LF/day did not meet the average rate of 400 LF/day.

Step 3 Adjust operations to ensure that the standard is met.

In Step 2 above, if the plan called for the job to be completed in five days, the crew leader would have to take action to ensure that this happens. If 300 LF/day is the actual average daily rate, it will have to be increased by 100 LF/day to meet the standard.

9.1.0 Materials Control

The crew leader's responsibility in materials control depends on the policies and procedures of the company. In general, the crew leader is responsible for ensuring on-time delivery, preventing waste, controlling delivery and storage, and preventing theft of materials.

9.1.1 Ensuring On-Time Delivery

It is essential that the materials required for each day's work be on the job site when needed. The crew leader should confirm in advance that all materials have been ordered and will be delivered on schedule. A week or so before the delivery date, follow-up is needed to make sure there will be no delayed deliveries or items on backorder.

To be effective in managing materials, the crew leader must be familiar with the plans and specifications to be used, as well as the activities to be performed. He or she can then determine how many and what types of materials are needed.

If other people are responsible for providing the materials for a job, the crew leader must follow up to make sure that the materials are available when they are needed. Otherwise, delays occur as crew members stand around waiting for the materials to be delivered.

9.1.2 Preventing Waste

Waste in construction can add up to loss of critical and costly materials and may result in job delays. The crew leader needs to ensure that every crew member knows how to use the materials efficiently. The crew should be monitored to make certain that no materials are wasted.

An example of waste is a carpenter who saws off a piece of lumber from a fresh piece, when the size needed could have been found in the scrap pile. Another example of waste involves installing a fixture or copper tube incorrectly. The time spent installing the item incorrectly is wasted because the task will need to be redone. In addition, the materials may need to be replaced if damaged during installation or removal.

Under LEED, waste control is very important. Credits are given for finding ways to reduce waste and for recycling waste products. Waste material should be separated for recycling if feasible (*Figure 22*).

Did you know?

Just-in-time (JIT) delivery is a strategy in which materials are delivered to the job site when they are needed. This means that the materials may be installed right off the truck. This method reduces the need for on-site storage and staging. It also reduces the risk of loss or damage as products are moved about the site. Other modern material management methods include the use of radio frequency identification tags (RFIDs) that make it easy to locate material in crowded staging areas.

9.1.3 Verifying Material Delivery

A crew leader may be responsible for the receipt of materials delivered to the work site. When this happens, the crew leader should require a copy of the shipping ticket and check each item on the shipping ticket against the actual materials to see that the correct amounts were received.

Figure 22 Waste material separated for recycling.

46101-11_F22.EPS

The crew leader should also check the condition of the materials to verify that nothing is defective before signing the shipping ticket. This can be difficult and time consuming because it means that cartons must be opened and their contents examined. However, it is necessary, because a signed shipping ticket indicates that all of the materials were received undamaged. If the crew leader signs for the materials without checking them, and then finds damage, no one will be able to prove that the materials came to the site in that condition.

Once the shipping ticket is checked and signed, the crew leader should give the original or a copy to the superintendent or project manager. The shipping ticket will then be filed for future reference because it serves as the only record the company has to check bills received from the supply house.

9.1.4 Controlling Delivery and Storage

Another very important element of materials control is where the materials will be stored on the job site. There are two factors in determining the appropriate storage location. The first is convenience. If possible, the materials should be stored near where they will be used. The time and effort saved by not having to carry the materials long distances will greatly reduce the installation costs.

Next, the materials must be stored in a secure area where they will not be damaged. It is important that the storage area suit the materials being stored. For instance, materials that are sensitive to temperature, such as chemicals or paints, should be stored in climate-controlled areas. Otherwise, waste may occur.

9.1.5 Preventing Theft and Vandalism

Theft and vandalism of construction materials increase costs because these materials are needed to complete the job. The replacement of materials and the time lost because the needed materials are missing can add significantly to the cost. In addition, the insurance that the contractor purchases will increase in cost as the theft and vandalism rate grows.

The best way to avoid theft and vandalism is a secure job site. At the end of each work day, store unused materials and tools in a secure location, such as a locked construction trailer. If the job site is fenced or the building can be locked, the materials can be stored within. Many sites have security cameras and/or intrusion alarms to help minimize theft and vandalism.

9.2.0 Equipment Control

The crew leader may not be responsible for long-term equipment control. However, the equipment required for a specific job is often the crew leader's responsibility. The first step is to identify when the required equipment must be transported from the shop or rental yard. The crew leader is responsible for informing the shop where it is being used and seeing that it is returned to the shop when the job is done.

It is common for equipment to lay idle at a job site because the job has not been properly planned and the equipment arrived early. For example, if wire-pulling equipment arrives at a job site before the conduit is in place, this equipment will be out of service while awaiting the conduit installation. In addition, it is possible that this unused equipment could be damaged, lost, or stolen.

The crew leader needs to control equipment use, ensure that the equipment is operated in accordance with its design, and that it is being used within time and cost guidelines. The crew leader must also ensure that equipment is maintained and repaired as indicated by the preventive maintenance schedule. Delaying maintenance and repairs can lead to costly equipment failures. The crew leader must also ensure that the equipment operators have the necessary credentials to operate the equipment, including applicable licenses.

The crew leader is responsible for the proper operation of all other equipment resources, including cars and trucks. Reckless or unsafe operation of vehicles will likely result in damaged equipment and a delayed or unproductive job. This, in turn, could affect the crew leader's job security.

The crew leader should also ensure that all equipment is secured at the close of each day's work in an effort to prevent theft. If the equipment is still being used for the job, the crew leader should ensure that it is locked in a safe place; otherwise, it should be returned to the shop.

9.3.0 Tool Control

Among companies, various policies govern who provides hand and power tools to employees. Some companies provide all the tools, while others furnish only the larger power tools. The crew leader should find out about and enforce any company policies related to tools.

Tool control is a twofold process. First, the crew leader must control the issue, use, and maintenance of all tools provided by the company. Next, the crew leader must control how the tools are being used to do the job. This applies to tools that are issued by the company as well as tools that belong to the workers.

Using the proper tools correctly saves time and energy. In addition, proper tool use reduces the chance of damage to the tool being used. Proper use also reduces injury to the worker using the tool, and to nearby workers.

Tools must be adequately maintained and properly stored. Making sure that tools are cleaned, dried, and lubricated prevents rust and ensures that the tools are in the proper working order.

In the event that tools are damaged, it is essential that they be repaired or replaced promptly. Otherwise, an accident or injury could result.

> **NOTE**
>
> Regardless of whether a tool is owned by a worker or the company, OSHA will hold the company responsible for it when it is used on a job site. The company can be held accountable if an employee is injured by a defective tool. Therefore, the crew leader needs to be aware of any defects in the tools the crew members are using.

Company-issued tools should be taken care of as if they are the property of the user. Workers should not abuse tools simply because they are not their own.

One of the major causes of time lost on a job is the time spent searching for a tool. To prevent this from occurring, a storage location for company-issued tools and equipment should be established. The crew leader should make sure that all company-issued tools and equipment are returned to this designated location after use. Similarly, workers should make sure that their personal toolboxes are organized so that they can readily find the appropriate tools and return their tools to their toolboxes when they are finished using them.

Studies have shown that the key to an effective tool control system lies in:

- Limiting the number of people allowed access to stored tools
- Limiting the number of people held responsible for tools
- Controlling the ways in which a tool can be returned to storage
- Making sure tools are available when needed

9.4.0 Labor Control

Labor typically represents more than half the cost of a project, and therefore has an enormous impact on profitability. For that reason, it is essential to manage a crew and their work environment in a way that maximizes their productivity. One of the ways to do that is to minimize the time spent on unproductive activities such as:

- Engaging in bull sessions
- Correcting drawing errors
- Retrieving tools, equipment, and materials
- Waiting for other workers to finish

If crew members are habitually goofing off, it is up to the crew leader to counsel those workers. The counseling should be documented in the crew leader's daily diary. Repeated violations will need to be referred to the attention of higher management as guided by company policy.

Errors will occur and will need to be corrected. Some errors, such as mistakes on drawings, may be outside of the crew leader's control. However, some drawing errors can be detected by carefully examining the drawings before work begins.

If crew members are making mistakes due to inexperience, the crew leader can help avoid these errors by providing on-the-spot training and by checking on inexperienced workers more often.

The availability and location of tools, equipment, and materials can have a profound effect on a crew's productivity. If the crew has to wait for these things, or travel a distance to get them, it reduces their productivity. The key to minimizing such problems is proactive management of these resources. As discussed earlier, practices such as checking in advance to be sure equipment and materials will be available when scheduled and placing materials close to the work site will help minimize unproductive time.

Delays caused by others can be avoided by carefully tracking the project schedule. By doing so, crew leaders can anticipate delays that will affect the work of their crews and either take action to prevent the delay or redirect the crew to another task.

Participant Exercise G

1. List the methods your company uses to minimize waste.

_____

_____

_____

_____

2. List the methods your company uses to control small tools on the job.

_____

_____

_____

_____

3. List five ways that you feel your company could control labor to maximize productivity.

_____

_____

_____

_____

_____

 Fundamentals of Crew Leadership

10.0.0 PRODUCTION AND PRODUCTIVITY

Production is the amount of construction put in place. It is the quantity of materials installed on a job, such as 1,000 linear feet of waste pipe installed in a given day. On the other hand, productivity depends on the level of efficiency of the work. It is the amount of work done per hour or day by one worker or a crew.

Production levels are set during the estimating stage. The estimator determines the total amount of materials to be put in place from the plans and specifications. After the job is complete, the actual amount of materials installed can be assessed, and the actual production can be compared to the estimated production.

Productivity relates to the amount of materials put in place by the crew over a certain time period. The estimator uses company records during the estimating stage to determine how much time and labor it will take to place a certain quantity of materials. From this information, the estimator calculates the productivity necessary to complete the job on time.

For example, it might take a crew of two people ten days to paint 5,000 square feet. The crew's productivity per day is obtained by dividing 5,000 square feet by ten days. The result is 500 square feet per day. The crew leader can compare the daily production of any crew of two painters doing similar work with this average, as discussed previously.

Planning is essential to productivity. The crew must be available to perform the work, and have all of the required materials, tools, and equipment in place when the job begins.

The time on the job should be for business, not for taking care of personal problems. Anything not work-related should be handled after hours, away from the job site. Planning after-work activities, arranging social functions, or running personal errands should be handled after work or during breaks.

Organizing field work can save time. The key to effectively using time is to work smarter, not necessarily harder. For example, most construction projects require that the contractor submit a set of as-built plans at the completion of the work. These plans describe how the materials were actually installed. The best way to prepare these plans is to mark a set of working plans as the work is in progress. That way, pertinent details will not be forgotten and time will not be wasted trying to remember how the work was done.

The amount of material actually used should not exceed the estimated amount. If it does, either the estimator has made a mistake, undocumented changes have occurred, or rework has caused the need for additional materials. Whatever the case, the crew leader should use effective control techniques to ensure the efficient use of materials.

When bidding a job, most companies calculate the cost per labor hour. For example, a ten-day job might convert to 160 labor hours (two painters for ten days at eight hours per day). If the company charges a labor rate of $30/hour, the labor cost would be $4,800. The estimator then adds the cost of materials, equipment, and tools, along with overhead costs and a profit factor, to determine the price of the job.

After a job has been completed, information gathered through field reporting allows the home office to calculate actual productivity and compare it to the estimated figures. This helps to identify productivity issues and improves the accuracy of future estimates.

The following labor-related practices can help to ensure productivity:

- Ensure that all workers have the required resources when needed.
- Ensure that all personnel know where to go and what to do after each task is completed.
- Make reassignments as needed.
- Ensure that all workers have completed their work properly.

1. Which of these activities occurs during the development phase of a project?
   a. Architect/engineer sketches are prepared and a preliminary budget is developed.
   b. Government agencies give a final inspection of the design, adherence to codes, and materials used.
   c. Project drawings and specifications are prepared.
   d. Contracts for the project are awarded.

2. The type of contract in which the client pays the contractor for their actual labor and material expenses they incur is known as a _____.
   a. firm fixed-price contract
   b. time-spent contract
   c. cost-reimbursable contract
   d. performance-based contract

3. On-site changes in the original design that are made during construction are recorded in the _____.
   a. as-built plans
   b. takeoff sheet
   c. project schedule
   d. job specifications

4. On a design-build project, _____.
   a. the owner is responsible for providing the design
   b. the architect does the design and the general contractor builds the project
   c. the same contractor is responsible for both design and construction
   d. a construction manager is hired to oversee the project

5. One example of a direct cost when bidding a job is _____.
   a. office rent
   b. labor
   c. accounting
   d. utilities

6. The control method that a crew leader uses to plan a few weeks in advance is a _____.
   a. network schedule
   b. bar chart schedule
   c. daily diary
   d. look-ahead schedule

7. A job diary should typically indicate _____.
   a. items such as job interruptions and visits
   b. changes needed to project drawings
   c. the estimated time for each job task related to a particular project
   d. the crew leader's ideas for improving employee morale

8. Gantt charts can help crew leaders in the field by _____.
   a. offering a comparison of actual production to estimated production
   b. providing short-term and long-term schedule information
   c. stating the equipment and materials necessary to complete a task
   d. providing the information needed to develop an estimate or an estimate breakdown

9. What is the crew leader's responsibility with regard to cost control?
   a. Cost control is outside the scope of a crew leader's responsibility.
   b. The crew leader is responsible only for minimizing material waste.
   c. The crew leader must ensure that all team members are aware of project costs and how to control them.
   d. The crew leader typically prepares the company's cost estimate and is therefore responsible for cost performance.

10. Which of the following is a correct statement regarding project cost?
    a. Cost is handled by the accounting department and is not a concern of the crew leader.
    b. A company's profit on a project is affected by the difference between the estimated cost and the actual cost.
    c. Wasted material is factored into the estimate, and is therefore not a concern.
    d. The contractor's overhead costs are not included in the cost estimate.

11. The crew leader is responsible for ensuring that equipment used by his or her crew is properly maintained.

   a. True
   b. False

12. Who is responsible if a defect in an employee's tool results in an accident?

   a. The employee
   b. The company
   c. The crew leader
   d. The tool manufacturer

13. Productivity is defined as the amount of work accomplished.

   a. True
   b. False

14. If a crew of masons is needed to lay 1,000 concrete blocks, and the estimator determined that two masons could complete the job in one eight-hour day, what is the estimated productivity rate?

   a. 125 blocks per hour
   b. 62.5 blocks per hour
   c. 31.25 blocks per hour
   d. 16 blocks per hour

Additional Resources

This module presents thorough resources for task training. The following resource material is suggested for further study.

Aging Workforce News, www.agingworkforce-news.com.

American Society for Training and Development (ASTD), www.astd.org.

Architecture, Engineering, and Construction Industry (AEC), www.aecinfo.com.

CIT Group, www.citgroup.com.

Equal Employment Opportunity Commission (EEOC), www.eeoc.gov.

National Association of Women in Construction (NAWIC), www.nawic.org.

National Census of Fatal Occupational Injuries (NCFOI), www.bls.gov.

National Center for Construction Education and Research, www.nccer.org.

National Institute of Occupational Safety and Health (NIOSH), www.cdc.gov/niosh.

National Safety Council, www.nsc.org.

NCCER Publications:
Your Role in the Green Environment
Sustainable Construction Supervisor

Occupational Safety and Health Administration (OSHA), www.osha.gov.

Society for Human Resources Management (SHRM), www.shrm.org.

United States Census Bureau, www.census.gov.

United States Department of Labor, www.dol.gov.

USA Today, www.usatoday.com.

Figure Credits

NCCER CURRICULA — USER UPDATE

NCCER makes every effort to keep its textbooks up-to-date and free of technical errors. We appreciate your help in this process. If you find an error, a typographical mistake, or an inaccuracy in NCCER's curricula, please fill out this form (or a photocopy), or complete the online form at **www.nccer.org/olf**. Be sure to include the exact module ID number, page number, a detailed description, and your recommended correction. Your input will be brought to the attention of the Authoring Team. Thank you for your assistance.

Instructors – If you have an idea for improving this textbook, or have found that additional materials were necessary to teach this module effectively, please let us know so that we may present your suggestions to the Authoring Team.

NCCER Product Development and Revision

13614 Progress Blvd., Alachua, FL 32615

Email: curriculum@nccer.org
Online: www.nccer.org/olf

❏ Trainee Guide   ❏ Lesson Plans   ❏ Exam   ❏ PowerPoints   Other _____

Craft / Level: _____   Copyright Date: _____

Module ID Number / Title: _____

Section Number(s): _____

Description: _____

_____

_____

_____

_____

Recommended Correction: _____

_____

_____

_____

_____

Your Name: _____

Address: _____

_____

Email: _____   Phone: _____

02403-14

Water Pressure Booster and Recirculation Systems

Overview

Water pressure booster systems boost pressure up to the level required by a building's plumbing system. The components of booster systems include storage tanks, pumps, controls, and water hammer arresters. Recirculation systems are special installations that constantly circulate hot water throughout a building's water supply piping. The location of the water heater in the building determines the method used to circulate water back to the heater. Large recirculation systems often require more than one water heater or storage tank.

Module Three

Trainees with successful module completions may be eligible for credentialing through the NCCER Registry. To learn more, go to **www.nccer.org** or contact us at **1.888.622.3720**. Our website has information on the latest product releases and training, as well as online versions of our *Cornerstone* magazine and Pearson's product catalog.

Your feedback is welcome. You may email your comments to **curriculum@nccer.org**, send general comments and inquiries to **info@nccer.org**, or fill in the User Update form at the back of this module.

This information is general in nature and intended for training purposes only. Actual performance of activities described in this manual requires compliance with all applicable operating, service, maintenance, and safety procedures under the direction of qualified personnel. References in this manual to patented or proprietary devices do not constitute a recommendation of their use.

Objectives

When you have completed this module, you will be able to do the following:

1. Describe the characteristics of a water pressure booster system and identify its components.
   a. Describe the types of water pressure booster systems.
   b. Describe the components of water pressure booster systems.
   c. Describe the design of water pressure booster systems.
   d. Describe how to install and maintain water pressure booster systems.
2. Describe the characteristics of a recirculation system and identify its components.
   a. Describe the types of recirculation systems.
   b. Describe the components of recirculation systems.
   c. Describe how to install and maintain recirculation systems.

Performance Tasks

Under the supervision of your instructor, you should be able to do the following:

1. Install the basic components of a water pressure booster system.
2. Install the basic components of a recirculation system.

Trade Terms

Average flow	Elevated-tank system	Maximum flow	Upfeed system
Balancing valves	Expansion joint	Maximum probable flow	Variable-capacity system
Circuit Setter™	Expansion tank	Pressurized air-bladder	Vertical turbine pump
Combined upfeed and	Forced circulation system	storage tank	Vibration isolators
downfeed system	Gravity return system	Recirculation system	Volute
Constant-speed system	High-limit aquastat	Single-suction impeller	Water pressure booster
Continuous demand	Hydropneumatic tank	Stage	system
Double-suction impeller	system	Tempering mixing valve	
Downfeed system	Intermittent demand	Thermal purge valve	

Industry-Recognized Credentials

If you're training through an NCCER-accredited sponsor, you may be eligible for credentials from NCCER's Registry. The ID number for this module is 02403-14. Note that this module may have been used in other NCCER curricula and may apply to other level completions. Contact NCCER's Registry at 888.622.3720 or go to **www.nccer.org** for more information.

Code Note

Codes vary among jurisdictions. Because of the variations in code, consult the applicable code whenever regulations are in question. Referring to an incorrect set of codes can cause as much trouble as failing to reference codes altogether. Obtain, review, and familiarize yourself with your local adopted code.

Contents

Topics to be presented in this module include:

Figures and Tables

1.0.0 WATER PRESSURE BOOSTER SYSTEMS

Objective

Describe the characteristics of a water pressure booster system and identify its components.

a. Describe the types of water pressure booster systems.
b. Describe the components of water pressure booster systems.
c. Describe the design of water pressure booster systems.
d. Describe how to install and maintain water pressure booster systems.

Performance Task

Install the basic components of a water pressure booster system.

Trade Terms

Average flow: The flow rate in a water pressure booster system under average operating conditions, used to calculate off-peak loads.

Constant-speed system: A water pressure booster system in which water circulates continuously through the system and pumps replacement water from the main.

Continuous demand: Demand for water caused by outlets, pumps, and other devices with a relatively constant flow.

Double-suction impeller: A pump impeller with two cavities, one each on opposite sides of the impeller. It is used to channel water in high-flow conditions in water pressure booster systems.

Elevated-tank system: The oldest and simplest type of water pressure booster system, featuring a water tank on the roof operated by the weight of the water column.

Hydropneumatic tank system: A water pressure booster system that uses air pressure to circulate the water.

Intermittent demand: Demand for water caused by fixtures that are used no more than about five minutes at a time.

Maximum flow: The theoretical total flow in a water pressure booster system if all outlets are simultaneously opened.

Maximum probable flow: The flow in a water pressure booster system during periods of peak demand.

Pressurized air-bladder storage tank: A tank used in a water pressure booster system that uses compressed air to maintain water pressure.

Single-suction impeller: A type of impeller used in centrifugal pumps that have a single-suction cavity. Single-suction impellers are used in low-flow water pressure booster systems.

Stage: In a vertical turbine pump, the arrangement of a single impeller and its water passage.

Thermal purge valve: A valve that allows heated water to drain from the recirculation system when no-flow conditions exist.

Variable-capacity system: A type of water pressure booster system that uses two or more pumps to provide water in response to peaks and lows in demand.

Vertical turbine pump: A pump used in water pressure booster systems where high-pressure and low-flow conditions exist.

Vibration isolators: Devices such as rubber pads and flexible line connectors that reduce the effects of pump vibration.

Volute: In centrifugal pumps, a geometrically curved outlet path.

Water pressure booster system: A plumbing installation that increases water pressure in the fresh water supply system.

Water supply systems sometimes need help in order to provide potable water at the right pressure. Many rural houses rely on deep wells for potable water, but the water pressure alone may not be enough to raise the water to the houses. Tall city office buildings can connect to the city water supply, but the city water pressure may not be enough to raise water to the top floor. Plumbers install water pressure booster systems to solve problems like these. Water pressure booster systems do exactly what their name suggests. They boost water pressure up to the level required by the plumbing system. Booster systems are designed to meet the specific needs of individual buildings.

1.1.0 Identifying Types of Water Pressure Booster Systems

There are four main types of water pressure booster systems:

- Elevated-tank systems
- Hydropneumatic tank systems
- Constant-speed systems
- Variable-capacity systems

Each system was designed as an improvement over the previous one, and now earlier systems are considered outdated because of new technology and standards.

1.1.1 Elevated-Tank Systems

The elevated-tank system is the oldest and simplest type of water pressure booster system. In an elevated-tank system, a water tank is installed on the top floor of a building. A small pump fills the tank. The weight of the water column provides the required pressure to the fixtures below (*Figure 1*). Elevated tanks are also called gravity tanks.

Elevated-tank systems are no longer installed. Many years ago, they were well suited for smaller buildings not connected to a city service. However, they are not effective in tall buildings or compatible with modern plumbing systems. In addition, the weight of a large water tank on a tall building can cause severe structural problems.

1.1.2 Hydropneumatic Tank Systems

The next major innovation in water pressure booster systems was the hydropneumatic tank system (*Figure 2*). These systems are used in smaller installations, such as in a floor of a building. Air pressure forces water from a tank to various parts of a building. These systems have several drawbacks. The constant contact between air and water can cause excessive corrosion, which can cause unsanitary conditions in the system. Over time, air diffuses into the water, which causes the tank to become waterlogged and reduces the system pressure. Water in the plumbing system is also at risk from contamination by compressor oil. These problems can plague older systems, but newer and more sophisticated systems are less likely to have them.

Install the largest hydropneumatic tank possible for the installation's needs. Select a pump that will be able to charge the tank fully. The tank should be able to maintain a minimum of a 10 percent water seal at full air volume. The system pressure will be highest when the tank is full, and it will be lowest when the tank is nearly empty. Set the pump to activate when the tank reaches a specified low pressure.

> **CAUTION**
>
> Because of contamination concerns, some jurisdictions prohibit hydropneumatic tank systems. Others may require special testing and certification to ensure the tanks' structural soundness. Check your local code for restrictions on these systems in your area.

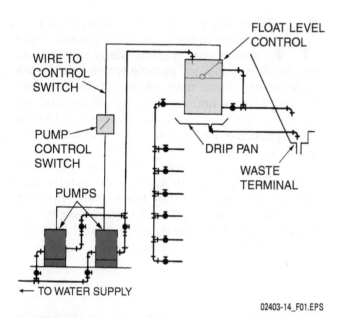

Figure 1 Schematic of an elevated-tank system.

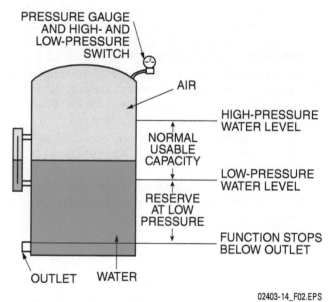

Figure 2 Hydropneumatic tank.

Consult the tank specifications for this information. The goal is to design a system that operates economically and efficiently. Select a variation that makes sense. Hydropneumatic tanks should only be used in nonpotable systems.

> **CAUTION**
>
> Hydropneumatic tanks are under high pressure. When working near a tank, avoid striking the tank with any hard or sharp object (for example, a hammer, chisel, or other tool). A rupture in the tank could cause severe damage and personal injury. Maintenance of hydropneumatic tank systems should be performed only by qualified technicians. Some jurisdictions require the installation of pressure-relief valves on hydropneumatic tanks.

1.1.3 Constant-Speed Systems

In a constant-speed system (*Figure 3*), water is pumped continuously through the plumbing installation. When water is used, the system pumps replacement water from the main. This system does not use air to create the water pressure, which eliminates the risk of corrosion that can affect hydropneumatic tank systems.

Constant-speed systems have some drawbacks, however. They require large pumps to maintain pressure during peak demand, but during periods of low demand, the pumps still do the same amount of work. This wastes a lot of electricity. Modern energy-use guidelines limit the installation of constant-speed systems, so review the local code before installing one of these systems.

1.1.4 Variable-Capacity Systems

The variable-capacity system was invented to correct the shortcomings of the constant-speed system. Variable-capacity systems use more than one pump (see *Figure 4*). The pumps engage one at a time or all together in response to pressure drops in the system. A plumber programs the pumps to activate at specific times, depending on the anticipated demand. This approach is more energy-efficient than having one pump operate all the time, which makes this system more efficient than constant-speed systems.

Before a variable-capacity system can be installed, you must establish the water-use cycle for an installation. This can take the form of a chart showing use over time. Then, select pumps that will provide enough water pressure to meet the projected demand at all times. Using the chart in *Figure 5*, for example, a plumber might choose to install two pumps. One would provide 33 percent of the system's capacity, and the other would provide 67 percent. During low-flow periods, the small pump operates. As demand increases, the second pump engages and the first pump shuts down. To meet the peak demand, both pumps operate together (see *Figure 6*).

Some variable-capacity systems use a pressurized air-bladder storage tank to replenish water in the system. This is a tank in which water pressure is kept constant by compressed air. Pressure builds up in the tank until it reaches the prescribed charge. A flexible bladder or diaphragm keeps the air and water apart. The system's pumps operate only to fill the tank. A pressure switch shuts off the system during periods of no flow (see *Figure 7*).

Pressurized tanks have several advantages. The bladder eliminates the threat of corrosion, and because the pumps do not have to cycle on and off as often, they last longer. A pressurized tank also eliminates the need for constant pump cycling during periods of no flow caused by minor leaks. A properly designed system with a pressurized tank is highly energy efficient. *Figure 8* illustrates a variable-capacity system.

1.2.0 Identifying Components of Water Pressure Booster Systems

When designing a water pressure booster system to meet the specific needs of a building or structure, you must consider the intended use, predicted capacity, and energy requirements. These factors will help determine what components are needed. Components of a water pressure booster system include the following:

- Storage tanks
- Pumps
- Pump motors
- Controls
- Water hammer arresters

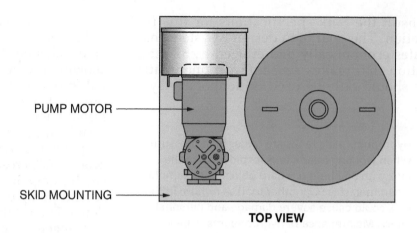

PUMP MOTOR

SKID MOUNTING

TOP VIEW

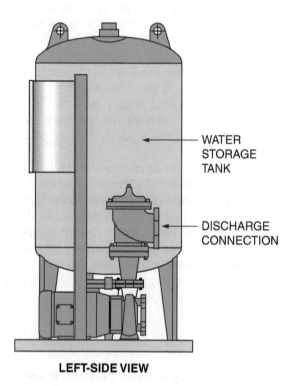

CONTROLS

WATER
STORAGE
TANK

DISCHARGE
CONNECTION

LEFT-SIDE VIEW

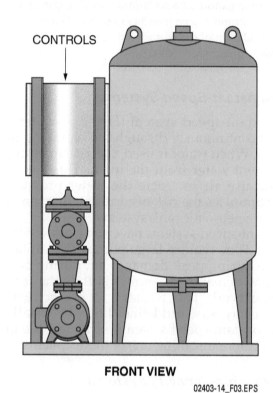

FRONT VIEW

02403-14_F03.EPS

Figure 3 Constant-speed system pump and storage tank.

Ensure that the components are suitable for the type of water system installation. Refer to the local code for specific requirements governing booster system components.

1.2.1 Storage Tanks

Booster system tanks are manufactured from materials such as galvanized steel, polyethylene, fiberglass-reinforced plastic, and even steel with a glass or cement lining. Tanks range in size from 2 gallons to more than 5,000 gallons. Tank installations may require flexible hoses, manual

or automatic float valves, manual shutoff valves, a sight gauge, and a drain. The tank can be purchased from the manufacturer as a kit or it can be built on site. Ensure that the tank materials and accessories are permitted by the local code.

1.2.2 Pumps

Centrifugal pumps (*Figure 9*) are the most widely used type of water pump in plumbing systems because of their mechanical simplicity and low cost. They are most efficient when pumping large amounts of water. Centrifugal pumps can

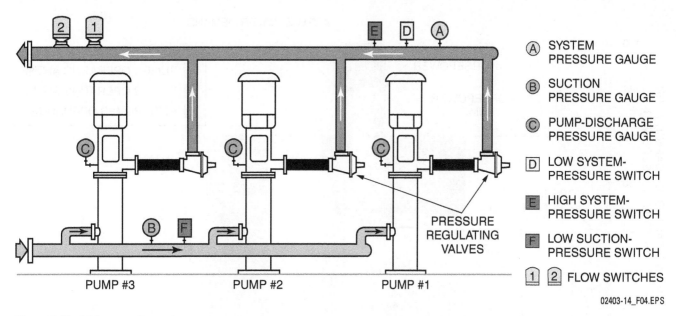

Figure 4 Variable-capacity system.

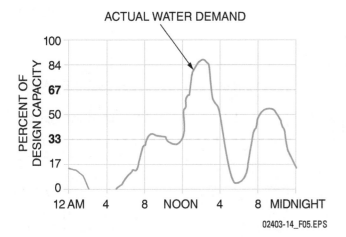

Figure 5 Water-use chart.

be classified according to how many suction cavities, or intakes, the impeller has. A single-suction impeller has a single suction cavity on one side of the impeller. A double-suction impeller has two cavities that are on opposite sides of the impeller. Both single- and double-suction impeller pumps can be used in booster systems. Single-suction impellers work well in low-flow conditions, while double-suction impellers are designed for high-flow conditions.

NOTE

Many local codes require the use of stainless steel pumps in water pressure booster systems that provide potable water.

Single-suction pumps can suffer from a high side-load thrust, which is an unbalanced load caused by water exiting from only one side of the impeller. Side-load thrust can deform the motor shaft over time. Pump seals and impellers will wear out if the condition is allowed to persist. Avoid excess wear of pump seals and impellers by installing a volute in the pump body (*Figure 10*). A volute is a geometrically curved outlet path that directs water from the impeller into the discharge pipe more efficiently than the standard pump-outlet design. Volutes improve pump efficiency and reduce wear.

Vertical turbine pumps (*Figure 11*) are also used in water pressure booster systems. They consist of a series of centrifugal impellers stacked one above each other in stages. A stage is a single impeller and its water passage. Water is drawn into the first impeller from the bottom. The water is pumped from one stage to the next. The pressure

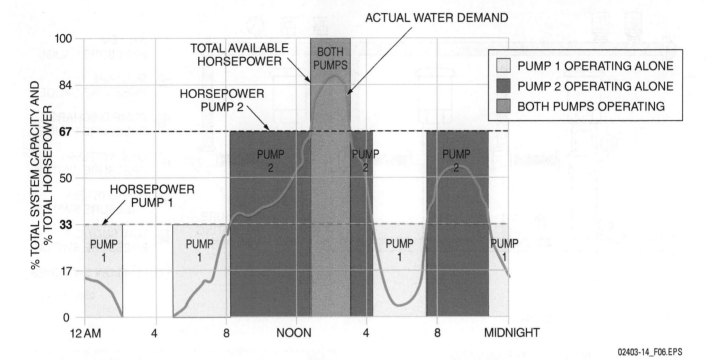

Figure 6 Variable-capacity system designed to meet water demand.

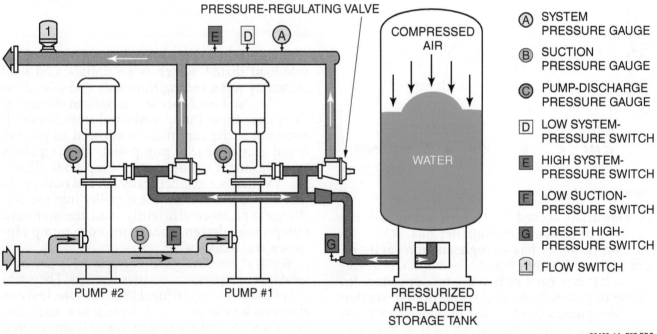

Figure 7 Variable-capacity booster system using a pressurized air-bladder storage tank.

NCCER – *Plumbing Level Four* 02403-14

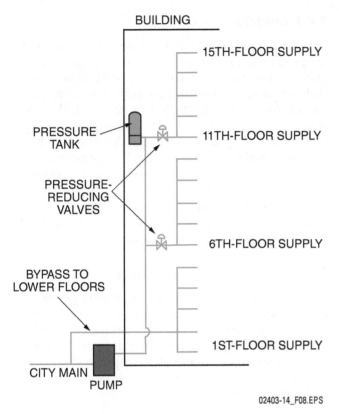

Figure 8 Variable-capacity booster system with an elevated pressurized tank in a multistory structure.

increases at each stage. The pressurized water is finally discharged after it passes through the last impeller. Vertical turbine pumps work well in high-pressure and low-flow conditions, such as in high-rise buildings. They do not experience side-load thrust problems.

1.2.3 Pump Motors

Many booster systems are designed to provide low flow at high pressure. Select a pump that offers the most efficient combination of water flow and pump speed. Note that doubling the pump capacity results in a fourfold increase in head. In other words, a pump that operates at 3,500 revolutions per minute (rpm) can pump twice the capacity of a pump that operates at 1,750 rpm, but the 3,500-rpm pump generates four times the head of the 1,750-rpm pump. Faster pumps operate more efficiently than slower ones.

Faster pumps often operate with more noise than slower pumps. To reduce noise, use a smaller motor with the same capacity. Install vibration isolators to cut down on the transfer of vibrations to the structure. These include rubber pads on the pump base and flexible connectors in the inlet and discharge lines.

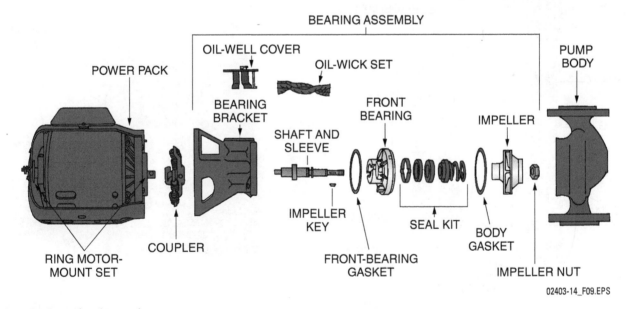

Figure 9 Centrifugal water booster pump.

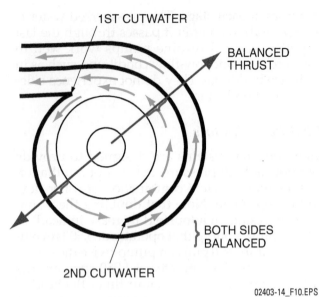

1ST CUTWATER

BALANCED THRUST

BOTH SIDES BALANCED

2ND CUTWATER

02403-14_F10.EPS

Figure 10 Booster-pump volute.

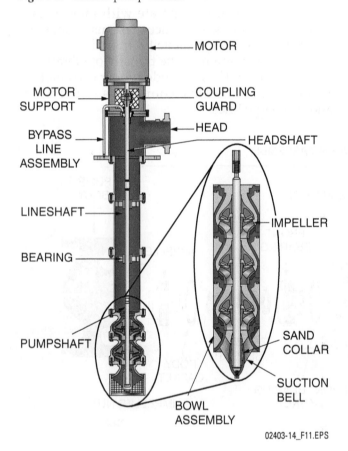

MOTOR

MOTOR SUPPORT

COUPLING GUARD

BYPASS LINE ASSEMBLY

HEAD

HEADSHAFT

LINESHAFT

IMPELLER

BEARING

PUMPSHAFT

SAND COLLAR

SUCTION BELL

BOWL ASSEMBLY

02403-14_F11.EPS

Figure 11 Vertical turbine pump.

1.2.4 *Controls*

Booster system controls include sensors for water pressure, temperature, and flow (*Figure 12*). A packaged booster system comes with its own controls, or you can use other controls. Use a pressure switch for pumps installed in residential water systems serviced by a well. Install flow sensors in a variable-capacity system to determine which pumps should be functioning. Temperature sensors, also called thermal sensors, shut the pumps down during no-flow periods. A temperature sensor detects the heat caused by a spinning impeller after flow stops, and then the sensor shuts off the pump until flow starts again.

1.2.5 *Water Hammer Arresters*

Water hammer can occur in water pressure booster systems. It occurs most often in higher-pressure systems, and can shorten the service life of the water supply system. Install water hammer arresters to protect the water supply system from damage. They will also help quiet the booster system's operation.

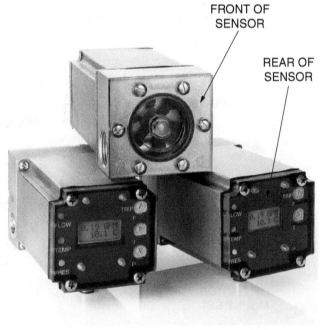

FRONT OF SENSOR

REAR OF SENSOR

02403-14_F12.EPS

Figure 12 Water pressure, temperature, and flow sensors.

1.3.0 Designing Water Pressure Booster Systems

A water pressure booster system must be selected and designed to meet the needs of the installation in order to work effectively and efficiently. For example, if the system is intended for a building with intermittent demand, such as a hotel or an office building, consider a variable-capacity system. This system is well-suited for a wide variation between peak and low demand. A constant-supply system might be appropriate for a hospital, where water is required at all times. The booster system's performance depends on correct estimates of several factors, including the following:

- Energy efficiency
- Type of pump
- Flow rates
- Anticipated water demand

The system should be designed so that it achieves the best balance of all these considerations. Each of them is discussed in more detail below.

1.3.1 Energy Efficiency

Inefficient booster systems cost more to operate, which is an important consideration with rising energy costs. Inefficient booster systems waste water and electricity. The difference in operational costs between a 10-horsepower (hp) pump motor and a 50-hp pump motor over a one-year period could amount to hundreds of dollars. Be sure to calculate the pumping needs of a system carefully.

Some water pressure booster systems have a thermal purge valve installed. These valves drain heated water out of the system when a no-flow condition exists, a procedure that wastes large amounts of water. Avoid installing thermal purge valves where operating cost is a concern.

1.3.2 Pump Selection

Selecting the right type of pump involves considering factors such as cost, ease of maintenance, reliability, and operational efficiency (see *Table 1*). Double-suction centrifugal pumps and vertical turbine pumps are very reliable, but they are also the most expensive. The cost of pump replacement parts is often less than the cost of the labor needed to install them. If replacement parts for a particular pump are not available locally, the cost of shipping replacement parts from a distant location is an addictional consideration.

1.3.3 Flow and Demand

Engineers are responsible for calculating flow and demand in a water pressure booster system (see *Table 2*). Average flow is considered to be the flow rate under average conditions. This rate cannot be used to determine maximum operating capability of a system, but can be important when calculating off-peak loads. Maximum flow is the total flow possible if all outlets are opened at the same time. Because it is highly unlikely that this would happen, this flow factor is unrealistic. The most important factor in sizing water pressure booster systems is maximum probable flow. This term refers to the flow under peak demand conditions.

Additional factors that deserve consideration are continuous demand and intermittent demand. Outlets such as hose bibbs, irrigation pumps, and other relatively constant-flowing devices are considered continuous-demand items. When these devices are used, the water flow is constant and makeup water must be at hand to replace it. Intermittent demand, as its name implies, refers to fixtures such as sinks, lavatories, showers, water closets, and similar devices with typical usage time of about five minutes or less. Continuous and intermittent demand within a system significantly affect the maximum probable-demand factor.

1.4.0 Installing and Maintaining Water Pressure Booster Systems

There are several elements involved in the process of installing and maintaining water pressure booster systems, including the system's piping materials, water tanks, motors, and pumps. Plumbers must be aware of the particular requirements for each of these components when installing a water pressure booster system. These components must also be taken into consideration when performing service or maintenance work on the system. The following sections provide guidance for installing and servicing each of these components. Refer to the manufacturer's instructions and the local applicable code before installing or servicing water pressure booster systems.

1.4.1 Installing and Maintaining Piping Materials

Ensure that the pipes and connections are suitable for use in a potable water system. Refer to your local code. Pipes must not corrode or degrade as a result of contact with potable water. Local codes also specify limits on lead content in pipes. Do

Table 1 Guidelines for Selecting an Appropriate Booster Pump

| | Single-Suction Centrifugal | | | Double-Suction Centrifugal | Vertical Turbine |
	Close-Mounted	Frame-Mounted	In-Line		
Efficiency	—Especially good in low-capacity range—			Good in high-capacity range	Good in all ranges
Reliability	Good	Good	Good	Best (except at shutoff)	Best
Suction-Lift Ability	Good	Good	Good	Best	Best*
Bearings to Maintain	Motor	Motor + 2	Motor	Motor + 2	Motor
Number of Shaft Seals	One	One	One	Two	One
Cost to Repair Seal	High	Moderate	High	Highest	Low
Hours to Repair Seal	2 to 4	1 to 3	2 to 4	4 to 8	1 to 2
Ability to Operate at Shutoff	Poor	Poor	Poor	Poor	Good
Pump Curve	Flat	Flat	Flat	Flat**	Steep**
Motor Availability	Fair	Good	Poor***	Good	Fair
Space Required	Least	Medium	Least	Most	Least
Original Price	Low	Moderate	Moderate	High	High

Notes:
* Can also tank-mount pumps to eliminate suction lift.
** Multistage pumps can be selected further to left or right on curve.
*** If specially designed shaft is used.

Table 2 Sample Table for Calculating Booster System Flow and Demand

Design Data			
System Capacity			_____ GPM
Pressure Required at Discharge Header			_____ PSIG
Pressure Drop Through Package (include discharge valve)			_____ PSIG
Minimum Suction Pressure			_____ PSIG

Pump Data	Pump No. 1	Pump No. 2	Pump No. 3 (Optional)
Gallons per Minute	_____	_____	_____
Pump TDH (feet)	_____	_____	_____
Header size _____ inch			
Discharge Valve:			
❏ PRV or ❏ Check Valve Size	_____	_____	_____
Series _____ Size	_____	_____	_____
Motor HP _____			
Motor RPM _____ Voltage _____	Hertz _____	Phase _____	

Configuration	Pump Orientation	Header Material	
	❏ Vertical	❏ Copper	
	❏ Horizontal	❏ Galvanized Steel	

not use any of the following joints or connections in a water supply:

- Cement or concrete
- Nonapproved fittings
- Solvent joints between different types of plastic pipe
- Saddle-type fittings

Permitted pipe materials include acrylonitrile-butadiene-styrene (ABS), polybutylene (PB), cross-linked polyethylene (PEX), polyvinyl chloride (PVC), and chlorinated polyvinyl chloride (CPVC) plastic; asbestos-cement; brass; gray and ductile iron; copper; and steel. When connecting pipes made of different materials, use proper joining methods in accordance with the local applicable code.

1.4.2 Installing and Maintaining Water Tanks

Depending on the design, the plumber may be responsible for locating the water tank. Be sure to provide access for tank maintenance and tank removal. When designing a booster system for a multistory building, remember that installing the tank higher allows a greater head. This not only increases the service pressure but reduces the pumping requirement.

Provide adequate support for the tank. Ensure that all underground tanks are covered to prevent unauthorized access, contamination, and infestation. Do not locate tanks or manholes for potable water pressure tanks underneath soil- or waste-piping, or any other possible source of contamination. Install a valved drainpipe at the lowest point of each tank; refer to *Table 3* for drainpipe sizing requirements. Provide a pressure-relief valve at the top of the tank (pressurized air-bladder storage tanks may not require a relief valve). On hydropneumatic systems, install a pressure-relief valve on the supply pipe. Ensure that the relief valve is equal to the tank's rating.

Leaks in a pressure tank can cause excessive pump cycling and low water pressure. Test for

Table 3 Size of Drainpipes for Water Tanks

Tank Capacity (Gallons)	Drainpipe (Inches)
Up to 750	1
751 to 1,500	1½
1,501 to 3,000	2
3,001 to 5,000	2½
5,001 to 7,500	3
More than 7,500	4

leaks by applying a soap and water solution to the exterior of the tank above the waterline. Bubbles will appear where air is escaping from the tank. Repair the leak using methods approved for potable water systems.

1.4.3 Installing and Maintaining Motors and Pumps

Install a low-pressure cutoff on booster system pumps to prevent back siphonage caused by negative pressure. Prevent tank overflow by installing ball cocks on inlet piping. Follow the manufacturers' instructions when installing pumps. Ensure the pump is primed before testing, if required.

Knowing how to diagnose pump problems quickly and accurately will not only save you time, but will also save the owner money. Whether testing a new pump or repairing an existing one, the most common problems include the following:

- Pump will not start
- Pump motor has overheated
- Pump cycles excessively
- Pump will not turn off
- Pump delivers little or no water

Review the troubleshooting tips that follow for each type of problem before answering a service call. The more experience you have in the field, the more familiar you will become with pump problems and solutions. Ask other experienced plumbers for advice. Some problems may be more common in your area due to climate and local codes.

Fresh Water Supply in America's First Modern Hotels

Boston's Tremont Hotel opened in 1829. It has been called America's first modern hotel. The four-story building featured metal bathtubs in the basement and indoor toilets on the ground floor—an unheard-of luxury in those days. The water for these fixtures came from a metal water tank on the roof. A steam engine pumped the water from ground level to the tank. Five years later, the Tremont's designer, Isaiah Rogers, was commissioned to build the six-story Astor House in New York City. This hotel had 300 guest rooms. Rogers equipped 17 rooms on the upper floors with water closets and baths for the guests. The Tremont Hotel and Astor House were the first large modern buildings in the United States equipped from the outset with their own plumbing systems.

Pump Not Starting

If a pump motor will not start, examine the pump's electrical system. If there is a blown fuse or if the motor is short-circuiting, contact an electrician to secure or replace the wiring. This work should be done only by a licensed electrician. Mechanical problems may also keep a pump motor from starting. Examine the booster system controls. Reset improperly adjusted temperature, pressure, and flow switches. Ensure that the tubing to the pressure switch is not plugged. Test the impeller by first turning the pump off, then attempting to turn the impeller with a screwdriver. If the impeller does not rotate, it may be blocked. Remove the pump casing and inspect the impeller for blockage.

Pump Motor Overheating

Pump motors have overload protectors to trip the motor off in the event of overheating. If a motor overheats, use a voltmeter to test the line voltage. If the voltage is below the required level, contact an electrician to ensure that the wiring from the electrical main to the motor is the correct size. The electrician should also ensure that the motor is wired according to the manufacturer's instructions. Only those licensed to correct electrical problems should attempt to do so.

High air temperature and inadequate venting can also cause overheating. Ensure that the pump is adequately vented and that the surrounding air temperature is within the pump's specified limits. Another cause of overheating is prolonged operation at low water pressure. If this is the case, install a globe valve on the water discharge line, and throttle it to provide the desired pressure.

Problems with the Pump Cycling Excessively

Excessive cycling may be caused by leaks in various locations in the system. Seals are the most frequently required repairs for domestic water pumps. Inspect all seals for leaks, and replace if necessary. To test for leaks on the discharge side, shut off all fixtures in the system. Inspect the discharge piping while listening for the sound of running water, which will indicate a leak. Check all valves, especially ball cocks, for leaks and repair or replace any valves that are leaking. Use a pressure gauge to find a leak on the suction side of a shallow well system. To identify a leak on the suction side of a deep well system, use the following steps:

Step 1 Attach a pressure gauge to the pump.

Step 2 Close the discharge line valve.

Step 3 Using a compressor or hand pump, apply a pressure of about 30 pounds per square inch (psi) to the system.

Step 4 Once this pressure has been reached, stop pumping and read the pressure gauge. If the pressure drops, there is a leak somewhere in the suction side. Inspect all pipes and valves.

Pump Turning Off

If the pump will not turn off, check the pressure switch. It may be set too high for the actual operating conditions, or it may have "drifted" to an incorrect setting. An electrician can inspect the pressure switch for defects caused by arcing in the electrical system. Clear the pressure-switch tube. Ensure that the pump is correctly primed. Inspect the pump ejector and remove any blockage.

> **WARNING!**
>
> You must follow safe procedures when immersing pump motors in water. When doing this type of work, you are at risk of electrocution, which can cause serious injury and even death. Before working on a pump motor, ensure that it is turned off and power to the pump motor is disconnected. Dry the pump motor thoroughly before operating. Wear appropriate personal protective equipment, and follow proper installation and maintenance procedures.

Problems with the Pump Delivering Little or No Water

Electrical and mechanical problems can cause a pump to deliver little or no water. Only a licensed electrician can check the line voltage and the priming. To clear an air lock, turn off the pump for about one minute and then restart it. Repeat this several times until the air is cleared from the pump. Ensure that the piping to the pump is correctly sized and free of leaks. Inspect the air volume control for leaks, and repair or replace the control, if required. Check the setting of pressure-regulating valves against the manufacturer's instructions. Ensure that the pump ejector is correctly sized. Remove any blockages in the intake, impeller, and ejector. Inspect the pump for worn parts and install appropriate replacements.

Additional Resources

Efficient Building Design Series, Volume 3: *Water and Plumbing*. 1999. Ifte Choudhury and J. Trost. New York: Prentice Hall.

International Plumbing Code®, Latest Edition. Falls Church, VA: International Code Council.

Plumbing: Cold Water Supplies, Drainage, and Sanitation. 1994. Fred Hall. New York: Longman.

1.0.0 Section Review

1. Because their pumps operate at the same rate during periods of low and high demand, modern energy-use guidelines limit the installation of _____.

   a. elevated-tank systems
   b. variable-capacity systems
   c. underground tank systems
   d. constant-speed systems

2. Doubling the capacity of a pump results in _____.

   a. a fourfold increase in head
   b. a twofold increase in head
   c. a fourfold decrease in head
   d. a twofold decrease in head

3. Average flow can be used to determine a system's _____.

   a. operating pressure range
   b. pump sizing
   c. maximum operating capability
   d. off-peak loads

4. Install a valved drainpipe at a water tank's _____.

   a. highest point
   b. lowest point
   c. center line
   d. maximum fill line

2.0.0 RECIRCULATION SYSTEMS AND COMPONENTS

Objective

Describe the characteristics of a recirculation system and identify its components.

a. Describe the types of recirculation systems.
b. Describe the components of recirculation systems.
c. Describe how to install and maintain recirculation systems.

Performance Task

Install the basic components of a recirculation system.

Trade Terms

Balancing valves: Valves in a recirculation system that ensure steady water flow.

Circuit Setter™: A type of balancing valve used to regulate the flow on the return of an entire circuit, with a dial permitting fine adjustments.

Combined upfeed and downfeed system: A recirculation system in which hot water is supplied in the upward and downward flow through the system.

Downfeed system: A recirculation system that supplies hot water as it travels down the system.

Expansion joint: A mechanical device installed on a pipe that allows the pipe to expand and flex as hot water flows through it.

Expansion tank: A tank attached to the supply pipes in a recirculation system that allows water to cool and slow down as it flows.

Forced circulation system: A recirculation system that uses a pump to circulate hot water through the system.

Gravity return system: A recirculation system in which the force of gravity circulates hot water throughout the system.

High-limit aquastat: A safety device that deactivates a hot water boiler when the water reaches the maximum temperature.

Recirculation system: A plumbing installation that circulates hot water within a building, providing customers with hot water on demand.

Tempering mixing valve: A valve that adds cold water to the hot water flow to control water temperature.

Upfeed system: A type of recirculation system that supplies hot water as it travels up the system.

When you take a shower at home, you may have to let the water run for a little while before it gets hot. However, in a hotel, the water is usually hot as soon as you turn on the tap. This is because the hotel uses a recirculation system. This is a special installation that constantly circulates hot water throughout a building's water supply piping. Recirculation systems can make hot water more readily available to any fixture in the system.

Recirculation systems work best in larger buildings, such as hotels, apartments, schools, factories, and restaurants. They are less effective in buildings where the farthest hot water tap is less than 100 feet from the water heater. Most recirculation systems are designed by building engineers. However, plumbers need to know about the different types of recirculation systems, their components and layouts, and the basics of installation.

> **NOTE**
>
> Hot water systems with piping under 2 inches require thermal expansion protection.

2.1.0 Identifying Types of Recirculation Systems

There are three types of recirculation systems. In an upfeed system, hot water is supplied to fixtures as the water travels up the system (*Figure 13*). In a downfeed system, hot water is supplied as the water travels down the system (*Figure 14*). In a combined upfeed and downfeed system, hot water is supplied on both the upward flow and downward flow (*Figure 15*). If required by the design, each of these systems may include hot water storage tanks. In all three types of recirculation systems, the water heater and storage tank can be located at either the top or the bottom of the installation.

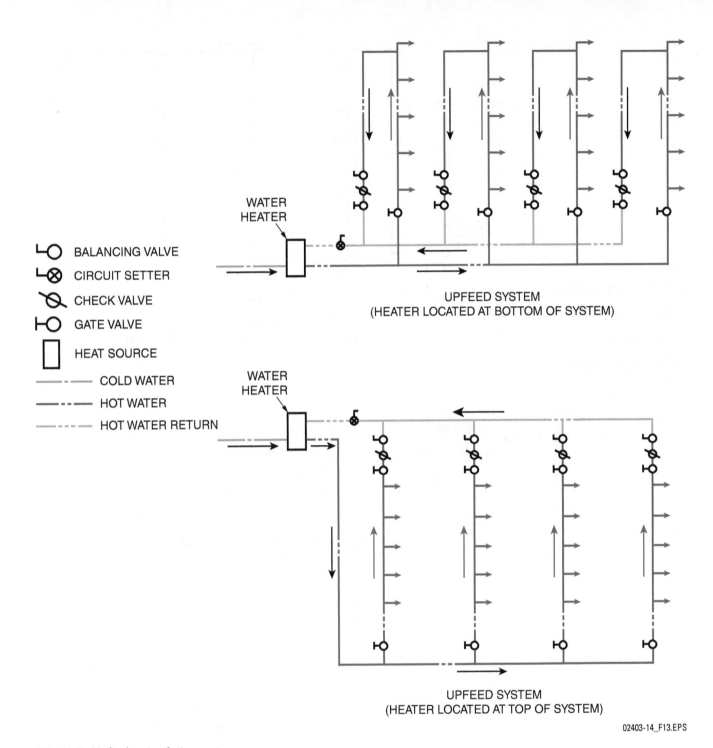

KEY:

- ⌐O BALANCING VALVE
- ⌐⊗ CIRCUIT SETTER
- ⊗ CHECK VALVE
- ⊢O GATE VALVE
- ☐ HEAT SOURCE
- —·— COLD WATER
- —·— HOT WATER
- —·— HOT WATER RETURN

WATER HEATER

UPFEED SYSTEM
(HEATER LOCATED AT BOTTOM OF SYSTEM)

WATER HEATER

UPFEED SYSTEM
(HEATER LOCATED AT TOP OF SYSTEM)

02403-14_F13.EPS

Figure 13 Upfeed recirculation system.

When selecting a recirculation system, there are several things you should consider. These include whether it is an upfeed or a downfeed system, and whether the water heater is positioned at the top or the bottom of the building. The various ways these systems can be configured require more or less piping. For example, as shown in *Figure 13*, an upfeed system with the water heater on top requires less piping than an upfeed system with the water heater on the bottom. Conversely, in a downfeed system, (shown in *Figure 14*), less

piping is required when the water heater is located at the bottom, and more piping is required when the water heater is located at the top. Regardless of which system you select, be aware that the potential for leaks increases according to the amount of pipe used, not according to the location of the water heater.

Systems that have the heater and storage tank at the top usually require a pump to circulate the water back to the heater. These are called forced circulation systems. Today, most hot water sys-

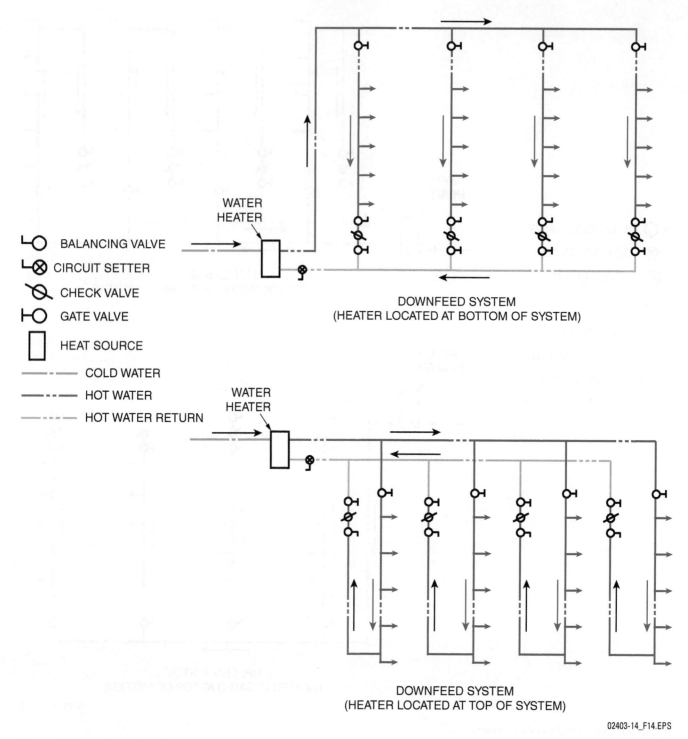

BALANCING VALVE

CIRCUIT SETTER

CHECK VALVE

GATE VALVE

HEAT SOURCE

— · — COLD WATER

— · — HOT WATER

— · — HOT WATER RETURN

WATER HEATER

DOWNFEED SYSTEM
(HEATER LOCATED AT BOTTOM OF SYSTEM)

WATER HEATER

DOWNFEED SYSTEM
(HEATER LOCATED AT TOP OF SYSTEM)

02403-14_F14.EPS

Figure 14 Downfeed recirculation system.

tems use forced circulation. A smaller number of systems use the force of gravity to return water to the heater. These are referred to as gravity return systems. Gravity return systems can be found mostly in smaller and older buildings. Plumbers need to know how to install and maintain both types of hot water recirculation systems.

2.1.1 Laying Out a Gravity Return System

Gravity return systems (*Figure 16*) are fairly small systems that have the water heater and storage tanks installed below the system. That way, water can flow back into the heater without mechanical help. Water in a gravity return system moves more

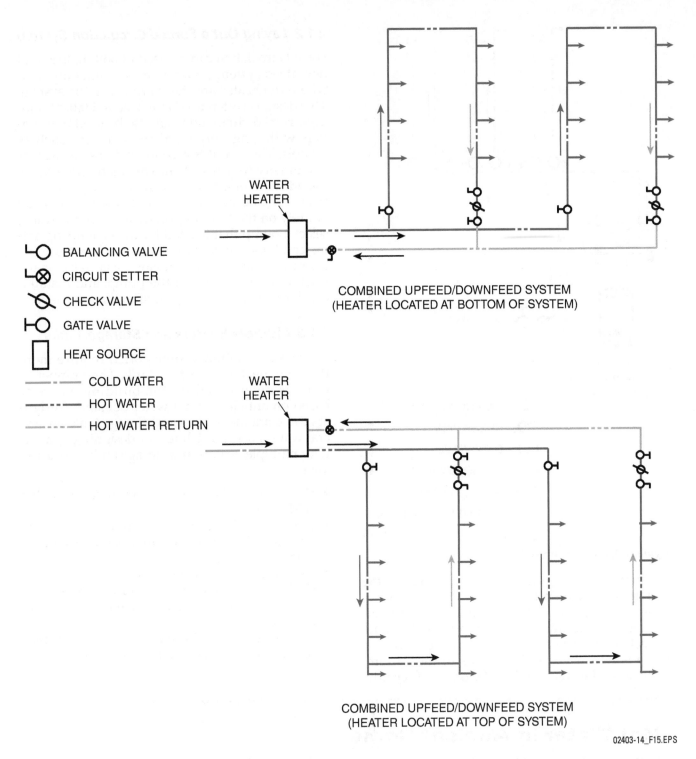

BALANCING VALVE

CIRCUIT SETTER

CHECK VALVE

GATE VALVE

HEAT SOURCE

COLD WATER

HOT WATER

HOT WATER RETURN

WATER HEATER

COMBINED UPFEED/DOWNFEED SYSTEM
(HEATER LOCATED AT BOTTOM OF SYSTEM)

WATER HEATER

COMBINED UPFEED/DOWNFEED SYSTEM
(HEATER LOCATED AT TOP OF SYSTEM)

02403-14_F15.EPS

Figure 15 Combined upfeed and downfeed recirculation system.

slowly than water in a forced circulation system, but the water flow is constant. Because of this, gravity return systems are used only in smaller buildings where speed is not as important. They can also be found in older buildings that have not been upgraded.

Proper grade is important in a gravity return system. The system uses a sloped supply pipe.

Locate the supply pipe so that it either inclines upward or rises straight up to the top of the building. That way, the cold water forces the hot water up into the system. Return pipes should also have an appropriate degree of slope. Grade is usually specified in construction drawings or in the local code, or obtain the information from the local plumbing inspector.

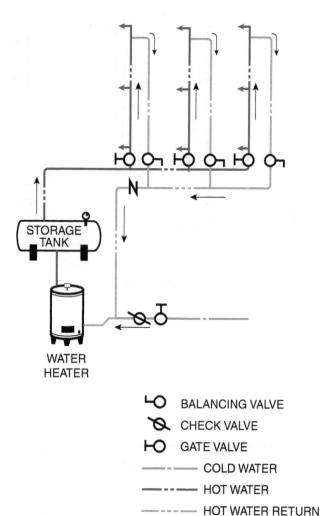

BALANCING VALVE

CHECK VALVE

GATE VALVE

—···— COLD WATER

—··—— HOT WATER

—·—·— HOT WATER RETURN

02403-14_F16.EPS

Figure 16 Gravity return system.

Circulation in a gravity system depends on the difference between the weight of cold and hot water. In a system with more than one circulating loop, ensure that the differential is the same in each loop. Use balancing valves to ensure that there is no differential variance in the system. Otherwise the system will not work as efficiently.

2.1.2 *Laying Out a Forced Circulation System*

Forced circulation can be used in both upfeed and downfeed systems, but it is most commonly used where the heater and tank are above the system. This design is often used in newer and taller buildings. Forced circulation can also be used in buildings with long horizontal runs of pipe, such as schools. The layout is essentially the same as that of a gravity return system. The main difference is the addition of a pump to circulate the hot water (*Figure 17*). Because forced circulation systems do not rely on the force of gravity to move the water, select a pump that is rated for the system's operating conditions. Install the pump on the return line before the water heater. Provide a gate valve on the inlet and outlet sides of the pump, and install a check valve on the inlet pipe after the gate valve.

2.1.3 *Multiple Heaters and Storage Tanks*

Large recirculation systems often require more than one heater or storage tank. The number of units is determined by the building engineer based on anticipated hot water demand. Designers can combine heaters and storage tanks in several ways. Basic guidelines for designing systems with multiple heaters and storage tanks are as follows:

- Do not allow hot water to circulate through a cold heater.
- Arrange piping and valves so that the heaters and tanks can function independently or together.
- Use as few valves as possible.
- On long pipe runs from heaters to storage tanks, ensure that the pipe runs have the proper degree of slope.
- Connect two water heaters to the discharge pipe using 45-degree wyes in the direction of flow.

Hot Water in Ancient Rome

Long before the development of recirculation systems, ancient Rome used hot water in its public and private baths. Public baths served as community centers where people gathered to hear news, talk with friends, and relax. The bathhouses were huge, ornate structures. The indoor baths of the Emperor Diocletian, for example, could seat 3,000 people. Most bathhouses had separate baths with cold, warm, and hot water. Water for the baths was heated by an underground furnace that also heated the air in the baths. The walls were covered in mosaic tiles. Water poured out from silver spigots shaped like fish and mythical creatures.

To supply the baths, Roman engineers accomplished some of the most complex engineering feats of their day. Water was channeled to the city's baths through 220 miles of gravity-powered aqueducts and tunnels. At its height, this network provided about 300 gallons of water for each citizen per day. By 400 AD, Rome had 11 public baths, 1,352 public fountains and cisterns, and 856 private baths.

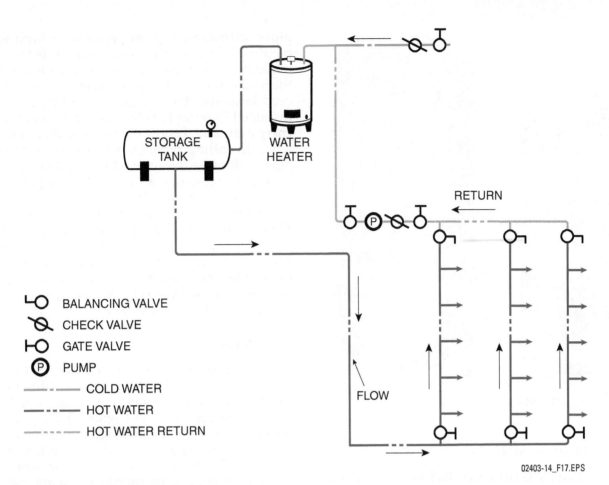

BALANCING VALVE
CHECK VALVE
GATE VALVE
PUMP
COLD WATER
HOT WATER
HOT WATER RETURN

Figure 17 Forced circulation in an upfeed system.

These guidelines are general recommendations only. Consult the plumbing drawings for specific details.

2.2.0 Identifying Components of Recirculation Systems

In addition to water heaters, hot water recirculation systems use storage tanks, pumps, valves, and temperature controls to manage the flow of hot water. Standard pipes and fittings carry the water. Building engineers choose the components that suit the needs of each installation. In this section, you will learn about the components of a recirculation system. Consult your local code for guidelines that are specific to your area.

2.2.1 Storage Tanks

Some systems use automatic storage tank water heaters (*Figure 18*) which heat and store water at a controlled temperature and deliver it on demand. Other systems require a separate storage tank. Storage tanks can be used with most electric and gas-fired water heaters. Do not install storage

tanks in systems that use indirect (instantaneous) water heaters. Storage tanks must be pressurized; install pressure-relief valves as required. The size of the storage tanks depends on the size of the facility. The engineers will estimate the size of the tank. Plumbers should consult the construction drawing to determine the tank size and refer to the local code for specific requirements.

> **WARNING!**
>
> Do not install valves on temperature and pressure (T/P) relief lines. Valves on a relief line could cause the line to block, which could damage the water heater or even cause it to explode. A malfunctioning water heater can result in water damage to property. An explosion can cause injury and even death.

2.2.2 Pumps

Forced circulation systems use centrifugal pumps to circulate water through the hot water lines. Use forced circulation when water cannot circulate by

02403-14_F18.EPS

Figure 18 Hot water storage tank.

gravity. Pumps increase the efficiency of a recirculation system. They can be designed to operate one of two ways. One option is to have the pump on constantly, which allows the water to circulate constantly through the system. The other option is to operate the pump by a temperature control, which allows the pump to switch on when the water cools below a preset minimum temperature. The choice depends on the type of installation and the demand for hot water. Some codes require that pumps be shut off when the system is not operating.

Use stainless steel or bronze pumps to handle potable water. These metals do not rust, which would contaminate water and may clog pipes. Consult local codes for additional requirements for recirculation pumps.

> **WARNING!**
>
> The maximum operating pressure of a pump is listed on its nameplate. Do not exceed this pressure, or the system could fail. This could not only damage the system, but could also cause injury or death.

2.2.3 Pipes

Piping materials used in hot water recirculation systems are generally the same as those that are used with cold water. Building engineers specify their size and type. Codes require that hot water pipes withstand a given pressure at high temperatures. Select only valves and joints that are made from compatible materials, and ensure that the lead content of the pipe material does not exceed local standards. Pipes should be properly insulated in order to help make the system more energy efficient. Follow the building plans closely when installing pipe. Undersized or oversized pipes will cause the system to malfunction.

High water temperatures will cause pipes and joints to expand. Install expansion joints in hot water piping as directed by the building plans. These will allow the pipes to expand and flex when hot water flows through them. Expansion joints can be installed several ways. Consult your local code. The building plan may call for expansion loops made from runs of pipe; specifications will be provided in the plans.

2.2.4 Valves

You may already be familiar with a type of tempering mixing valve. They are used in single-handle faucets. Hot water recirculation systems also use tempering valves (*Figure 19*). They control water temperature by adding cold water to the hot water flow. This ensures that the water in the system will remain at the right temperature. Install tempering mixing valves in systems that use indirect heaters.

One use for tempering mixing valves is in installations that require dual hot water systems. For example, a commercial kitchen might need two systems to provide different hot water temperatures for dishwashers and hand sinks.

Check valves are automatic valves that permit liquids to flow in only one direction. They protect against backflow caused by back pressure and back siphonage. Check valves are widely used in a variety of plumbing installations. They are available in horizontal and vertical styles (*Figure 20*).

Cold-Weather Tip

A recent estimate shows that homes can waste anywhere from 27 to more than 100 gallons a day waiting for running water to get hot. That means an efficient hot water recirculation system can save between 9,800 and 38,000 gallons per year in just one home. Depending on the installation, a recirculation system could represent significant cost savings to the customer, and reduced wastewater in the community.

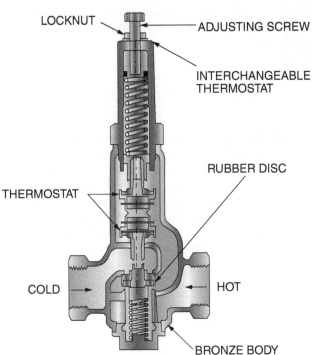

LOCKNUT — ADJUSTING SCREW

INTERCHANGEABLE THERMOSTAT

RUBBER DISC

THERMOSTAT

COLD → ← HOT

BRONZE BODY

02403-14_F19.EPS

Figure 19 Tempering mixing valve.

Install either a manual or an automatic air-relief valve at the highest point in the system in order to allow air in the pipes to be purged. An adjustable balancing valve is installed at the base of each water line to set the most efficient flow (refer to *Figure 13*, *Figure 14*, and *Figure 15*). Balancing valves compensate for pipe friction and other irregularities, ensuring a smooth and steady flow.

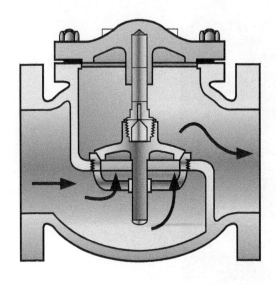

HORIZONTAL

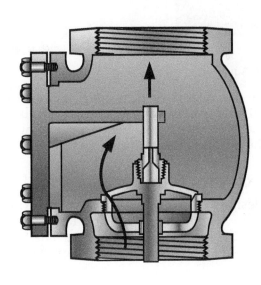

VERTICAL

02403-14_F20.EPS

Figure 20 Check valves.

On the return line near the water heater, install a special type of balancing valve with a dial that allows for fine adjustments to the water flow, such as the Bell & Gossett Circuit Setter™, to permit balancing of the entire circuit.

2.2.5 Aquastats

An aquastat (*Figure 21*) is a thermostat that regulates the temperature of hot water in a boiler. It can be used to control the water heater's operation. Install aquastats on or near the water heater or at high

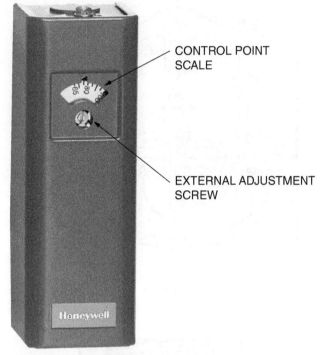

CONTROL POINT SCALE

EXTERNAL ADJUSTMENT SCREW

02403-14_F21.EPS

Figure 21 Aquastat.

points in the water lines. High-limit aquastats are a special kind of aquastat. They turn off the boiler when the water inside reaches maximum temperature. These aquastats act as safety devices.

2.2.6 Expansion Tanks

Some designs require the installation of expansion tanks (*Figure 22*). Expansion tanks are attached to the hot water supply pipes. The tanks shown in *Figure 22* use a flexible diaphragm, or

bladder, that separates system water from pressurized air. Pressurized expansion tanks often come precharged with air, but the pressure can be adjusted through a charge valve, if required.

Install expansion tanks above the system where the hottest water collects. Review the building plans to determine this location. Use pipes or a combination of pipes and a tank to allow the desired expansion. Expansion tanks can be installed in both gravity return and forced circulation systems. Check your local code for expansion tank requirements for your jurisdiction.

> **WARNING!**
>
> Hot water under pressure can escape with explosive force, and it can cause dangerous scalding. At 125°F (52°C), it takes about 1½ to 2 minutes for hot water to cause scalding. At 155°F (68°C), it takes about one second. Wear appropriate personal protective equipment. Review and follow all construction drawings closely.

2.3.0 Installing and Maintaining Recirculation Systems

Plumbers install and repair hot water recirculation systems. Many of the materials and the methods used to install recirculation systems are common to other types of plumbing installations. You will need to read the construction drawings before constructing the system. Your knowledge and experience will be helpful when you build a recirculation system, but you also need to be

(A)

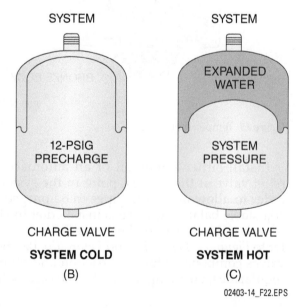

SYSTEM

SYSTEM

EXPANDED WATER

12-PSIG PRECHARGE

SYSTEM PRESSURE

CHARGE VALVE

CHARGE VALVE

SYSTEM COLD

SYSTEM HOT

(B)

(C)

02403-14_F22.EPS

Figure 22 Expansion tanks.

aware of some problems that are unique to recirculation systems. These problems are discussed below.

2.3.1 Installing a Hot Water Recirculation System

Depending on the design of the recirculation system, you may need to connect more than one riser to the hot water supply main. Ensure that the connection to each riser is correct. Otherwise, problems may develop with the hot water flow in some of the risers. Incorrect connections in a gravity system can keep it from working properly. Correct connections are also important in systems that have been altered by previous repairs. Each riser may require a slightly different connection because of the connection's location and the riser's length and slope. Four different connections can be used:

- 45-degree connection (the most typical)
- Horizontal connection
- Vertical connection
- Inverted 45-degree connection

Knowing when to install each type of connection comes from practical experience. For example, a short riser might need a vertical connection to draw as much water as a longer one with a 45-degree connection. The longest riser might need a horizontal connection so that it will not draw more water than the others. Consult your local code for guidelines, and also seek advice from experienced plumbers.

When installing more than one water heater or storage tank in a high-rise building, install check valves on the return risers. These valves will prevent the backflow of cold water into the hot water supply system. Install check valves near the water heater to prevent hot water from re-entering the cold water system. Check your local codes for specifications. Tempering mixing valves can be located in several places in the system. One effective location is outside the heat exchanger on the discharge line.

Install centrifugal pumps on the hot water return line. Locate them as close as possible to the water heater (refer to *Figure 17*). Install pumps with a bypass or a union. In some high-rise buildings, it may be necessary to install several pumps on different floors. If so, follow the plans and specifications closely.

2.3.2 Troubleshooting and Maintaining a Recirculation System

Make sure that water pressure booster systems are designed and installed correctly. A malfunctioning booster pump usually means an entire water supply system must be shut down until the pump can be repaired. In a home this can be a nuisance, but in a hospital it can cost lives. Quick and efficient troubleshooting and repair or replacement is essential.

Recirculation systems are made up of components that you are familiar with. You have learned how to troubleshoot problems with water tanks, water heaters, pipes, and valves. This knowledge will enable you to troubleshoot a recirculation system. In some cases, a manufacturer's service technician may be required to do this work.

Consider the use of balancing valves to prevent pressure differentials between the stacks in a recirculation system with more than one stack. Balancing valves compensate for different lengths and heights of pipe in various parts of the system. Open and close the individual valves until the pressure is uniform throughout the system.

Scale buildup is a common occurrence in a recirculation system. Scale buildup is caused by deposits that can create problems such as clogged pipes. Clogged pipes will need to be cleaned or replaced.

Solar Hot Water Systems

New city, state, and federal government programs are encouraging homeowners and businesses to use solar hot water heating. Solar energy is a renewable resource and creates few greenhouse gas by-products. Since 1992, the federal government has issued tax credits to businesses that use solar power. Many states offer similar incentives for homeowners and businesses. Some systems convert sunlight into electricity. Others collect solar energy and use it to heat fresh water supplies and swimming pools directly. These direct systems, called solar thermal systems, are the least expensive form of solar energy for applications connected to power grids. Solar power is gaining popularity as people become more environmentally conscious. Plumbers can expect to install an increasing number of solar hot water systems in the near future.

Additional Resources

The Hot Water Handbook: An Advanced Primer on Domestic Hot Water. 1998. George Lanthier and Robert Suffredini. Arlington, MA: Firedragon Enterprises.

Plumbing: Hot Water Supply and Heating Systems. 1994. Fred Hall. New York: Longman.

Water Quality & Systems: A Guide for Facility Managers. 1996. Robert N. Reid. New York: UpWord Publishing.

2.0.0 Section Review

1. When laying out a forced circulation system, install the pump on the _____.

   a. supply line before the water heater
   b. supply line after the water heater
   c. return line before the water heater
   d. return line after the water heater

2. Tempering mixing valves must be installed in systems that use _____.

   a. tankless heaters
   b. indirect heaters
   c. demand heaters
   d. storage heaters

3. Balancing valves are designed to compensate for different _____.

   a. pipe lengths and heights
   b. pipe diameters and fixtures
   c. water demands among fixtures
   d. operating pressures in a single system

SUMMARY

Water pressure boosters and hot water recirculation systems provide hot and cold water at the correct pressures and temperatures. Plumbers install and maintain these systems using techniques and tools that are similar to those used to install other types of water supply systems.

Water pressure booster systems increase water pressure to the required level. The four types of booster systems are elevated tank, hydropneumatic tank, constant speed, and variable capacity. Consider several factors when designing a booster system, including the intended use, the predicted capacity, and the energy requirements. Pumps, motors, and tanks require periodic maintenance.

Recirculation systems distribute hot water throughout a building. They work best in larger buildings. There are three types of recirculation systems: upfeed, downfeed, and combined upfeed and downfeed systems. These systems can be installed with water heaters and storage tanks above or below the system. Systems with heaters and tanks at the top, called forced circulation systems, usually require a pump to circulate the water. These systems are used in newer and taller buildings, and can also be used in buildings with long horizontal runs of pipe, such as schools. Systems with heaters and tanks at the bottom rely on gravity to circulate the water. They are called gravity return systems. Read the construction drawings before installing a recirculation system. Always ensure that water systems will perform the way they were designed.

1. A booster system that uses air pressure to force water from a tank to various parts of a building is the _____.
    a. constant-capacity system
    b. hydropneumatic tank system
    c. variable-capacity system
    d. elevated-tank system

2. A shortcoming of the constant-speed system is _____.
    a. excessive electricity use
    b. variations in water pressure
    c. excessively noisy operation
    d. temperature fluctuations

3. A pressurized air-bladder storage tank is sometimes used with a(n) _____.
    a. hydropneumatic tank system
    b. elevated-tank system
    c. variable-capacity system
    d. constant-capacity system

4. The most widely used type of water pump used in plumbing systems is the _____.
    a. positive-displacement pump
    b. peristaltic pump
    c. hydrostatic pump
    d. centrifugal pump

5. A centrifugal pump can be classified by the _____.
    a. the number of vanes on the impeller
    b. impeller rotational speed
    c. number of intakes the impeller has
    d. outlet size

6. Excessive wear of single-suction pump seals and impellers is avoided by installing a _____.
    a. flow restrictor
    b. venturi
    c. volute
    d. backflow preventer

7. Booster systems are frequently designed to provide _____.
    a. low flow at high pressure
    b. low flow at low pressure
    c. high flow at high pressure
    d. high flow at low pressure

8. Pumps in a water pressure booster system supplied by a well should be provided with a _____.
    a. thermal sensor
    b. pressure switch
    c. flow sensor
    d. water hammer arrester

9. The most important factor in sizing water pressure booster systems is _____.
    a. average flow
    b. minimum probable flow
    c. maximum probable flow
    d. adjusted mean flow

10. Items classified as intermittent flow typically have a usage time shorter than _____.
    a. 2 minutes
    b. 5 minutes
    c. 7 minutes
    d. 10 minutes

11. To locate leaks in a water pressure booster system tank, use a _____.
    a. soap-and-water solution
    b. fluorescent indicator solution
    c. pressure tester
    d. leak detector kit

12. If a pump motor overheats, it will be turned off by a device called a(n) _____.
    a. heat switch
    b. safety cutout
    c. overload protector
    d. flow controller

13. Recirculation systems are less effective when the distance from the water heater to the farthest hot water tap is less than _____.

   a. 50 feet
   b. 100 feet
   c. 150 feet
   d. 200 feet

14. The need for multiple heaters and storage tanks in a large recirculation system is determined by the _____.

   a. building engineer
   b. architect
   c. building inspector
   d. plumbing contractor

15. To allow air to be purged from piping, a manual or automatic air-relief valve is installed _____.

   a. at the water-heater outlet
   b. at the lowest point in the system
   c. at the highest point in the system
   d. ahead of the circulating pump

Trade Terms Quiz

Fill in the blank with the correct term that you learned from your study of this module.

1. A safety device that deactivates a hot water boiler when the water reaches the maximum temperature is called a(n) _____.

2. A(n) _____ is a plumbing installation that circulates hot water within a building, providing customers with hot water on demand.

3. A mechanical device installed on a pipe that allows the pipe to expand and flex as hot water flows through it is called a(n) _____.

4. A(n) _____ is a water pressure booster system in which water circulates continuously through the system and pumps replacement water from the main.

5. A recirculation system that uses a pump to circulate hot water through the system is called a(n) _____.

6. A(n) _____ is a valve that allows heated water to drain from the recirculation system when no-flow conditions exist.

7. A pump impeller with two cavities on opposite sides of the impeller that is used in water pressure booster systems to channel water in high-flow conditions is called a(n) _____.

8. _____ are devices such as rubber pads and flexible line connectors that reduce the effects of pump vibration.

9. The flow rate in a water pressure booster system under average operating conditions, used to calculate off-peak loads is called a(n) _____.

10. _____ are valves in a recirculation system that ensure steady water flow.

11. A recirculation system in which hot water is supplied in the upward and downward flow through the system is called a(n) _____.

12. A(n) _____ is a pump used in water pressure booster systems where high-pressure and low-flow conditions exist.

13. A recirculation system that supplies hot water as it travels down the system is called a(n) _____.

14. A(n) _____ is a valve that adds cold water to the hot water flow to control water temperature.

15. Demand for water caused by outlets, pumps, and other devices with a relatively constant flow is called _____.

16. A(n) _____ is a tank attached to the supply pipes in a recirculation system that allows water to cool and slow down as it flows.

17. In centrifugal pumps, a geometrically curved outlet path is called a(n) _____.

18. A(n) _____ is a recirculation system in which the force of gravity circulates hot water throughout the system.

19. The theoretical total flow in a water pressure booster system if all outlets are simultaneously opened is called _____.

20. A(n) _____ is a type of recirculation system that supplies hot water as it travels up the system.

21. A tank used in a water pressure booster system that uses compressed air to maintain water pressure is called a(n) _____.

22. A(n) _____ is a plumbing installation that increases water pressure in the fresh water supply system.

23. Demand for water caused by fixtures that are used no more than about five minutes at a time is called _____.

24. A(n) _____ is the arrangement of a single impeller and its water passage in a vertical turbine pump.

25. A type of impeller used in centrifugal pumps that have a single-suction cavity is called a(n) _____.

26. A(n) _____ is a type of water pressure booster system that uses two or more pumps to provide water in response to peaks and lows in demand.

27. The oldest and simplest type of water pressure booster system, featuring a water tank on the roof operated by the weight of the water column, is called a(n) _____.

28. A(n) _____ is a water pressure booster system that uses air pressure to circulate the water.

29. The flow in a water pressure booster system during periods of peak demand is called _____.

30. A(n) _____ is a type of balancing valve used to regulate the flow on the return of an entire circuit, with a dial permitting fine adjustments.

Trade Terms

Average flow
Balancing valves
Circuit Setter™
Combined upfeed and
    downfeed system
Constant-speed system
Continuous demand
Double-suction impeller
Downfeed system

Elevated-tank system
Expansion joint
Expansion tank
Forced circulation system
Gravity return system
High-limit aquastat
Hydropneumatic tank
    system
Intermittent demand

Maximum flow
Maximum probable flow
Pressurized air-bladder
    storage tank
Recirculation system
Single-suction impeller
Stage
Tempering mixing valve
Thermal purge valve

Upfeed system
Variable-capacity system
Vertical turbine pump
Vibration isolators
Volute
Water pressure booster
    system

Trade Terms Introduced in This Module

Average flow: The flow rate in a water pressure booster system under average operating conditions, used to calculate off-peak loads.

Balancing valves: Valves in a recirculation system that ensure steady water flow.

Circuit Setter™: A type of balancing valve used to regulate the flow on the return of an entire circuit, with a dial permitting fine adjustments.

Combined upfeed and downfeed system: A recirculation system in which hot water is supplied in the upward and downward flow through the system.

Constant-speed system: A water pressure booster system in which water circulates continuously through the system and pumps replacement water from the main.

Continuous demand: Demand for water caused by outlets, pumps, and other devices with a relatively constant flow.

Double-suction impeller: A pump impeller with two cavities, one each on opposite sides of the impeller. It is used to channel water in high-flow conditions in water pressure booster systems.

Downfeed system: A recirculation system that supplies hot water as it travels down the system.

Elevated-tank system: The oldest and simplest type of water pressure booster system, featuring a water tank on the roof operated by the weight of the water column.

Expansion joint: A mechanical device installed on a pipe that allows the pipe to expand and flex as hot water flows through it.

Expansion tank: A tank attached to the supply pipes in a recirculation system that allows water to cool and slow down as it flows.

Forced circulation system: A recirculation system that uses a pump to circulate hot water through the system.

Gravity return system: A recirculation system in which the force of gravity circulates hot water throughout the system.

High-limit aquastat: A safety device that deactivates a hot water boiler when the water reaches the maximum temperature.

Hydropneumatic tank system: A water pressure booster system that uses air pressure to circulate the water.

Intermittent demand: Demand for water caused by fixtures that are used no more than about five minutes at a time.

Maximum flow: The theoretical total flow in a water pressure booster system if all outlets are simultaneously opened.

Maximum probable flow: The flow in a water pressure booster system during periods of peak demand.

Pressurized air-bladder storage tank: A tank used in a water pressure booster system that uses compressed air to maintain water pressure.

Recirculation system: A plumbing installation that circulates hot water within a building, providing customers with hot water on demand.

Single-suction impeller: A type of impeller used in centrifugal pumps that have a single-suction cavity. Single-suction impellers are used in low-flow water pressure booster systems.

Stage: In a vertical turbine pump, the arrangement of a single impeller and its water passage.

Tempering mixing valve: A valve that adds cold water to the hot water flow to control water temperature.

Thermal purge valve: A valve that allows heated water to drain from the recirculation system when no-flow conditions exist.

Upfeed system: A type of recirculation system that supplies hot water as it travels up the system.

Variable-capacity system: A type of water pressure booster system that uses two or more pumps to provide water in response to peaks and lows in demand.

Vertical turbine pump: A pump used in water pressure booster systems where high-pressure and low-flow conditions exist.

Vibration isolators: Devices such as rubber pads and flexible line connectors that reduce the effects of pump vibration.

Volute: In centrifugal pumps, a geometrically curved outlet path.

Water pressure booster system: A plumbing installation that increases water pressure in the fresh water supply system.

Additional Resources

This module presents thorough resources for task training. The following resource material is suggested for further study.

Efficient Building Design Series, Volume 3: Water and Plumbing. 1999. Ifte Choudhury and J. Trost. New York: Prentice Hall.

The Hot Water Handbook: An Advanced Primer on Domestic Hot Water. 1998. George Lanthier and Robert Suffredini. Arlington, MA: Firedragon Enterprises.

International Plumbing Code®, Latest Edition. Falls Church, VA: International Code Council.

Plumbing: Cold Water Supplies, Drainage, and Sanitation. 1994. Fred Hall. New York: Longman.

Plumbing: Hot Water Supply and Heating Systems. 1994. Fred Hall. New York: Longman.

Water Quality & Systems: A Guide for Facility Managers. 1996. Robert N. Reid. New York: UpWord Publishing.

Figure Credits

Section Review Answer Key

Answer	Section Reference	Objective
Section One		
1. d	1.1.3	1a
2. a	1.2.3	1b
3. d	1.3.3	1c
4. b	1.4.2	1d
Section Two		
1. c	2.1.2	2a
2. b	2.2.4	2b
3. a	2.3.2	2c

NCCER CURRICULA — USER UPDATE

NCCER makes every effort to keep its textbooks up-to-date and free of technical errors. We appreciate your help in this process. If you find an error, a typographical mistake, or an inaccuracy in NCCER's curricula, please fill out this form (or a photocopy), or complete the online form at **www.nccer.org/olf**. Be sure to include the exact module ID number, page number, a detailed description, and your recommended correction. Your input will be brought to the attention of the Authoring Team. Thank you for your assistance.

Instructors – If you have an idea for improving this textbook, or have found that additional materials were necessary to teach this module effectively, please let us know so that we may present your suggestions to the Authoring Team.

NCCER Product Development and Revision
13614 Progress Blvd., Alachua, FL 32615

Email: curriculum@nccer.org
Online: www.nccer.org/olf

❏ Trainee Guide   ❏ Lesson Plans   ❏ Exam   ❏ PowerPoints   Other _____

Craft / Level: _____   Copyright Date: _____

Module ID Number / Title: _____

Section Number(s): _____

Description: _____

_____

_____

_____

_____

Recommended Correction: _____

_____

_____

_____

_____

Your Name: _____

Address: _____

_____

Email: _____   Phone: _____

02404-14

Indirect and Special Waste

OVERVIEW

Indirect waste systems prevent sewer gas and waste from contaminating fixtures and appliances by enabling fixtures and appliances to drain through pipes that are not directly connected to the building waste system. Wastes that require treatment before being discharged into the DWV system are called special wastes. Interceptors are used to slow, separate, and contain special wastes before they enter the sewer system. Each type of interceptor has its own requirements, and codes may vary widely on installation specifications.

Module Four

Trainees with successful module completions may be eligible for credentialing through the NCCER Registry. To learn more, go to **www.nccer.org** or contact us at **1.888.622.3720**. Our website has information on the latest product releases and training, as well as online versions of our *Cornerstone* magazine and Pearson's product catalog.

Your feedback is welcome. You may email your comments to **curriculum@nccer.org**, send general comments and inquiries to **info@nccer.org**, or fill in the User Update form at the back of this module.

This information is general in nature and intended for training purposes only. Actual performance of activities described in this manual requires compliance with all applicable operating, service, maintenance, and safety procedures under the direction of qualified personnel. References in this manual to patented or proprietary devices do not constitute a recommendation of their use.

Objectives

When you have completed this module, you will be able to do the following:

1. Identify indirect waste systems.
   a. Identify the installation requirements for indirect waste piping.
2. Identify special waste systems.
   a. Identify the various types of special waste.
   b. Explain how to install interceptors.

Performance Tasks

Under the supervision of your instructor, you should be able to do the following:

1. Install an indirect waste system.
2. Size and install an interceptor using information provided by the instructor.

Trade Terms

Air break
Air gap
Automatic trap primer
Baffle
Catch basin
Draw-off hose
Flow control fitting
Gravity draw-off system
Indirect waste

Indirect waste system
Interceptor
Manual draw-off system
Mechanical grease interceptor
Modified waste line
Open-hub waste receptor
Receptor
Sediment bucket
Special waste

Industry-Recognized Credentials

If you're training through an NCCER-accredited sponsor, you may be eligible for credentials from NCCER's Registry. The ID number for this module is 02404-14. Note that this module may have been used in other NCCER curricula and may apply to other level completions. Contact NCCER's Registry at 888.622.3720 or go to **www.nccer.org** for more information.

Code Note

Codes vary among jurisdictions. Because of the variations in code, consult the applicable code whenever regulations are in question. Referring to an incorrect set of codes can cause as much trouble as failing to reference codes altogether. Obtain, review, and familiarize yourself with your local adopted code.

Contents

Topics to be presented in this module include:

Figures and Table

SECTION ONE

1.0.0 INDIRECT WASTE SYSTEMS

Objective

Identify indirect waste systems.
  a. Identify the installation requirements for indirect waste piping.

Performance Task

Install an indirect waste system.

Trade Terms

Air break: A backflow preventer in which a smaller pipe drains into a larger pipe.

Air gap: A simple backflow preventer that consists of a space between the indirect waste system and the building waste system. The gap must be two times the diameter of the indirect waste disposal pipe and not less than 1 inch.

Automatic trap primer: A device that feeds water into a low-use trap, allowing the trap to maintain a constant seal. Automatic trap primers may be designed to close automatically (also called automatic dosing systems), or to operate electronically or under the pressure of gravity.

Indirect waste: Wastewater that flows through an indirect waste pipe.

Indirect waste system: A drainpipe attached to a fixture or appliance that is separated from the regular drainage system by a backflow preventer.

Modified waste line: A vertical pipe at the terminal of a waste line that allows indirect waste to mix with air.

Open-hub waste receptor: A waste drainpipe or pipe hub that extends above the floor line.

Receptor: A basin designed to hold liquids.

Sediment bucket: A removable basin in a drain that prevents solid wastes from entering a sanitary system.

All fixtures and appliances drain away waste products and thus become unsanitary. Fixtures are designed to be cleaned. Certain public-use fixtures need extra protection from sewage backup. If a backup occurred in a restaurant ice bin, for example, it could cause illness and even death. One way to protect fixtures and appliances is to have them drain through pipes that are not directly connected to the building waste system. Such pipes are called indirect waste systems, and the water that flows through them is called indirect waste. Indirect waste systems are used to keep sewer wastes from contaminating fixtures if drains back up. An air gap is commonly used to protect against indirect waste contamination. An air gap is simply a space between the indirect waste disposal pipe and the building waste system (see *Figure 1*). The gap must be two times the diameter of the indirect waste pipe, but not less than 1 inch.

The air break is another type of indirect waste device. In an air break, a smaller-diameter pipe drains into a larger-diameter pipe. The air break prevents the water in the larger pipe from returning to the smaller pipe. Note that the waste line must be below the flood rim of the receptor, but above the weir of the trap or the seal. Use air breaks where there is no possibility of siphonage, such as in a laundry tray. Washing machines are often installed with air breaks. Codes vary on the requirement for air gaps and air breaks. Always

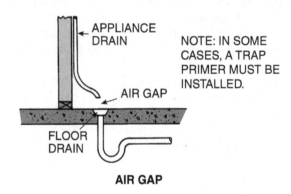

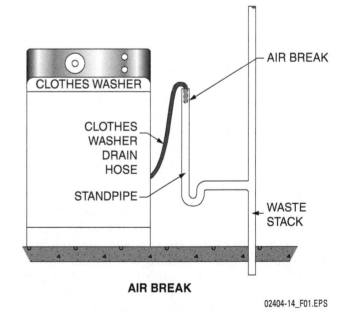

Figure 1 Air gap and air break.

consult your local code before installing either one.

Indirect wastes are produced by appliances and fixtures in the following locations:

• Commercial food-preparation facilities
• Food storage areas
• Medical and dental offices
• Air conditioning condensate and other potable and nonpotable clear-water waste installations

Indirect waste systems have many features in common. They also have their own special requirements. The following section discusses the requirements for the installation of piping for indirect waste systems. Always refer to the local applicable code for the requirements that apply to your area.

1.1.0 Installing Indirect Waste Piping

The air gap between an appliance waste pipe and the drain must be at least twice the diameter of the appliance waste pipe. For example, a waste pipe of 2 inches in diameter must have an air gap of at least 4 inches. Some indirect waste pipes empty into a sink basin. Measure the air gap from the flood level of the sink to the terminal of the waste pipe.

To prevent splashing, install a receptor on the inlet to the waste system. Receptors can be funnels, floor drains, trapped and vented slop sinks, or other similar fixtures. Ensure that the receptor has a removable strainer or basket that will keep solids out of the waste pipe. Do not install receptors in bathrooms or in unventilated spaces, such as storerooms. Make sure that receptors are easily accessible for cleaning. An open-hub waste receptor is a pipe or pipe hub extending at least 1 inch above the floor. This receptor does not require a strainer.

Another way to prevent splashing is to install a modified waste line. Using a sanitary tee, attach a vertical pipe to the terminal of the waste line (see Figure 2). The pipe should be open at both ends. A modified waste line allows the indirect waste to mix with air as it exits the pipe. This helps the wastewater to flow smoothly into the drain.

Many indirect waste systems are used only intermittently. They may drain only small amounts of waste at a time. As a result, trap seals can dry up if not periodically refreshed. If that happens, sewer gases can escape into the surrounding air, causing odors and presenting a potential health hazard. To prevent the buildup of odors from sewer gases, install a trap primer. Use an automatic trap primer to maintain a constant seal in

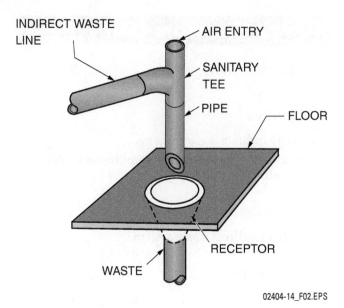

INDIRECT WASTE LINE — AIR ENTRY — SANITARY TEE — PIPE — FLOOR — RECEPTOR — WASTE

02404-14_F02.EPS

Figure 2 Modified indirect waste line.

a low-use trap (see *Figure 3*). Automatic trap primers may be designed to dose automatically (also called automatic dosing systems) due to pressure differences, or to operate electronically or by gravity. The type of automatic trap primer will be specified in the project specifications. Ensure that indirect waste installations can be easily flushed. A hose can be used to flush pipes not equipped with a hot water connection.

Do not connect steam lines to a plumbing system. Steam or other wastes that are hotter than 140°F should be discharged into a cooling sump or basin. After the wastewater has cooled, it is discharged into the sewer system.

1.1.1 Food Preparation, Handling, and Storage Installations

Codes require indirect waste lines with air gaps for walk-in refrigerators and freezers, icemakers, and ice compartments. Wastewater from multiple ice compartments and icemakers can be collected at one point, which can be a floor sink or a floor drain (*Figure 4*). This reduces the amount of piping required to drain multiple fixtures. Many codes require indirect waste lines longer than 2 feet to have traps.

Some codes allow commercial dishwashers and dish-washing sinks to attach directly to the building drain. If your code does not permit this, use an indirect waste system. Because large quantities of grease may be discharged from dish-washing fixtures, install a grease interceptor on the indirect waste line before the air gap. Refer to your local plumbing code for the requirements in your area.

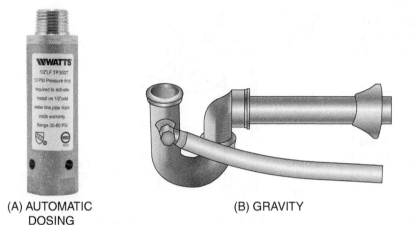
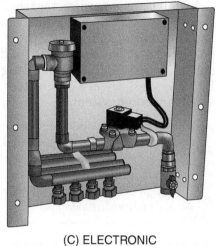

(A) AUTOMATIC
DOSING

(B) GRAVITY

(C) ELECTRONIC

02404-14_F03.EPS

Figure 3 Automatic trap primers.

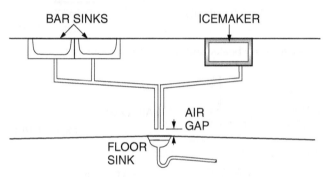

02404-14_F04.EPS

Figure 4 Indirect waste piping for bar sinks and icemakers.

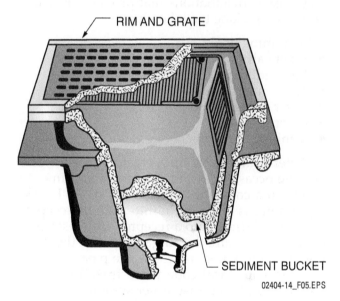

02404-14_F05.EPS

Figure 5 Floor sink with sediment bucket.

Many commercial kitchen appliances, such as potato peelers, create solid wastes. Install floor sinks with sediment buckets to trap solid wastes (*Figure 5*). A sediment bucket is a removable basin inside a drain that holds solid waste but allows water to drain.

Floor and area drains are often set before the concrete floor is poured. Ensure that the elevation of the drain is correct. Locate the drain at the lowest point in the floor or area. Refer to the floor plan, plumbing drawings, or specifications for the drain's elevation, the horizontal distances from walls, and the floor finishes. Floor sinks may used in installations that require a high degree of sanitation, such as hospitals, laboratories, cafeterias, and restaurants. The basic difference between a floor drain and a floor sink is that the floor sink is coated with a porcelain or other easily cleaned finish. Floor sinks reduce the amount of contaminant that can collect in the grate and drain body. Consult the local applicable code or the project specifications to determine if the installation of a reduced pressure-zone principle (RPZ) backflow preventer assembly is required on the same line as the floor sink.

1.1.2 Health Care Installations

Health care facilities need sanitary conditions to protect the health of their patients. Effective waste disposal helps to limit the spread of disease. Indirect waste systems are used in a wide variety of health care facilities, including the following:

- Surgical and dental theaters
- Clinics and infirmaries
- Pharmaceutical and research laboratories
- Youth and elder care institutions

Consult your local code before installing health care waste systems.

Install traps on indirect waste lines for bedpan steamers. Batteries of fixtures, such as sterilizers, can drain into a single receptor. A receptor is a basin that catches and holds solids. Attach clinical sinks directly to the building waste system. Bedpan washers are connected directly to the waste and vent system.

> **WARNING!**
>
> Wastewater from medical and dental facilities is hazardous. It can contain bacteria and viruses that can cause illness and death. Ensure that the design of the DWV system adheres strictly to code before installing it.

1.1.3 Clear-Water Waste Installations

Many other fixtures and appliances create indirect wastes. Installations that produce clear-water wastes include the following:

- Swimming pools
- Relief valves
- Air conditioner condensate drains
- Refrigerators
- Water storage tanks
- Water heaters
- Drinking fountains

Water-heater relief lines create indirect waste only on occasion. Install an automatic primer to maintain a constant trap seal (*Figure 6*). Drinking fountains often drain into a single receptor and should drain into a receptor with a strainer (*Figure 7*). Measure the air gap between the flood line of the receptor and the waste pipe terminal. Refer to your local code when installing these and other indirect wastewater systems.

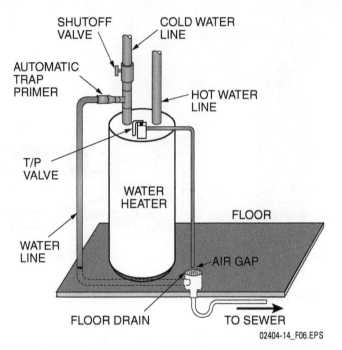

02404-14_F06.EPS

Figure 6 Water heater unit installed in a basement.

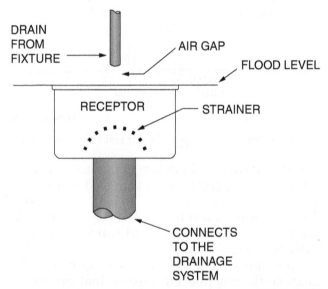

02404-14_F07.EPS

Figure 7 Drinking fountain draining into receptor with a strainer.

Additional Resources

Introduction to Wastewater Treatment Processes. 2014. Bruce Jefferson, Simon Parsons, and Elise Cartmell. New York: Wiley-Blackwell.

Water, Sanitary, and Waste Services for Buildings, Fifth Edition. 2001. Alan F. E. Wise and J. A. Swaffield. Boston, MA: Routledge.

1.0.0 Section Review

1. Many codes require indirect waste lines to have traps if their length exceeds _____.
   a. 4 feet
   b. 3 feet
   c. 2 feet
   d. 1 foot

2.0.0 SPECIAL WASTE SYSTEMS

Objective

Identify special waste systems.
a. Identify the various types of special waste.
b. Explain how to install interceptors.

Performance Task

Size and install an interceptor using information provided by the instructor.

Trade Terms

Baffle: Partition inside the body of an interceptor. Baffles are designed to reduce turbulence caused by incoming wastewater and to block wastes from escaping through the outlet.

Catch basin: A reservoir installed before the sanitary piping. It allows sediment to settle out of wastewater.

Draw-off hose: A hose used to siphon grease and oil from interceptors.

Flow control fitting: A device that regulates the flow of wastewater into grease and oil interceptors.

Gravity draw-off system: A system that automatically drains oil from an interceptor into a storage tank.

Interceptor: A device that separates special wastes before they can enter the sewer, septic, or stormwater system.

Manual draw-off system: A system that drains oil out of an interceptor through a valve-operated draw-off hose.

Mechanical grease interceptor: A grease interceptor fitted with a timer-activated self-cleaning capability.

Special waste: Waste that must be removed, diluted, or neutralized before it can be discharged into a sanitary system.

Indirect waste is not the only kind of waste that requires special drain installations. Special wastes must be treated before they can be discharged safely into the system, due to one or more of the following conditions:

- Their temperature is too high for the septic or sewer system, such as water above 140°F.

- They contain solid and partially solid matter that will settle out.
- They contain acids that must be neutralized.
- They contain harmful or potentially harmful substances, such as grease, oil, and gasoline.

Some special wastes are treated by evaporation, while others must be separated from the wastewater. Still others have to be diluted or chemically neutralized first.

2.1.0 Identifying Types of Special Waste

Special wastes include grease, oil, and gasoline; solids, such as sediment, glass, and hair; corrosive wastes; and high-temperature wastes. Grease or animal fat is generated in large quantities where food is prepared and served commercially. Grease will clog sewer and septic pipes if it is allowed to enter the system. Oil and gasoline are volatile petroleum products. They are used to operate, clean, repair, and store motor vehicles. These fluids could ignite and damage sewer, septic, or stormwater drainage systems. Sediment can clog drains and pipes. Corrosive chemicals like acids can damage pipes and drains, generate toxic fumes, and interfere with the waste treatment process. High-temperature wastes can cause pipes to expand and contract. This weakens pipe joints and causes leaks.

Treatments vary for each type of waste. Improper treatment could damage the sewer or septic system. It could also contaminate fresh water supplies and cause sickness or even death. Review the proper treatment method before treating wastes. Your local code will provide specific guidelines.

Health care facilities use special waste systems. As with indirect waste systems in health care facilities, install traps on special waste lines. Batteries of fixtures can be drained into a single receptor. An interceptor is often required in special waste piping depending on the types and concentrations of the waste materials in the system. In some installations, the waste may be treated prior to its entry into the waste system. Corrosive liquids or harmful chemicals must be diluted, neutralized, or treated using a code-approved neutralizing device. Always refer to the local applicable code before installing special waste systems in health care facilities.

2.2.0 Installing Interceptors

Interceptors separate special wastes before they can enter a sewer, septic, or stormwater system. Small interceptors are about the size of an ordinary P-trap. Some custom-built underground interceptors are large enough for a person to stand in. No matter what size they are, most interceptors operate the same basic way.

The internal volume of an interceptor is larger than that of the drainage pipes. This causes wastewater to slow down as it enters, allowing the waste materials time to settle, float, or evaporate out of the wastewater. Baffles are partitions in the interceptor that reduce turbulence caused by flowing wastewater. Baffles also keep wastes from escaping through the outlet. Free of the contaminants, the wastewater discharges into the sewer system by gravity.

Interceptors work a lot like fixture traps. Both are designed to seal one part of a plumbing installation from waste products in another part. However, interceptors violate one of the basic principles of traps: interceptors are not self-cleaning and must be manually cleaned.

Site plans specify where large interceptors should be installed. An interceptor can be above or below floor level. The installation's design and the available space determine the location. Provide enough space to allow cleaning equipment, such as pumps and augers, to fit. Ensure that covers and sediment buckets can be removed easily.

Locate the interceptor close to the fixture because this will help prevent the inlet piping from clogging. Install exterior interceptors below the frost line to keep them from freezing. Support large interceptors with a solid foundation because the weight of aboveground interceptors can cause structural problems in some buildings. Even interceptors installed below grade need support. Otherwise, pipe joints may fail and damage the tank. Packed sand over undisturbed earth is a good foundation.

If the waste includes toxic gases, steam, high-pressure wastes, or chemicals, vent the interceptor through a roof vent. Ensure that the inlet and outlet piping is installed properly, and seal all piping joints tightly to prevent liquid or gas leaks. For concrete and masonry interceptors, carefully grout the joints between the piping and the tank.

2.2.1 Grease Interceptors

For small amounts of grease, install a small interceptor inside the building (*Figure 8*). They are typically made of steel and have a bolt-on lid. Ensure that the interceptor is easy to reach for cleaning, and locate it as close to the sink or appliance as possible. Otherwise foul odors may occur. Odors are the biggest problem associated with indoor grease interceptors. Proper venting will help reduce odors. Vent grease interceptors using the guidelines for venting a trap. Cleaning can be a problem when lid removal is required.

One cleaning option is to use a draw-off hose (*Figure 9*). Draw-off hoses are used to siphon the grease out of an interior grease interceptor. They can be either gravity fed or connected to a pump. Ensure that a draw-off valve is installed at the outlet of the interceptor. Closing the valve allows the water level in the interceptor to rise, which forces the grease to flow into the draw-off hose. Consult your local code for the proper procedure.

Excessive flow into an interceptor can cause turbulence, and reduces the amount of time grease has to coagulate and float to the top for draw-off. These conditions will permit grease to escape through the outlet. Install a flow control fitting on the inlet pipe to slow the flow of wastewater into an interceptor (see *Figure 10*). The flow control's inlet is carefully sized, limiting the amount of wastewater that can flow into the interceptor. Flow control devices with removable or adjustable parts are not permitted in grease interceptors.

Large interceptors are located outside and underground. Precast concrete tanks, similar to sep-

The Trap "Trap"

Interceptors are sometimes called traps, such as a grease trap. You have already learned that the term *trap* also applies to installations that maintain water seals in fixtures. Both types of traps are vital components in DWV systems because they protect people from exposure to noxious and harmful substances. However, they operate very differently. Interceptors and traps have their own unique requirements. Do not confuse the two! The results could be dangerous. Take the time to review your local code requirements for both interceptor and trap installations.

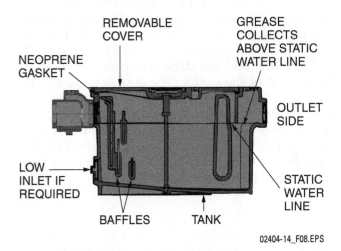

02404-14_F08.EPS

Figure 8 Typical interior grease interceptor.

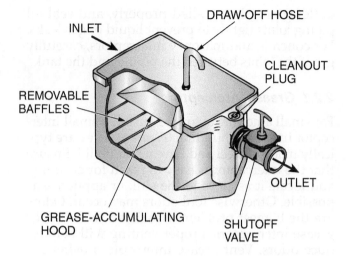

INLET

DRAW-OFF HOSE

CLEANOUT PLUG

REMOVABLE BAFFLES

OUTLET

GREASE-ACCUMULATING HOOD

SHUTOFF VALVE

02404-14_F09.EPS

Figure 9 Draw-off hose installed on a grease interceptor.

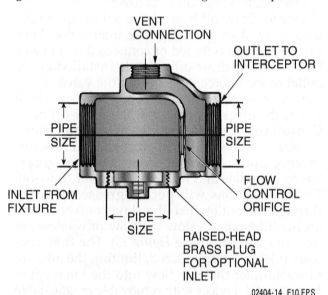

VENT CONNECTION

OUTLET TO INTERCEPTOR

PIPE SIZE

PIPE SIZE

FLOW CONTROL ORIFICE

INLET FROM FIXTURE

PIPE SIZE

RAISED-HEAD BRASS PLUG FOR OPTIONAL INLET

02404-14_F10.EPS

Figure 10 Flow control fitting.

tic tanks, can be used as large grease interceptors. They can be site-built from concrete, masonry, and pipe fittings (see *Figure 11*). The dimensions can be varied to meet size requirements. Ensure that the inlet is positioned correctly with regard to the outlet and the static waterline. Increasing the height of the outlet pipe allows more solids to collect. This reduces the number of times the interceptor has to be serviced. Access to the interceptor is provided by a manhole or through an opening for a draw-off hose. Two- and three-chamber grease interceptors (*Figure 12*) are more efficient than single-chamber designs.

Maintaining Special Waste Interceptors

Special waste interceptors require regular service. Provide the building managers or occupants with a maintenance schedule. Explain that regular servicing protects the health of the public and helps protect the owners from fines, penalties, and lawsuits.

WARNING!

Climb-in interceptors may contain corrosive, flammable, or odorous substances. Personal injury could result in poisoning or an infection, such as hepatitis. Obtain thorough confined-space training before servicing climb-in interceptors. Use appropriate personal protective equipment, and wear hand protection while working.

Codes provide formulas to allow you to calculate the right size for grease traps. The formulas in *Table 1* are based on the Environmental Protection Agency (EPA)-2 model; always refer to your

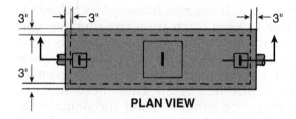

3" 3" 3"

3"

PLAN VIEW

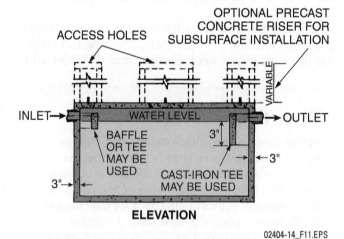

ACCESS HOLES

OPTIONAL PRECAST CONCRETE RISER FOR SUBSURFACE INSTALLATION

VARIABLE

INLET

WATER LEVEL

OUTLET

BAFFLE OR TEE MAY BE USED

3"

3"

CAST-IRON TEE MAY BE USED

3"

ELEVATION

02404-14_F11.EPS

Figure 11 Typical large site-built grease interceptor.

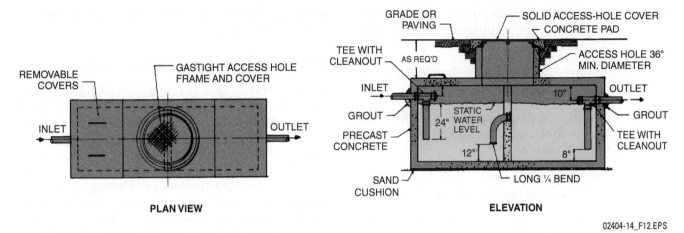

PLAN VIEW   ELEVATION

02404-14_F12.EPS

Figure 12 Two-chamber grease interceptor.

local code. First, determine the total maximum drainage flow from each fixture. This number is estimated in gallons per minute (gpm). Next, factor in the estimated load factors. Multiply the total drainage flow by the load factor. Multiply that number by 60 (minutes) to get the maximum flow in one hour. Finally, multiply the maximum flow by 2 (representing the two-hour retention time specified in the code). This number gives you the total required volume of the grease trap in gallons.

Grease interceptors may also be sized using the capacity method. This is done by calculating the flow rate of each sink that drains into the interceptor. Follow these steps to calculate grease interceptor capacity using flow-rate calculations:

Step 1 Determine the capacity of each sink in cubic inches:

$$length \times width \times height = capacity\ in\ cubic\ inches$$

Table 1 Grease-Interceptor Sizing Formula for a Restaurant

A. Determine maximum drainage flow from fixtures:		
Type of Fixture	**Flow Rate**	**Amount**
Restaurant kitchen sink	15 gpm	_____
Single-compartment sink	20 gpm	_____
Double-compartment sink	25 gpm	_____
2 single-compartment sinks	25 gpm	_____
2 double-compartment sinks	35 gpm	_____
Triple sink, 1½-inch drain	35 gpm	_____
Triple sink, 2-inch drain	35 gpm	_____
30-gallon dishwasher	15 gpm	_____
50-gallon dishwasher	25 gpm	_____
50- to 100-gallon dishwasher	40 gpm	_____
B. Total		_____
C. Estimate Loading Factors:		
Restaurant type	Fast food/paper delivery =	.50
	Low volume =	.50
	Medium volume =	.75
	High volume =	1.00
D. Step B × Step C:	_____	(Subtotal)
E. Subtotal × 60 min. × max. flow for 1 hour		_____
F. Step E × 2 hours' retention time = vol. of trap (gal.)		_____

Step 2 Convert the capacity from cubic inches into gallons per minute (gpm) using the following formula:

$$\text{capacity in cubic inches}/231 = \text{capacity in gpm}$$

Step 3 Account for displacement, which is the sink's actual usable volume, using the following formula:

$$\text{capacity in gpm} \times 0.75 = \text{actual capacity in gpm}$$

Step 4 Add the actual capacities of the sinks together. This is the total capacity that the grease interceptor should be able to handle in one minute.

Note that if drain-down time is not a critical consideration, the project specifications may call for an interceptor that is sized for as much as half the calculated flow rate.

To clean small grease interceptors, use the following procedure:

Step 1 Remove the lid.

Step 2 Scoop the grease out of the interceptor and place into a watertight container (wear proper protective equipment for the hands and face).

Step 3 Thoroughly clean the sides and inside lid of the interceptor.

Step 4 Make sure the container is tightly sealed before properly disposing of it according to local code requirements.

Step 5 Re-install and seal the lid.

Grease disposal varies from city to city. Some areas forbid placing grease in landfills. Other codes allow some grease to be placed in garbage cans in sealed containers. Professional contractors perform service and repair on exterior grease interceptors. They use large pumper trucks to suck the grease out of the interceptor. Some recycling companies collect grease and convert it into fuel and industrial products. Rendering companies also collect grease and use it in the manufacture of consumer products, such as soap and perfume. As a plumber, you may be called on to recommend contractors, recyclers, and rendering companies. Only recommend those that follow environmental standards.

Ensure that food-waste grinders and garbage disposals do not discharge into a grease trap. Most codes place restrictions on garbage-disposal wastes entering sewer systems. Check your local code for the requirements.

2.2.2 Mechanical Grease Interceptors

Mechanical grease interceptors (*Figure 13*), also called automatic grease removal devices (GRDs), are used in systems where grease, oils, and fats are in excess of the amount that small grease interceptors can handle. As with a regular grease interceptor, a mechanical grease interceptor works by passing wastewater through a strainer in order to separate fats, oils, and grease from the wastewater. The wastewater is then allowed to drain through the interceptor's outlet, while the greasy wastes are stored until they are removed for recycling. However, mechanical grease interceptors are also designed to be self-cleaning. At regular intervals, an internal heater warms the interceptor to 115°F–130°F, which causes any remaining grease to turn into liquids that then float to the top of the interceptor's retention tank. A skimming wheel collects the floating liquid waste, and a wiper blade slides the wastes off the wheel into a special outlet that is connected to a separate waste container (see *Figure 14*). Mechanical grease interceptors are designed to operate automatically. They only require attention when removing or replacing the waste container, or for maintenance.

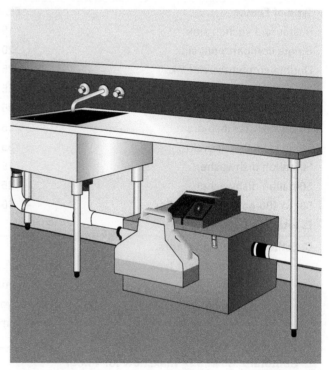

02404-14_F13.EPS

Figure 13 Mechanical grease interceptor.

NCCER – *Plumbing Level Four*   02404-14

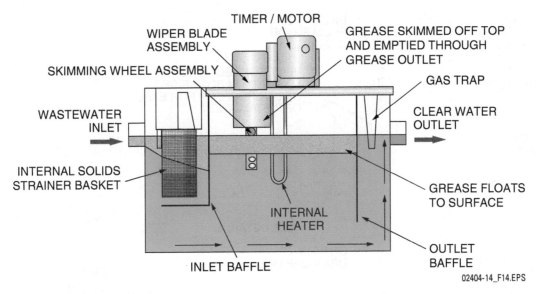

TIMER / MOTOR

WIPER BLADE
ASSEMBLY

GREASE SKIMMED OFF TOP
AND EMPTIED THROUGH
GREASE OUTLET

SKIMMING WHEEL ASSEMBLY

GAS TRAP

WASTEWATER
INLET

CLEAR WATER
OUTLET

INTERNAL SOLIDS
STRAINER BASKET

GREASE FLOATS
TO SURFACE

INTERNAL
HEATER

OUTLET
BAFFLE

INLET BAFFLE

02404-14_F14.EPS

Figure 14 Cut-away illustration of a mechanical grease interceptor showing the self-cleaning process in operation.

American National Standards Institute (ANSI) standard 112.14, *Grease Interceptors*, covers the sizing, design, installation, and testing requirements for mechanical grease interceptors. Ensure that the interceptor has been correctly sized for the calculated wastewater flows for all connected fixtures. Refer to the manufacturer's instructions and the local applicable code for sizing requirements. Install the interceptor so that it can be accessed easily for inspection and maintenance.

Install mechanical grease interceptors downstream from the fixtures to which they are connected, as close to the fixtures as possible. Use flexible-sleeve pipe couplings to connect the interceptor inlet and outlet to the DWV (drain, waste, and vent) system piping. Ensure that the unit has been filled with water before starting the first self-cleaning cycle. Otherwise, the heating element can be damaged. To prevent the risk of back siphonage, install an outlet vent or vacuum breaker as close as possible to the interceptor outlet. The vent or vacuum breaker should be at least half the diameter of the system outlet connection. Where permitted by code, air admittance vents may also be used to vent the interceptor.

> **CAUTION**
>
> Do not install mechanical grease interceptors on waste lines connected to food waste disposers, potato peelers, or solid-waste grinders. The solid waste can clog the interceptor and cause the motor to burn out. Also, the self-cleaning cycle could cause the solid waste to catch fire.

When the installation has been completed, set the interceptor's self-cleaning timer for an appropriate interval as specified in the manufacturer's instructions. Typically, a single cleaning cycle should not be allowed to exceed 2 hours. The internal strainer and external grease container should both be emptied at least daily. The skimming wheel and wiper blade should be inspected at least weekly to ensure it is free of buildup. The manufacturer's instructions will specify other regular and periodic maintenance requirements that operators should observe.

2.2.3 Oil Interceptors

The design of an oil interceptor is similar to that of a grease interceptor. However, oil and gasoline are harder to separate from wastewater than grease is. Codes vary widely on how to install oil interceptors. You may need to consult a local code authority.

Use draw-off hoses to remove oil from an interceptor. Otherwise, the oil will have to be skimmed out by hand—a messy operation. There are two ways to use draw-off hoses. Gravity draw-off systems (*Figure 15*) drain oil into a storage tank. The oil drains out when it reaches a certain height within the interceptor. The height is determined by an adjustable draw-off sleeve inside the interceptor. For gravity systems, set the terminal of the sleeve to ⅛ inch above the normal water operating line. In manual draw-off systems (*Figure 15*), oil drains out of the interceptor by operating a draw-off valve. The terminal of the adjustable draw-off sleeve must be set to ¼ inch below the static wa-

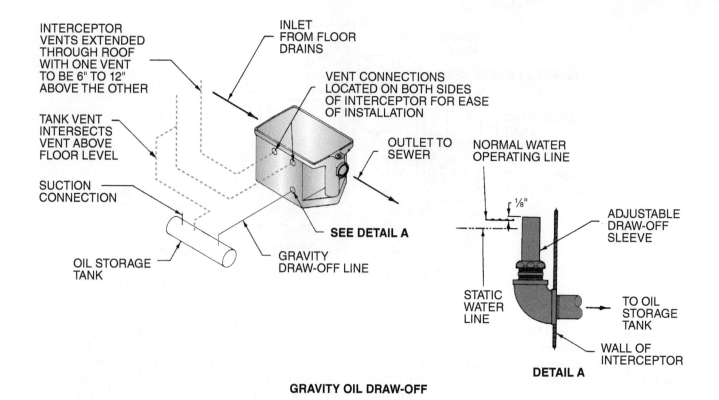

GRAVITY OIL DRAW-OFF

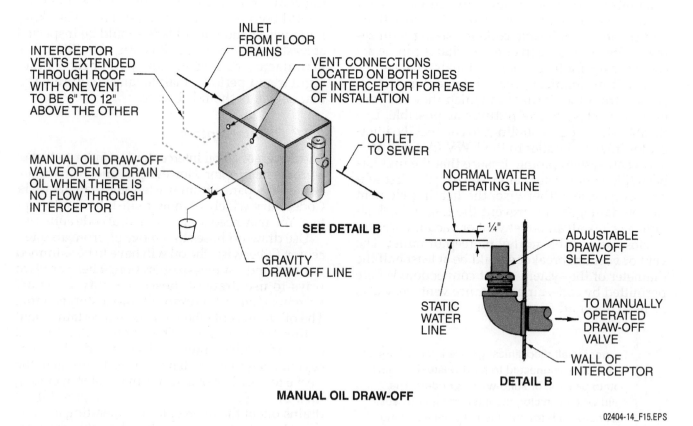

MANUAL OIL DRAW-OFF

02404-14_F15.EPS

Figure 15 Gravity and manual draw-off oil interceptors.

terline. Be sure to close the flow control valve before operating the draw-off valve. Both draw-off systems use sediment buckets or similar devices to catch dirt and grit before they enter the drain lines.

Install vents on oil interceptors. Vents allow flammable and noxious gases to escape harmlessly and prevent back pressure from building up in the interceptor. Oil interceptors require two vents; terminate one vent 6 to 12 inches above the other. Oil storage tanks on gravity draw-off systems must be vented.

A number of businesses produce special wastes that must be handled according to the local applicable health and plumbing codes. For example, dentists' offices require the installation of sediment buckets on waste lines connected to suction devices. Wastes produced by the processing of X-ray plates must be disposed of separately from regular wastewater. Some codes treat the wastes produced by funeral homes during the embalming process as special wastes. Always refer to the local applicable health and plumbing codes when installing DWV systems in facilities that produce unusual wastes.

2.2.4 Sediment Interceptors

Sediment interceptors (*Figure 16*) prevent sand, dirt, and grit from entering the waste system. Small interceptors are installed in place of the P-trap. The top and bottom of sediment interceptors are designed to be removed. This allows for regular cleaning no matter how it is installed. Large sediment interceptors can be installed below, on,

or above the floor (*Figure 17*). They can be either precast or site built (*Figure 18*). Larger inlets may require the installation of a vent.

Specially designed sediment interceptors can also be installed in a variety of other facilities, including the following:

- Car washes
- Parking garages
- Swimming pools
- Commercial laundries (*Figure 19*)
- Jewelry shops
- Computer-chip manufacturing plants

Each type of interceptor has its own special requirements. Be sure to talk with an experienced plumber before you put in a special interceptor. Review your local code as well.

Provide enough space for a drain pan to fit below the interceptor. This will prevent liquid wastes from spilling when the interceptor is cleaned. On small sediment interceptors, the lower opening should be the inlet. Otherwise, the interceptor will not hold enough liquid to form a trap seal. In that case, sewer gas can escape and special wastes can enter the sewer or septic system.

2.2.5 Catch Basins

As you have seen, sediment buckets installed in floor drains intercept sand and grit deposited on the floor. Catch basins perform a similar function, but on a larger scale. Catch basins are reservoirs that allow sediment to settle before wastewater discharges into the drain system. Install catch basins in parking lots, area drains, and yard drains. Catch basins can be customized to match installation needs. Review local codes for applicable requirements.

Catch basins are also used to cool high-temperature wastes before they enter the waste system. Do not install catch basins to intercept sanitary sewage. Catch basins do not require traps if they drain into a storm sewer or drainage field. Ensure that cover plates on the catch basin are easy to access so that built-up sediment can be removed.

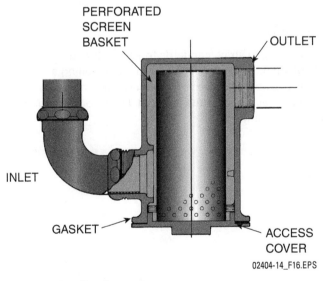

PERFORATED SCREEN BASKET

OUTLET

INLET

GASKET

ACCESS COVER

02404-14_F16.EPS

Figure 16 Small sediment interceptor.

> **CAUTION**
>
> Do not use catch basins or sediment interceptors to trap sanitary wastes. Install traps on all interceptors that discharge into sanitary sewer systems. This will allow the interceptor to drain properly.

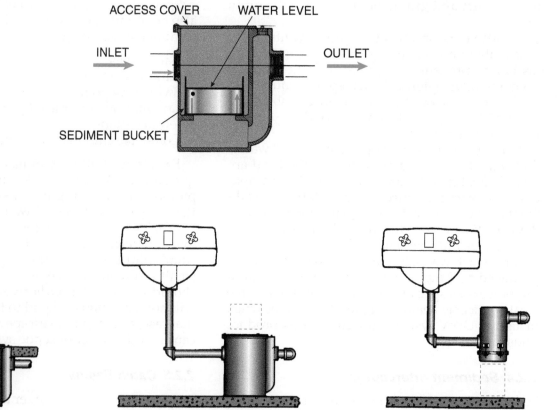

ACCESS COVER WATER LEVEL

INLET OUTLET

SEDIMENT BUCKET

FLUSH-WITH-
FLOOR INSTALLATION

ON-THE-FLOOR
INSTALLATION

SUSPENDED
INSTALLATION

02404-14_F17.EPS

Figure 17 Large sediment interceptors installed on wall-hung hand sinks.

Best Practices for Water Disposal

Special waste drainage systems must be designed and built according to strict standards to ensure that the environment is protected from accidental contamination and pollution. For example, a contractor installing a stormwater interceptor must be able to show how stormwater is cleaned, the types of special wastes removed, and where the wastewater is drained. In effect, stormwater is treated as a hazardous material. Local codes require these systems to be documented, particularly those that dispose of sediment.

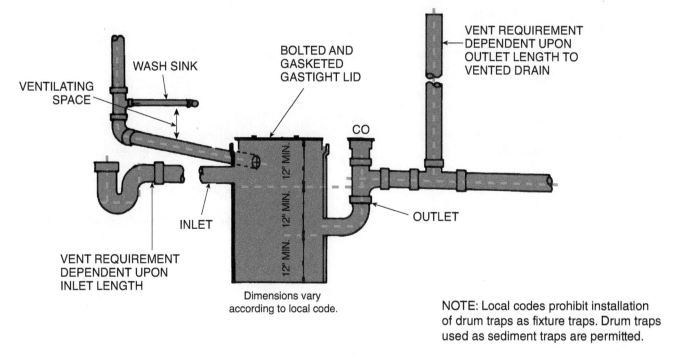

NOTE: Local codes prohibit installation of drum traps as fixture traps. Drum traps used as sediment traps are permitted.

02404-14_F18.EPS

Figure 18 Example of a site-built sediment interceptor with typical information provided in design drawings.

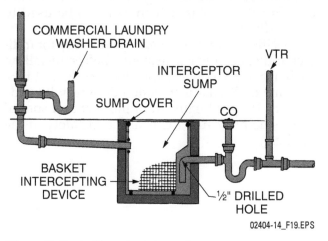

02404-14_F19.EPS

Figure 19 Typical laundry interceptor.

Neutralization Sumps

Chemical waste systems are a type of special waste. Corrosive liquids, spent acids, and toxic chemicals pose serious health risks. They can't discharge directly into the drainage system. They must first be diluted, chemically treated, or neutralized. Neutralization sumps, or tanks, are used to treat these types of wastes. Only then can these wastes discharge into the sewer. Consult your local code for specific requirements.

Some wastes must pass through a neutralizing agent before they can be discharged. Neutralizing agents react chemically with special wastes. The reaction turns the wastes into substances that can be discharged safely. One of the most commonly used agents is limestone.

Additional Resources

ANSI Standard 112.14, *Grease Interceptors*, Latest Edition. Washington, DC: ANSI.

2.0.0 Section Review

1. The type of special waste that can cause pipes to expand and contract is _____.

   a. corrosive waste
   b. high-temperature waste
   c. volatile waste
   d. solid waste

2. When installing an outlet vent or vacuum breaker on a mechanical grease interceptor, it should be sized at least _____.

   a. twice the diameter of the system inlet connection
   b. ½ the diameter of the system inlet connection
   c. twice the diameter of the system outlet connection
   d. ½ the diameter of the system outlet connection

SUMMARY

Indirect waste systems protect food preparation, handling, and storage areas; health care facilities; and residential appliances from contamination by sewer wastes. Indirect waste connections with air gaps are installed on sinks, refrigerators and freezers, and medical and dental equipment.

Special wastes must be treated before they can enter the waste system. Interceptors prevent special wastes from entering the sewer, septic, or stormwater system. Chemicals are diluted and neutralized in sumps. High-temperature wastes cool in catch basins, which can also be used to trap sediment.

Specific installation requirements vary for indirect and special waste systems. Before installing such a system, consult your local code and take the time to discuss installation designs with code experts. Proper treatment of indirect and special wastes helps ensure the health and safety of the public.

1. Systems used to keep sewer wastes from contaminating fixtures if drains back up are known as _____.

   a. safety drain systems
   b. indirect waste systems
   c. sanitary drain systems
   d. isolated waste systems

2. In an air break, _____.

   a. a minimum 1-inch gap is required
   b. a larger pipe drains into a smaller pipe
   c. a partial vacuum is formed
   d. a smaller pipe drains into a larger pipe

3. An open-hub waste receptor must extend above the floor a distance of at least _____.

   a. 1 inch
   b. 3 inches
   c. 6 inches
   d. 12 inches

4. To prevent the escape of sewer gases from a low-use trap, install a(n) _____.

   a. adjustable-level valve
   b. continuous drip system
   c. automatic trap primer
   d. antiventing sealer

5. A grease interceptor must be installed in the indirect waste line serving a _____.

   a. commercial dishwasher
   b. vegetable-preparation sink
   c. commercial clothes washer
   d. medical laboratory

6. A receptor is _____.

   a. any pipe or container into which an indirect waste line empties
   b. another name for a grease trap
   c. a type of overflow preventer
   d. a basin that catches and holds solids

7. Treatment methods for special wastes include evaporation, separation, chemical neutralization, and _____.

   a. encapsulation
   b. dilution
   c. condensation
   d. ablation

8. A device or structure that slows wastewater flow so that waste materials can be separated by settling, floating, or evaporating is referred to as a(n) _____.

   a. isolator
   b. separator
   c. interceptor
   d. receptor

9. A draw-off hose is used to clean a _____.

   a. catch basin
   b. grease interceptor
   c. water filter
   d. sediment interceptor

10. The automatic cleaning cycle for a mechanical grease interceptor should be limited to a maximum of _____.

    a. 30 minutes
    b. 1 hour
    c. 90 minutes
    d. 2 hours

Trade Terms Quiz

Fill in the blank with the correct term that you learned from your study of this module.

1. A hose used to siphon grease and oil from interceptors is called a(n) _____.

2. A(n) _____ is a simple backflow preventer that consists of a space between the indirect waste system and the building waste system.

3. A removable basin in a drain that prevents solid wastes from entering a sanitary system is called a(n) _____.

4. A(n) _____ is a drainpipe attached to a fixture or appliance that is separated from the regular drainage system by a backflow preventer.

5. The partition inside the body of an interceptor that is designed to reduce turbulence caused by incoming wastewater and to block wastes from escaping through the outlet is called a(n) _____.

6. A(n) _____ is a vertical pipe at the terminal of a waste line that allows indirect waste to mix with air.

7. A system that drains oil out of an interceptor through a valve-operated draw-off hose is called a(n) _____.

8. _____ is wastewater that flows through an indirect waste pipe.

9. A basin designed to hold liquids is called a(n) _____.

10. A(n) _____ is a system that automatically drains oil from an interceptor into a storage tank.

11. A device that feeds water into a low-use trap, either automatically, electronically, or under the pressure of gravity, to allow the trap to maintain a constant seal is called a(n) _____.

12. _____ is waste that must be removed, diluted, or neutralized before it can be discharged into a sanitary system.

13. A grease interceptor fitted with a timer-activated self-cleaning capability is called a(n) _____.

14. A(n) _____ is a device that separates special wastes before they can enter a sewer, septic, or stormwater system.

15. A waste drainpipe or pipe hub that extends above the floor line is called a(n) _____.

16. A(n) _____ is a backflow preventer in which a smaller pipe drains into a larger pipe.

17. A reservoir installed before the sanitary piping that allows sediment to settle out of wastewater is called a(n) _____.

18. A(n) _____ is a device that regulates the flow of wastewater into grease and oil interceptors.

Trade Terms

Air break	Draw-off hose	Interceptor	Open-hub waste receptor
Air gap	Flow control fitting	Manual draw-off system	Receptor
Automatic trap primer	Gravity draw-off system	Mechanical grease	Sediment bucket
Baffle	Indirect waste	interceptor	Special waste
Catch basin	Indirect waste system	Modified waste line	

Trade Terms Introduced in This Module

Air break: A backflow preventer in which a smaller pipe drains into a larger pipe.

Air gap: A simple backflow preventer that consists of a space between the indirect waste system and the building waste system. The gap must be two times the diameter of the indirect waste disposal pipe and not less than 1 inch.

Automatic trap primer: A device that feeds water into a low-use trap, allowing the trap to maintain a constant seal. Automatic trap primers may be designed to close automatically (also called automatic dosing systems), or to operate electronically or under the pressure of gravity.

Baffle: Partition inside the body of an interceptor. Baffles are designed to reduce turbulence caused by incoming wastewater and to block wastes from escaping through the outlet.

Catch basin: A reservoir installed before the sanitary piping. It allows sediment to settle out of wastewater.

Draw-off hose: A hose used to siphon grease and oil from interceptors.

Flow control fitting: A device that regulates the flow of wastewater into grease and oil interceptors.

Gravity draw-off system: A system that automatically drains oil from an interceptor into a storage tank.

Indirect waste: Wastewater that flows through an indirect waste pipe.

Indirect waste system: A drainpipe attached to a fixture or appliance that is separated from the regular drainage system by a backflow preventer.

Interceptor: A device that separates special wastes before they can enter the sewer, septic, or stormwater system.

Manual draw-off system: A system that drains oil out of an interceptor through a valve-operated draw-off hose.

Mechanical grease interceptor: A grease interceptor fitted with a timer-activated self-cleaning capability.

Modified waste line: A vertical pipe at the terminal of a waste line that allows indirect waste to mix with air.

Open-hub waste receptor: A waste drainpipe or pipe hub that extends above the floor line.

Receptor: A basin designed to hold liquids.

Sediment bucket: A removable basin in a drain that prevents solid wastes from entering a sanitary system.

Special waste: Waste that must be removed, diluted, or neutralized before it can be discharged into a sanitary system.

Additional Resources

This module presents thorough resources for task training. The following resource material is suggested for further study.

ANSI Standard 112.14, *Grease Interceptors*, Latest Edition. Washington, DC: ANSI.

Introduction to Wastewater Treatment Processes. 2014. Bruce Jefferson, Simon Parsons, and Elise Cartmell. New York: Wiley-Blackwell.

Water, Sanitary, and Waste Services for Buildings, Fifth Edition. 2001. Alan F. E. Wise and J. A. Swaffield. Boston, MA: Routledge.

Figure Credits

Courtesy of Watts, Figure 3a
Wade Division, Tyler Pipe, Figures 8–10, Figures 15–17

Section Review Answer Key

Answer	Section Reference	Objective
Section One		
1. c	1.1.1	1a
Section Two		
1. b	2.1.0	2a
2. d	2.2.2	2b

NCCER CURRICULA — USER UPDATE

NCCER makes every effort to keep its textbooks up-to-date and free of technical errors. We appreciate your help in this process. If you find an error, a typographical mistake, or an inaccuracy in NCCER's curricula, please fill out this form (or a photocopy), or complete the online form at **www.nccer.org/olf**. Be sure to include the exact module ID number, page number, a detailed description, and your recommended correction. Your input will be brought to the attention of the Authoring Team. Thank you for your assistance.

Instructors – If you have an idea for improving this textbook, or have found that additional materials were necessary to teach this module effectively, please let us know so that we may present your suggestions to the Authoring Team.

NCCER Product Development and Revision

13614 Progress Blvd., Alachua, FL 32615

Email: curriculum@nccer.org
Online: www.nccer.org/olf

❏ Trainee Guide   ❏ Lesson Plans   ❏ Exam   ❏ PowerPoints   Other _____

Craft / Level: _____ Copyright Date: _____

Module ID Number / Title: _____

Section Number(s): _____

Description: _____

_____

_____

_____

_____

Recommended Correction: _____

_____

_____

_____

_____

Your Name: _____

Address: _____

_____

Email: _____ Phone: _____

02405-14

Hydronic and Solar Heating Systems

Overview

Hydronic systems use heated water or steam to heat rooms or an entire building through the use of radiant or forced heat. Most hydronic systems are powered by electricity or gas; solar heating systems, are powered by the sun's heat. Hydronic and solar heating systems typically consist of a water heater or boiler (or, in solar heating systems, a collector), a circulation system, radiators, and controls. When installing these systems, plumbers first rough-in the piping, then install the heater, boiler or collector, and controls, and finally test, balance, and start the system.

Module Five

Trainees with successful module completions may be eligible for credentialing through the NCCER Registry. To learn more, go to **www.nccer.org** or contact us at **1.888.622.3720**. Our website has information on the latest product releases and training, as well as online versions of our *Cornerstone* magazine and Pearson's product catalog.

Your feedback is welcome. You may email your comments to **curriculum@nccer.org**, send general comments and inquiries to **info@nccer.org**, or fill in the User Update form at the back of this module.

This information is general in nature and intended for training purposes only. Actual performance of activities described in this manual requires compliance with all applicable operating, service, maintenance, and safety procedures under the direction of qualified personnel. References in this manual to patented or proprietary devices do not constitute a recommendation of their use.

Objectives

When you have completed this module, you will be able to do the following:

1. Describe the principles of hydronic and solar heating systems.
   a. Describe water heaters, boilers, and collectors.
   b. Describe hot water circulation.
   c. Describe radiators and radiant loops.
2. Describe the basic types of hydronic and solar heating systems and their components.
   a. Describe the basic types of hydronic systems.
   b. Describe the basic types of solar heating systems.
3. Describe the procedures for roughing-in, installing, and testing hydronic and solar heating system piping.
   a. Describe how to rough-in the piping.
   b. Describe how to install boilers and heating units.
   c. Describe how to install controls.
   d. Describe how to test and balance the system.
   e. Describe corrosion prevention techniques.

Performance Task

Under the supervision of your instructor, you should be able to do the following:

1. Lay out a hydronic or solar heating system.

Trade Terms

Active system
Airtrol valve
Anodic inhibitor
Batch system
Boiler
Closed loop system
Collector
Convection

Differential
    thermostat
Direct return system
Direct system
Diverter tee
Drain-down system
Ethylene glycol
Forced heat system
Heat exchanger

Hydronic system
Indirect system
Limit switch
One-pipe system
Open loop system
Parallel
Passive system
Primary-secondary
    system

Radiant loop
Radiant system
Radiation
Radiator
Reverse return
    system
Reverse
    thermosiphoning
Series

Solar heating system
Specific heat
Steam dome
Thermistor
Thermosiphon
    system
Thermosiphoning
Two-pipe system
Zone control valve

Industry-Recognized Credentials

If you're training through an NCCER-accredited sponsor, you may be eligible for credentials from NCCER's Registry. The ID number for this module is 02405-14. Note that this module may have been used in other NCCER curricula and may apply to other level completions. Contact NCCER's Registry at 888.622.3720 or go to **www.nccer.org** for more information.

Code Note

Codes vary among jurisdictions. Because of the variations in code, consult the applicable code whenever regulations are in question. Referring to an incorrect set of codes can cause as much trouble as failing to reference codes altogether. Obtain, review, and familiarize yourself with your local adopted code.

Contents

Topics to be presented in this module include:

Figures

1.0.0 PRINCIPLES OF HYDRONIC AND SOLAR HEATING SYSTEMS

Objective

Describe the principles of hydronic and solar heating systems.

   a. Describe water heaters, boilers, and collectors.
   b. Describe hot water circulation.
   c. Describe radiators and radiant loops.

Trade Terms

Boiler: In a hydronic system, a device used to heat or boil water. Boilers differ from typical water heaters in that they have a greater capacity, operate at higher temperatures and pressures, and have a greater heat output.

Collector: A device used in a solar heating system to receive heat from sunlight and transfer it to water or another liquid.

Convection: Circulation caused by the sinking of dense cold water and the rising of hot water.

Forced heat system: A type of hydronic system that uses fans and ducts to circulate air heated by passing it over coils containing hot water.

Hydronic system: A plumbing system that uses water or steam to heat a building. Also called a radiant system.

Radiant loop: A heating unit that consists of a network of hot water pipes in the floor or ceiling.

Radiant system: Another name for a hydronic system.

Radiation: The transfer of heat between bodies through space.

Radiator: A heating unit that emits heat from water via a series of metal fins.

Solar heating system: A hydronic system that uses the sun to heat water.

Specific heat: A measure of how much heat a liquid can hold per pound, measured in British thermal units (Btus).

Plumbing systems that use water or steam to heat a room or an entire building are called hydronic systems. While most hydronic systems use electricity or gas to heat the water in the system, solar heating systems use the sun to heat water. Hydronic and solar heating systems can be divided into two broad categories: radiant systems and forced heat systems. Radiant hydronic systems transfer the heat from hot water to objects in a room, utilizing a principle called radiant heat. In forced heat systems, also called water-sourced heat pumps, blowers circulate air over coils containing hot water in order to transfer the heat from the coils to the air. The warmed air is then circulated throughout the building using air handlers and ducts, much like a conventional heating, ventilating, and air conditioning (HVAC) system. Forced heat systems are hybrid systems that are used mostly in large buildings. Hydronic and solar heating systems offer many advantages over forced heat systems, including the following:

- Eliminate drafts, cold spots, and dry air
- Do not spread dust, mold, or airborne diseases
- Heat all the objects in a room, resulting in greater heat retention
- Are more cost-effective to operate

Hydronic and solar heating systems work somewhat like recirculation systems. The major difference is that hydronic and solar fixtures provide heat instead of water. Hot water recirculation systems must be insulated to prevent heat loss, but in hydronic and solar heating systems, heat loss at the correct locations is precisely the goal. The heat from the water in the system is used to heat the rooms in a building.

Hydronic and solar heating systems use the heat given off by cooling water or steam to heat rooms. Heating occurs because water has a higher specific heat than almost any other liquid. Specific heat is a measure of how much heat a liquid can hold per pound. Specific heat is measured in British thermal units (Btus).

Remember that 1 Btu is the amount of energy required to raise one pound of water 1°F. When one pound of hot water cools from 190°F to 150°F, it gives off 40 Btus of heat. Steam is even more efficient than water. When one pound of steam condenses back into water, it gives off 970 Btus.

The components of typical hydronic systems are illustrated in *Figure 1*. A basic hydronic heating system is able to accomplish the following:

- Create hot water or steam
- Distribute hot water or steam
- Transfer the heat from the hot water or steam into the room

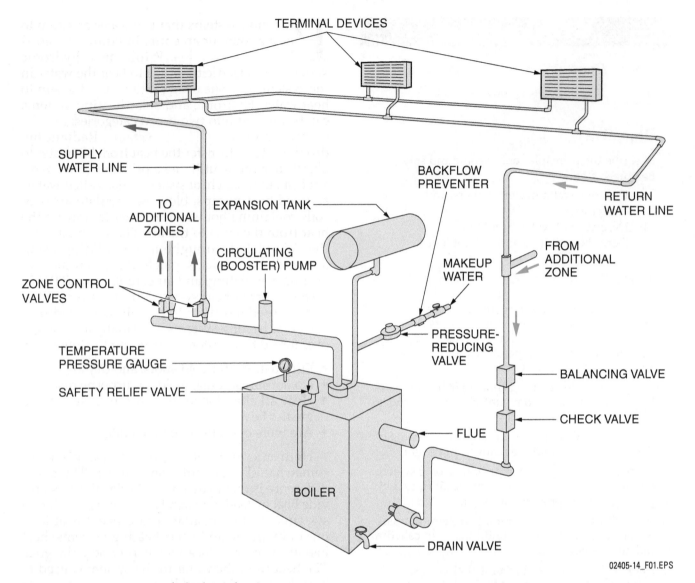

TERMINAL DEVICES

SUPPLY
WATER LINE

TO
ADDITIONAL
ZONES

EXPANSION TANK

BACKFLOW
PREVENTER

RETURN
WATER LINE

ZONE CONTROL
VALVES

CIRCULATING
(BOOSTER) PUMP

MAKEUP
WATER

FROM
ADDITIONAL
ZONE

TEMPERATURE
PRESSURE GAUGE

PRESSURE-
REDUCING
VALVE

BALANCING VALVE

SAFETY RELIEF VALVE

CHECK VALVE

FLUE

BOILER

DRAIN VALVE

02405-14_F01.EPS

Figure 1 Basic components of a hydronic heating system.

1.1.0 Understanding How Water Heaters, Boilers, and Collectors Work

Most hydronic systems use potable water from the water supply system. The system is topped off as needed when water is lost through evaporation. Water in hydronic systems typically circulates at a temperature between 180°F and 210°F. Steam systems usually operate between 212°F and 220°F. When steam cools, it turns back into water; this water is called condensate. Condensate returns to the heating system and is reheated into steam. Steam systems can usually be found in houses that were built before World War II.

Some small hydronic systems use water heaters to supply the hot water, but most systems use a boiler to provide the hot water or steam. Boilers have a larger capacity than water heaters; they also generate more heat. Often, the terms *water heater* and *boiler* are used to mean the same thing,

Early Hot-Water and Steam Hydronic Systems

The earliest hydronic systems used steam to provide radiant heat. The famous Roman baths relied on steam heat, for example. By the mid-nineteenth century, steam systems were in common use in homes and commercial buildings. Because many early steam systems operated under pressure, their boilers often exploded. Hot water systems were introduced into the United States from Canada in the late 1880s. Hot water was safer to use than steam because hot water systems operated at normal air pressure. Hot water systems were also easier to maintain and operate. By the end of World War II, hot water replaced steam as the heating medium of choice.

The Franklin Stove

In 1741, Benjamin Franklin invented a metal-lined stove that produced more heat and less smoke than other types of stoves in use at the time. The stove uses the principle of radiant heat to warm a room. Stoves based on Franklin's design are still widely used today. The Franklin stove is considered to be the first radiant heat system.

but they have very different requirements and applications. According to many state and local plumbing codes, for example, a boiler has the following characteristics:

- A capacity of at least 120 gallons
- Water temperature at or above 210°F
- Pressure of at least 150 pounds per square inch gauge (psig)
- Heat input of at least 200,000 Btus per hour

As a professional plumber, you need to know how to use these terms correctly.

Another way to heat water is to use solar energy. In a solar heating system, water is pumped through a collector, where it is heated by sunlight. A typical collector is a narrow rectangular box with a metal plate, a water pipe, and a transparent cover (see *Figure 2*). The collector is set up so that sunlight enters through the transparent cover. Sunlight heats the metal plate inside the collector. The heat is transferred to the water pipe, which

runs through the collector in a series of S-curves. The transparent cover acts as a reflector to keep the heat from radiating back out. The hot water is then pumped throughout the system. Some systems use a liquid other than water to absorb the sun's heat.

Hydronic and solar heating systems require the installation of an expansion tank. The tank absorbs excess pressure caused by hot water expansion. Expansion tanks prevent hot water from being wasted through the temperature and pressure (T/P) relief valve. The size of the expansion tank will vary based on the capacity of the system. Some hydronic systems may use a bladder-type expansion tank.

Guidelines for Water Heaters and Boilers

In many jurisdictions water heaters and boilers are governed by separate sets of guidelines. The 2003 collection of model building codes published by the International Code Council (ICC) reflects this separate treatment. The *International Plumbing Code®* (IPC) covers water heaters but not boilers, while the *International Mechanical Code®* (IMC) covers boilers in some detail. Many state and local codes are based on the ICC codes. Some jurisdictions require a special permit for installation of boilers.

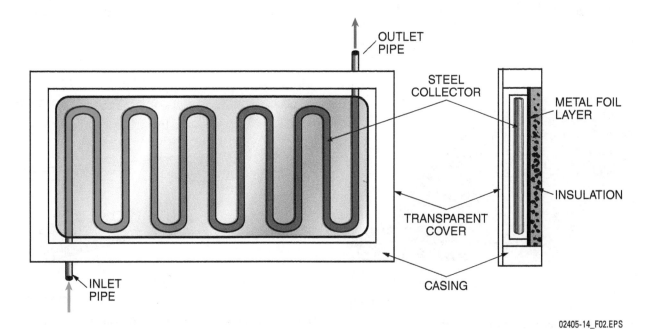

OUTLET PIPE

STEEL COLLECTOR

METAL FOIL LAYER

INSULATION

TRANSPARENT COVER

CASING

INLET PIPE

02405-14_F02.EPS

Figure 2 Typical solar collector.

1.2.0 Understanding How Hot Water Circulation Works

In many systems, once the water is heated, a pump circulates the water throughout the system. Centrifugal pumps are the most commonly used type of circulating pump. Use a stainless steel or bronze pump to circulate potable water. These metals do not rust; rust will contaminate the water and may clog the pipes. Consult local codes for additional requirements about circulating pumps.

Some hydronic and solar heating systems do not use circulating pumps. These systems rely on convection to circulate the water. Convection is circulation caused by the different densities of hot and cold water. Hot water rises, and cold water sinks. The system is designed so that the heavier cold water flows back to the tank or collector. This forces the hot water in the tank or collector to circulate.

1.3.0 Understanding How Radiators and Radiant Loops Work

Sometimes when you pick up a cup of hot coffee, the cup is too hot to hold. That transfer of heat from the coffee cup to your hand is called radiation. Radiation is the discharge of heat through space from one body to another. Radiation does not heat the air between bodies very much. Hydronic systems transfer the heat from hot water to objects in a room in the same way that the coffee cup warms your hand. Because hydronic systems use radiation to distribute heat, they are often called radiant systems.

Hydronic systems use radiators and radiant loops to transfer heat from the hot water or steam into a room. Radiators emit heat from a series of metal fins. Water circulating through the hydronic system heats the fins (see *Figure 3*).

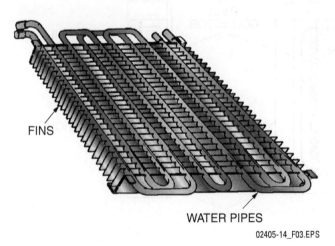

FINS

WATER PIPES

02405-14_F03.EPS

Figure 3 Components of a typical radiator.

Objects in the room, including people, then absorb this heat. The fins also set up natural convection air currents, which draw cool air from the floor of the room, heat the air as it passes between the fins, and discharge the heated air into the room.

Radiators, the most common method of heat transfer in a hydronic system, can be built into the baseboards of a room. They can also be installed as freestanding units (*Figure 4*). Most radiators use copper tubing and aluminum fins. Steel is sometimes used for the pipe and fins. Radiators are the most common method of heat transfer in a hydronic system. They are relatively easy to install and service.

A radiant loop is a network of hot water pipes embedded in the ceiling or floor (*Figure 5*). Radiant loops turn the floor or ceiling into a radiator.

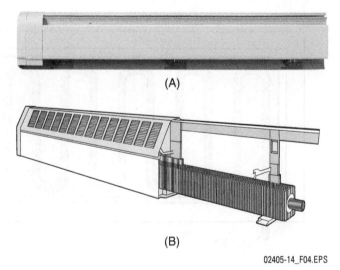

(A)

(B)

02405-14_F04.EPS

Figure 4 Examples of freestanding radiators.

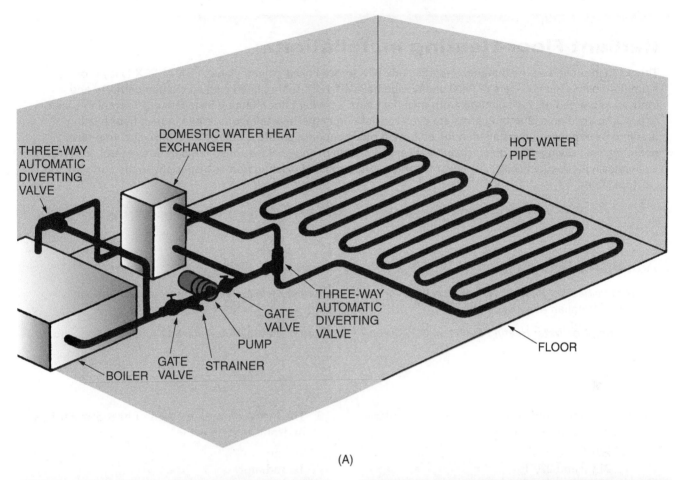

(A)

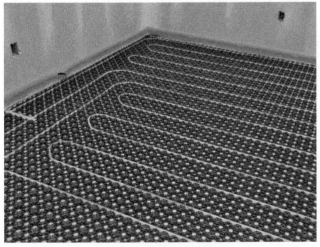

(B) MATS OVER A
CONCRETE SUBFLOOR

(C) STAPLE-DOWN INSTALLATION OVER A
FOAM-INSULATED SUBFLOOR

02405-14_F05.EPS

Figure 5 Radiant loops installed in floors.

Pipe in a radiant loop can be made from copper, CPVC (chlorinated polyvinyl chloride), polyethylene, or PEX (cross-linked polyethylene). Tubing for the loop can be laid in a concrete slab or hung underneath a wood floor. Because they are located in the floor or ceiling, radiant loops do not get in the way when the room is painted or cleaned. However, they are more complex to design and install than a radiator system. They also require more piping.

Radiant Floor-Heating Installations

Thanks to improved piping materials—particularly PEX (cross-linked polyethylene), PEX-AL-PEX (a layer of aluminum between two layers of PEX) and Wirsbo hePEX™ (PEX tubing lined with an oxygen barrier)—and innovative pre-engineered solutions from manufacturers — radiant floor-heating installation is becoming easier and less costly. Some plumbing firms are even specializing in these installations. While radiant floor heating is still more expensive to install than forced heat, it offers superior comfort, energy efficiency, and long-term operating cost savings. It is quiet, clean, and easy to maintain, and enables zoned temperature control, making it an increasingly frequent choice for custom homes and retrofits of existing homes, especially in the northern and western states.

Additional Resources

Hydronic Radiant Heating: A Practical Guide for the Nonengineer Installer. 1998. Dan Holohan. Bethpage, NY: Dan Holohan Associates.

Repairing Hot Water Heating Systems. 2012. Robert VanNorden. Seattle, WA: Amazon Digital Services, Inc.

1.0.0 Section Review

1. Steam in hydronic systems typically circulates at a temperature between _____.

    a. 220°F and 228°F
    b. 212°F and 220°F
    c. 180°F and 210°F
    d. 150°F and 180°F

2. Hydronic and solar heating systems that do not use circulating pumps rely instead on the principle of _____.

    a. radiation
    b. evaporation
    c. convection
    d. conduction

3. The most common method of heat transfer in a hydronic system is the _____.

    a. fan
    b. radiator
    c. heat pump
    d. thermosiphon

SECTION TWO

2.0.0 TYPES AND COMPONENTS OF HYDRONIC AND SOLAR HEATING SYSTEMS

Objective

Describe the basic types of hydronic and solar heating systems and their components.

a. Describe the basic types of hydronic systems.
b. Describe the basic types of solar heating systems.

Trade Terms

Active system: A solar heating system that uses a pump and controls to circulate hot water through the system.

Batch system: A solar heating system in which water is both heated and stored in the collector. The system uses the principle of convection to circulate water.

Closed loop system: Another name for an indirect system.

Direct return system: A two-pipe system in which the return water flows in the opposite direction from the flow in the supply pipe.

Direct system: A solar heating system that uses a pump and temperature controls to control the flow of hot water. It is also called an open loop system.

Drain-down system: A direct system in which water drains out of the collector when the pump shuts off.

Ethylene glycol: An alcohol-based liquid used in an indirect system to transfer heat from a collector to a heat exchanger.

Heat exchanger: A device used in an indirect system to transfer heat from one liquid to another in such a way that the two liquids do not come into contact with each other.

Indirect system: An active system that uses a liquid other than water to distribute heat from the collector. It is also called a closed loop system.

One-pipe system: A hydronic system in which water or steam is fed to the heating units and returned to the boiler through a single run of pipe.

Open loop system: Another name for a direct system.

Parallel: A one-pipe system in which each heating unit draws only the amount of water it requires from the supply line.

Passive system: A solar heating system that relies on convection to circulate water.

Primary-secondary system: A one-pipe system with a parallel arrangement.

Reverse return system: A two-pipe system in which the return water flows in the same direction as the flow in the supply pipe.

Reverse thermosiphoning: In a thermosiphon system, a condition in which water flows backward through the system.

Series: A one-pipe system in which all the water in the supply loop goes through each heating unit.

Thermosiphon system: A passive system in which hot water is stored in a tank located higher than the solar collector.

Thermosiphoning: In a thermosiphon system, the process whereby cold water flows back to the collector to be reheated.

Two-pipe system: A hydronic system that uses separate supply and return lines.

Hydronic and solar heating systems can be designed to meet the heating needs of almost any building. The same basic components can be assembled to heat a three-bedroom home or a multistory office complex. Your skills as a plumber will allow you to design a system that suits both the type of building and the client's needs. In this section, you will learn about the different types of hydronic and solar heating systems. Each of these systems is designed for specific applications. Local codes specify the heating systems that can be installed in your area.

2.1.0 Types of Hydronic Systems

Some hydronic systems supply water or steam to the heating units and return the water to the boiler through a single run of pipe. They are called one-pipe systems. Other systems use separate supply and return lines. They are called two-pipe systems. Each of these systems can be built in two different ways, depending on the type of building and its heating needs. At first, these configurations may seem confusing. Take the time to review the different designs and become familiar with them. Each has advantages for different applications.

2.1.1 One-Pipe Systems

Heating units in a one-pipe system can be installed in either series or parallel. In a series arrangement, all the water in the loop goes through each heating unit (see *Figure 6*). Series systems are simple to install and relatively inexpensive. However, series arrangements have several drawbacks. Because all the water flows through each heating unit, each unit must be sized to carry the total flow. The water temperature drops as the water flows through each heating unit. This means that heating units farther down the line cannot provide as much heat as units closer to the boiler. This type of hydronic system is practical only for the smallest applications.

In a parallel arrangement, not all the water in the loop flows through each heating unit. Each heating unit draws only what is required from the main supply (see *Figure 7*). One-pipe systems with a parallel arrangement are also called primary-secondary systems. Parallel systems distribute heat more evenly throughout the building than series systems. Heating units can be sized to take less water.

For one-pipe steam systems, supply pipes should be at least 2 inches in diameter. This allows steam to flow to the heating units while the condensate drains back to the boiler down the same pipe. This two-way flow often makes steam systems noisier than one-pipe water systems.

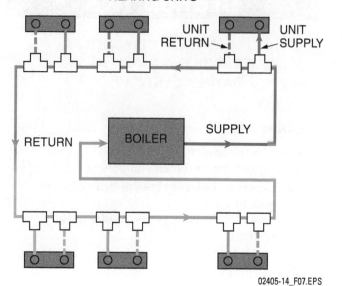

02405-14_F07.EPS

Figure 7 Parallel arrangement of heating units in a one-pipe hydronic system.

2.1.2 Two-Pipe Systems

Two-pipe systems have separate supply and return lines. This results in better temperature control than one-pipe systems. In a two-pipe system, the diameter of the supply pipe decreases after it passes each heating unit. Likewise, the return pipe's diameter increases after each unit. The heating units are sized only for their individual load. Because they require more pipe and fittings, two-pipe systems are more expensive to install.

There are two types of two-pipe systems. The difference has to do with the direction of the return flow. In a direct return system, the return pipe is installed parallel to the supply pipe. The return water or condensate flows back to the boiler in the opposite direction from the flow in the supply pipe (see *Figure 8*). This means that the water from the first heating unit in the system is the first to return to the boiler. In a reverse return system, the return and supply pipes are also parallel. However, the return water flows in the same direction as the flow in the supply pipe (see *Figure 9*). In a reverse return system, water from the last heating unit is the first to return to the boiler.

2.2.0 Types of Solar Heating Systems

For buildings that receive direct sunlight year-round, solar power is a cost-effective heating method. In a solar heating system, the fuel—sunlight—is free. This can offset the added expense of installing special solar heating equipment. A

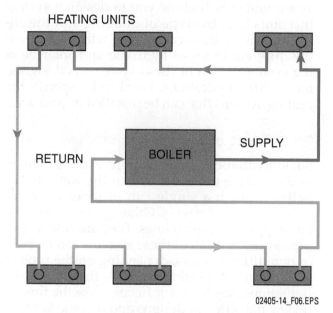

02405-14_F06.EPS

Figure 6 Series arrangement of heating units in a one-pipe hydronic system.

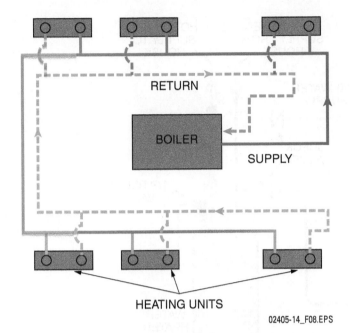

RETURN

BOILER

SUPPLY

HEATING UNITS

02405-14_F08.EPS

Figure 8 Direct return arrangement in a two-pipe system.

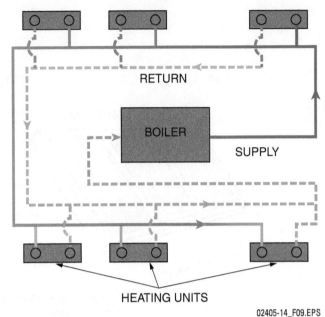

RETURN

BOILER

SUPPLY

HEATING UNITS

02405-14_F09.EPS

Figure 9 Reverse return arrangement in a two-pipe system.

boiler or water heater is often used to supplement solar heating so that a building can still be heated when there is little or no sunlight. Solar heating systems are available in many different designs. This module will focus on the systems you are most likely to encounter. Solar heating systems can be grouped into the following categories:

- Passive systems, which include batch systems and thermosiphon systems
- Active systems, which include direct systems and indirect systems

Passive systems rely on convection to move water through the system, whereas active systems use pumps and controls. Each of these systems is discussed in more detail in the following sections.

2.2.1 Batch Systems

A batch system is the simplest way to use solar energy to provide heat. A batch system consists of a solar collector with one or more water tanks inside it. The collector heats the water in the tanks. When hot water is drawn, water flows from the collector tanks. Cold water then flows into the collector to make up for the hot water that was used. A conventional water heater supplements the collector supply as needed (see *Figure 10*).

Batch systems are less effective than other solar heating systems for space heating. They can be built from supplies available at most hardware stores. Manufactured batch systems are also available. Because the water in the system does not move when there is no demand, batch systems are vulnerable to freezing. Ensure that pipes

are insulated thoroughly, and install a cover on the collector to prevent heat loss after sunset.

2.2.2 Thermosiphon Systems

Thermosiphon systems also use convection to move hot water. Hot water is stored in a tank located higher than the solar collector (see *Figure 11*). After water is heated in the collector, it flows into the top of the storage tank. As the water in the tank cools, it settles to the bottom of the tank. From there, the cool water flows back to the collector, where it is warmed again. This cycle is called thermosiphoning.

Thermosiphon systems operate by the difference in density of the water in the storage tank and the water in the collector. Cool water in the storage tank and hot water in the collector means rapid flow. Hot water in the storage tank and cool water in the collector means little or no flow. This condition could even lead to reverse thermosiphoning. Reverse thermosiphoning is a condition in which water flows backwards through the system. If this happens, the system will lose heat. Install check valves to prevent reverse thermosiphoning. Reverse thermosiphoning also can occur in other solar heating systems.

As with batch systems, ensure that thermosiphon systems are insulated against heat loss. Thermosiphon systems do not have temperature controls. Water temperature in the system will vary with changes in the weather. Thermosiphon systems do not provide 100 percent of the heating needs of a typical building. Backup heating is required for most installations.

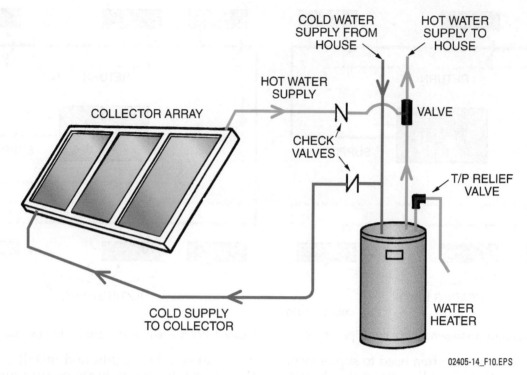

Figure 10 Typical batch system.

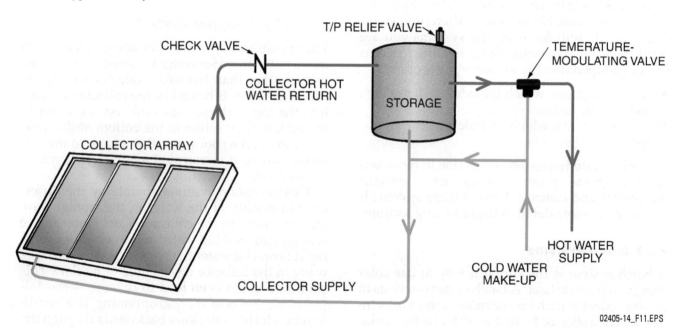

Figure 11 Typical thermosiphon system.

2.2.3 Direct Systems

Like thermosiphon systems, direct systems have a collector and a storage tank. Direct systems use pumps to circulate the hot water (see *Figure 12*). In a direct system, water temperature can be controlled. The pump turns on when the water in the collector reaches a preset temperature higher than the temperature of the water in the storage

tank. Direct systems are simple to construct. They are also very efficient to operate. Direct systems are also called open loop systems.

In some direct systems, water drains out of the collector when the pump shuts off. The system will not freeze when it is not in use. Such systems are called drain-down systems (see *Figure 13*). Drain-down systems often have valve problems if

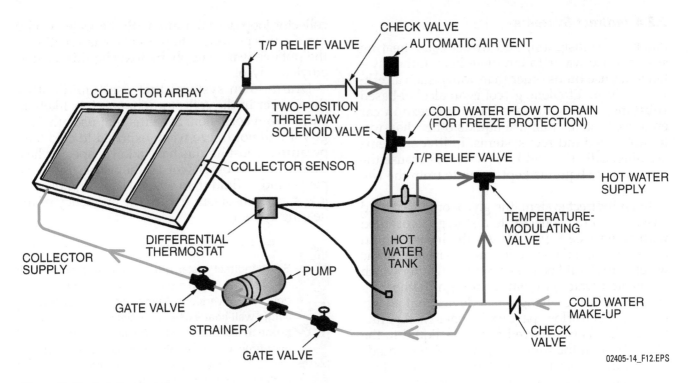

Figure 12 Typical direct system.

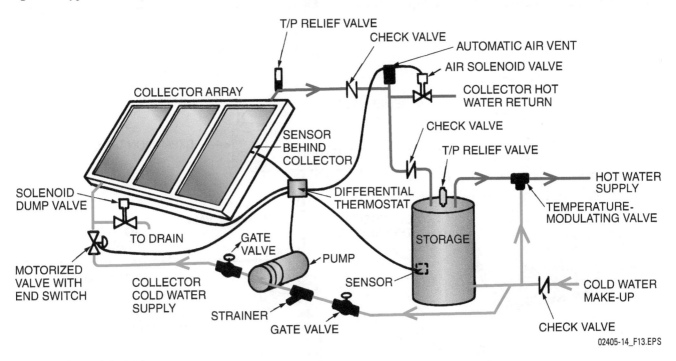

Figure 13 Typical drain-down system.

the valve is not used for a long time. Failure of the valve in a drain-down system could damage the collector. When water drains from the system, it is replaced with air. Repeated exposure to air and water can cause corrosion in the pipes.

Direct systems are vulnerable to scale buildup from minerals in domestic water. Many direct systems use aluminum collectors. Aluminum eventually corrodes when exposed to domestic water supplies. Corrosion can cause scale buildup in the pipes and other components exposed to the water. As a result, the system cannot operate as efficiently as before. These problems limit the applications of direct systems.

2.2.4 Indirect Systems

Batch, thermosiphon, and direct solar heating systems use water to circulate heat. Other systems can use fluids other than water, such as ethylene glycol. Ethylene glycol is an alcohol-based substance commonly used as antifreeze in car engines. Systems that use liquids other than water are called indirect systems. Indirect systems are often called closed loop systems because the heat-transfer liquid is kept separate from the water supply.

In an indirect system, the glycol or other liquid passes from the collector to a storage tank filled with water (see *Figure 14*). As the heated liquid passes through the tank, it transfers its heat to the water with the aid of a heat exchanger, after which the heated water circulates through the heating system. A heat exchanger is a device that allows two liquids to flow past each other in separate tubes. Heat is transferred from one liquid to the other through the tube walls without the liquids coming into contact with each other. Most heat exchangers are designed with a double wall (*Figure 15*). The double wall ensures that the liquid in the collector loop does not mix with the water. Codes require this protection to prevent contamination of the potable water supply by toxic liquids such as ethylene glycol.

Drain-down systems that use liquids other than water in the collector loop are less likely to suffer from corrosion. Systems that use ethylene glycol will not freeze when exposed to cold temperatures. Most liquids have a lower specific heat than water. In other words, they carry less heat per pound of fluid than water. Using liquids with low specific heat can reduce the efficiency of the heating system.

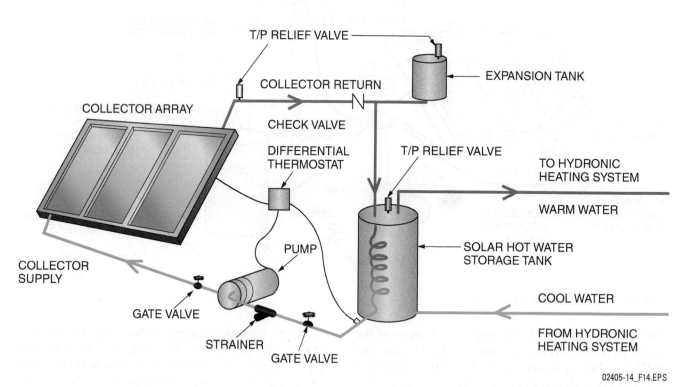

Figure 14 Typical indirect system.

02405-14_F14.EPS

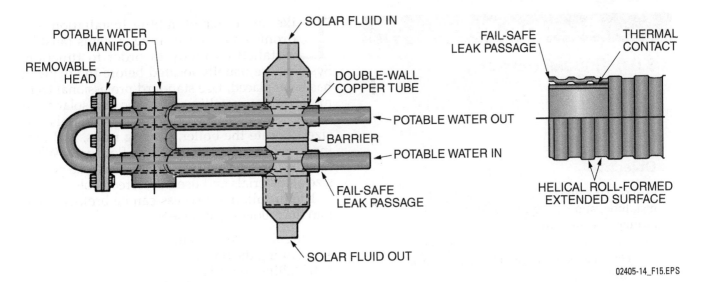

POTABLE WATER MANIFOLD

REMOVABLE HEAD

SOLAR FLUID IN

DOUBLE-WALL COPPER TUBE

POTABLE WATER OUT

BARRIER

POTABLE WATER IN

FAIL-SAFE LEAK PASSAGE

SOLAR FLUID OUT

FAIL-SAFE LEAK PASSAGE

THERMAL CONTACT

HELICAL ROLL-FORMED EXTENDED SURFACE

02405-14_F15.EPS

Figure 15 Double-walled heat exchanger.

Making Solar Heating Cost Effective

Solar heating systems often have high initial costs but lower operating costs than other types of heating systems. Operating costs can be reduced even further if other energy conservation steps are taken. Insulate piping, walls, and ceilings to reduce heat loss. Size the system to account for seasonal changes in demand, and available sunlight; the client should not expect a solar heating system to provide 100-percent of the building's heating needs. Install a backup system for cloudy days.

Federal tax incentives for solar heating were eliminated in the mid-1980s. Many cities and states have followed suit. Some jurisdictions still provide rebates and tax incentives for homes that use solar heating systems. Ask an accountant about applicable local laws. Explain all the variables and options to a client before you install a solar heating system.

Additional Resources

Quick & Basic Hydronic Controls: A Contractor's Easy Guide to Hydronic Controls, Wiring, and Wiring Diagrams. 2001. Carol Fey. Littleton, CO: P.I.G. Press.

Solar Hot Water Fundamentals: Siting, Design, and Installation. 2011. Peter Skinner, et al. Albany, NY: E2G Solar.

2.0.0 Section Review

1. Two-pipe systems offer better temperature control than one-pipe systems because they _____.

   a. require less insulation
   b. use two or more heat exchangers
   c. have separate supply and return lines
   d. allow for the installation of more radiators per line

2. After long periods of disuse, the devices in a drain-down system that will often experience problems are the _____.

   a. valves
   b. thermosiphons
   c. radiators
   d. heat exchangers

3.0.0 ROUGHING-IN, INSTALLING, AND TESTING HYDRONIC AND SOLAR HEATING SYSTEM PIPING

Objective

Describe the procedures for roughing-in, installing, and testing hydronic and solar heating system piping.

   a. Describe how to rough-in the piping.
   b. Describe how to install boilers and heating units.
   c. Describe how to install controls.
   d. Describe how to test and balance the system.
   e. Describe corrosion prevention techniques.

Performance Task

Lay out a hydronic or solar heating system.

Trade Terms

Airtrol valve: In a hydronic system, a valve that purges air from the system into the expansion tank.

Anodic inhibitor: A chemical applied to a pipe to prevent galvanic corrosion.

Differential thermostat: In a solar heating system, a control that activates a pump when water in the collector reaches a specified temperature above that of the water in the storage tank.

Diverter tee: In a one-pipe system, a tee that diverts the water flow by using an internal baffle to create a pressure drop.

Limit switch: In a solar heating system, a switch that activates when the water in the collector reaches a preset temperature.

Steam dome: In a steam hydronic system, the place in the top of the boiler where steam collects.

Thermistor: In a solar heating system, a control that adjusts a pump's speed according to changes in the water temperature.

Zone control valve: In a hydronic system, a valve that allows the temperature in various parts of the building to be preset.

Like any other plumbing installation, hydronic and solar heating systems must be installed correctly in order to function. Systems are usually located before the drywall has been placed. Use standard professional techniques when installing hydronic and solar heating systems. Use approved piping materials and sizes. Locate the boiler, heating units, and piping to achieve the most efficient flow path. When hanging and supporting pipes, select the proper size of materials and use the correct tools.

The installation process can be broken down into the following four steps:

- Roughing-in the piping
- Installing the boiler
- Installing controls
- Starting, testing, and balancing the system

Note that these instructions are general guidelines only. Follow the manufacturers' instructions when installing boilers, pumps, and controls. Always refer to the local code when installing and testing hydronic and solar heating systems.

As you work on a heating system installation, you need to anticipate potential problems and provide ways to either prevent them or allow for quick repair. Locate shut-off valves behind an access panel or in basements and crawl spaces. Do not run pipes along exterior walls or near air vents. Run pipe through the center of walls to avoid nail punctures.

Some codes require the installation of backflow prevention devices (*Figure 16*) on hydronic heating system supply lines to protect against backflow from the heated part of the system. Refer to your local applicable code to determine whether backflow preventers are required. If so, install the appropriate backflow prevention devices according to the manufacturer's instructions.

Press-Fitting Technology

Solderless copper joints are an increasingly popular method for joining fittings and connections. Advantages include much faster prep and installation time, and reliable water- and gastight connections. Press fittings can be installed even with water present in the system. Press fit connections are especially useful for maintenance and repair work because they significantly reduce system downtime.

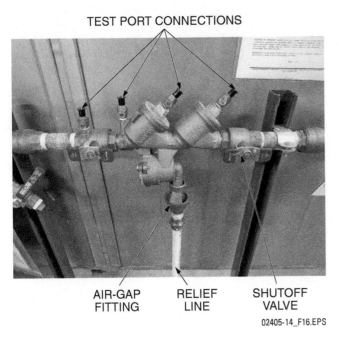

TEST PORT CONNECTIONS

AIR-GAP FITTING   RELIEF LINE   SHUTOFF VALVE

02405-14_F16.EPS

Figure 16 Backflow preventer installed on a supply line.

3.1.0 Roughing-In the Piping

Begin by determining the type of heating unit to be used. If radiators are to be installed, lay out the location of each heating unit. Determine the distance between units and the distance of each unit from the finished wall. Rough-in the entry points for the inlet and outlet piping in the wall. For a radiant loop, determine the location of the supply and return lines in the floor or ceiling. Refer to the manufacturer's instructions for installation requirements. Consult your local code for materials and other guidelines.

For a hydronic system, begin by establishing the location of the boiler. Plan the pipe runs between the boiler and the heating units. Inspect the planned path of the pipe runs for obstacles, structural members, and other concerns. Determine the location of different heating zones

---

Heating Unit Rough-In

Earlier modules in this training explained how to rough-in a fixture. Use this technique to locate the heating units in a hydronic or solar heating system. Measurements are available in the manufacturers' catalogs. Catalogs are available from plumbing distributors or on the internet. Note that critical dimensions may vary between styles of the same brand. Rough-in measurements tell you exactly where the piping should exit the walls or floors. Use these dimensions to run the water supply piping and to attach the fixtures to the floor or wall.

---

within the building. Locate zone control valves at each of these zones. These are valves that allow the owner to set the temperature in various parts of the building. Then determine the most efficient flow path. Ensure that there are no overlaps. A typical hydronic system installation with boiler, water heater, control valves, and piping is shown in *Figure 17*.

After you determine the layout of the supply and return piping, install the pipe runs. Install the mains first, followed by the zone valves. Drill holes through the wall for the supply and return pipes. For pipe that is not insulated, drill the hole one size larger than the pipe's outside diameter. Place a plastic isolator on the pipe to reduce the noise of pipe expansion. Then run branches from the top of the mains to and from the radiation units.

Pressurized, or diaphragm, expansion tanks are also in common use in closed systems. They contain a flexible diaphragm, or bladder, that separates the system water from the air. *Figure 18* shows the action of the diaphragm as the water expands. Note that the tank pressure must be adjusted and checked with the tank isolated from the water system; tank and system pressure will change as thermal expansion takes place. Pressurized expansion tanks often come precharged with air, but the pressure can be adjusted through the charge valve, if needed, to fit system design conditions. Pressurized expansion tanks must be installed according to the manufacturer's instructions. This is because the piping between the system and tank differs from that used with a standard expansion tank. It is important that any air in the system be purged to the atmosphere and not be allowed to enter the tank(s).

Pressure-reducing valves are used in hydronic systems anywhere it is necessary to reduce a higher water or steam pressure to a lower one for input to a device. All systems have a feedwater pressure-reducing valve installed in the boiler water make-up line (*Figure 19*). This device automatically replenishes any water lost through leaks in the system. It also reduces the pressure of the cold water supplied from the water utility to a pressure suitable for use with the boiler. The valve maintains the water supplied to the boiler at a pressure less than that of the boiler relief valve. Feedwater pressure-reducing valves have a built-in strainer and a low-inlet-pressure check valve.

The pressure-reducing valve typically used as a boiler feedwater line valve controls the output pressure and flow by keeping a balance between the pressure of an internal spring applied to one side of a diaphragm and the pressure of the delivered water applied to the other side of the dia-

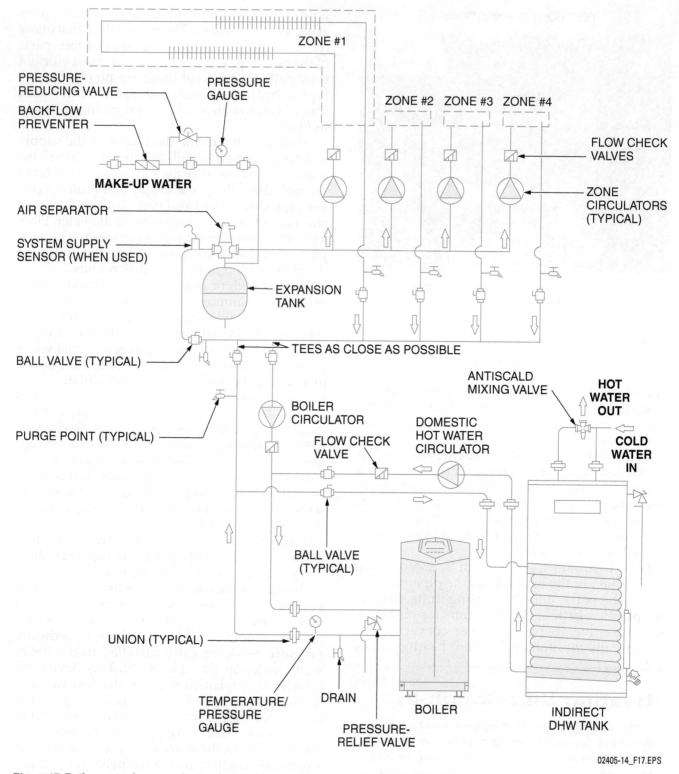

Figure 17 Boiler, water heater, valve, and piping installation for a typical hydronic system.

phragm. It allows water to pass through the valve whenever the pressure at its outlet side drops below its pressure setting. The spring pressure is manually adjusted to set the desired output pressure level. Typically, it is factory-set at 12 pounds per square inch (psi), but is adjustable between 10 and 25 psi. When adjusted, ensure that the setting provides the needed pressure at all elevated points in the system. The model shown in *Figure 19* has a lever that can be pulled up, allowing the water to bypass the regulator and flow through quickly to fill an empty system.

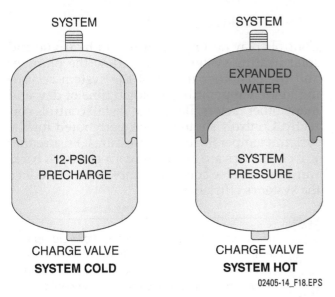

SYSTEM COLD

12-PSIG PRECHARGE

CHARGE VALVE
SYSTEM COLD

SYSTEM HOT

EXPANDED WATER

SYSTEM PRESSURE

CHARGE VALVE
SYSTEM HOT

02405-14_F18.EPS

Figure 18 Diaphragm tank operation.

02405-14_F19.EPS

Figure 19 Feedwater pressure-reducing valve.

A pressure-relief valve (*Figure 20*) is used to protect the boiler and the system from high pressures caused by either water thermal conditions or steam pressure conditions in the boiler. It does not operate unless an overpressure condition exists. The typical hot water boiler is constructed for a maximum working pressure of less than 30 psi, at which point the relief valve is designed to open fully (note that this is significantly less than the average household domestic water pressure). As a result, water fed to the boiler is admitted through a pressure-reducing

02405-14_F20.EPS

Figure 20 Pressure-relief valve.

valve, set at an appropriate pressure to accommodate proper system operation while protecting the boiler from damage. The safety relief valve is set to open at the same pressure as the boiler's specified maximum operating pressure, and discharge hot water via a waste pipe attached to its outlet port. Safety relief valves should always be installed as directed by the manufacturer and in conformance with the local applicable code.

Check valves (*Figure 21*) come in several types and are used to allow water or steam to flow through a pipe in one direction, but not in the other. The most common type of check valve has a swinging gate or flapper that swings open to allow flow in one direction, but closes if the flow is reversed. This type of valve must be installed horizontally with the hinge at the top of the valve. Some check valves are spring-loaded to help in closing the flapper. Another type, called a lift check valve, works so that the flapper lifts off the seat to allow flow.

Check valves are used anywhere in a system where it is necessary to prevent the reverse flow of water or steam. For example, when installed in the circulating pump line of a hot water system, the check valve allows hot water to flow only when the circulating pump is running. When

02405-14_F21.EPS

Figure 21 Check valve.

the pump is off, it prevents any backflow of water caused by gravity to be returned to the boiler. In hot-water and steam heating systems, check valves are used in the domestic water supply lines to the boiler. They are also commonly used in the condensate return lines of steam systems.

3.2.0 Installing the Boiler and Heating Units

For a hydronic system, install the boiler after you rough-in the heating units and install the piping. Locate the boiler according to the manufacturer's instructions and local code. Ensure that the boiler is plumb, level, and accessible. Allow room for access covers. For gas boilers, ensure that there is access for the flue and combustion air. For steam boilers, note that the steam supply line must be precisely located at or near the top of the boiler. The supply line feeds steam from the boiler's steam dome at the top of the boiler. The steam dome is the part of the boiler where steam collects after it is formed.

Shim the boiler to a level position and follow the manufacturer's instructions to hold the boiler in place. Then install the supply and return pipes using approved connectors. Connect the piping for the boiler feedwater, T/P relief valve, and expansion tank, if one is used. Ensure that the relief valve is piped to a drain. Install the pump according to the manufacturer's instructions. Install a purging valve on the boiler's return line.

After you install the boiler, install the heating units. Follow the manufacturers' instructions when installing heating units. Install air vents on each heating unit and install the covers on the units. Connect each unit to the supply and return risers.

3.3.0 Installing Controls

Controls play an important role in hydronic and solar heating systems. They allow the plumber and the customer to adjust the system according to the temperature, weather, time of day, and many other factors. Be sure to install controls correctly. Controls should be properly rated for the system's temperatures and pressures. Ensure that valves and tees are facing the correct way. A backward valve or tee could interrupt flow and reduce the system's efficiency.

> **WARNING!**
> The maximum operating pressure of a pump is stamped on its nameplate. Do not exceed this pressure or the system could fail. Failure could not only damage the system but also could result in injury or death.

> **CAUTION**
> When installing a pump in a hydronic or solar heating system, install a strainer on the inlet side; this will protect the pump from damage caused by solids entering the pump. Install a check valve on the discharge side.

Refer to your local code for specific guidelines for installing controls. Some types of controls and materials may be prohibited by code. Controls for hydronic and solar heating systems are covered in the following sections. Many of them are already familiar to you. Others are used only in hydronic or solar heating systems.

3.3.1 Installing Hydronic System Controls

Like many other plumbing installations, hydronic systems use various types of controls to ensure smooth operation. You are already familiar with many of these controls. Others are specially designed for use in hydronic systems.

A pressure gauge monitors the system's pressure. Use a T/P relief valve or safety valve to keep the system pressure below the boiler's maximum working pressure. Attach an airtrol valve to the expansion tank to purge air from the system into the expansion tank (airtrol is short for *air control*).

Flow control valves are check valves that allow hot water to flow only when the pump is on. The valve prevents the system from draining when the pump is turned off. This helps maintain high temperatures throughout the system. Install a pressure regulator valve (see *Figure 22*) in the cold water line. Pressure regulator valves ensure that the water pressure in the system remains steady. The valve will protect the heating equipment from damage caused by a surge in pressure. Install the circulating pump on the boiler's hot water outlet. Size the pump to overcome pipe friction in the system. Review the pump's specifications to ensure that the pump can supply the correct amount of water at the proper temperature to each heating unit.

> **CAUTION**
>
> In spite of a manufacturer's quality controls, controls may have defects. Be sure to examine controls for holes, cracks, or poor casting. Read and follow the manufacturer's installation instructions.

For one-pipe parallel systems, install a diverter tee (*Figure 23*) on each supply or return riser leading to the hot water supply pipe. Also refer to *Figure 7*, which shows heating units connected to a one-pipe system via diverter tees. Diverter tees, also called venturi tees, use an internal baffle to create a pressure drop. The pressure drop draws the water along the desired direction of flow. Diverter tees allow a single pipe to act as both a supply and a return main. Diverter tees are often referred to as monoflo tees, after the

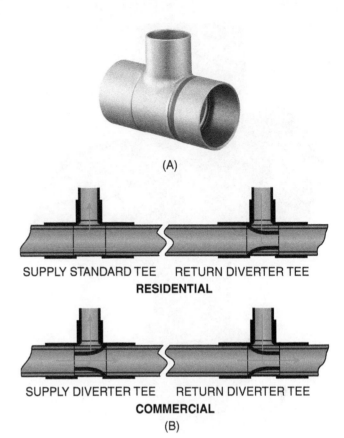

(A)

SUPPLY STANDARD TEE   RETURN DIVERTER TEE
RESIDENTIAL

SUPPLY DIVERTER TEE   RETURN DIVERTER TEE
COMMERCIAL

(B)

02405-14_F23.EPS

Figure 23 Diverter tees.

popular Monoflo® brand manufactured by Bell & Gossett (a Xylem brand).

Diverter tees can be installed on supply risers or return risers. Consult the heating-unit specifications for the proper location. Install a balancing valve (*Figure 24*) to set, balance, and control the flow rate. Balancing valves can be used to control flow for a single heating unit or an entire zone.

Proper Installation of Diverter Tees

Diverter tees allow a pipe to act as both a supply and a return line. If tees are not installed correctly, heat will not circulate efficiently throughout the system.

One of the most popular brands of diverter tees is the Monoflo® tee, invented by the Bell & Gossett Company (now part of Xylem) in the early 1920s. Monoflo® tees have a red band on one side. Install the tee so that the side with the band faces the heating unit. This will ensure proper flow. If a one-pipe system is experiencing heating problems, ensure that the tees are installed correctly.

02405-14_F22.EPS

Figure 22 Pressure regulator valve.

02405-14_F24.EPS

Figure 24 Balancing valve.

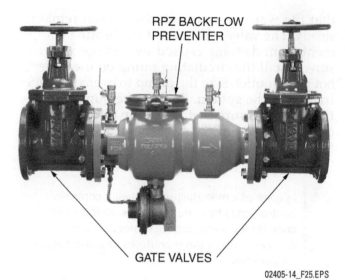

02405-14_F25.EPS

Figure 25 A reduced pressure-zone principle backflow preventer assembly.

02405-14_F26.EPS

Figure 26 Zone control valve.

Some codes require the installation of reduced pressure-zone principle backflow preventers (RPZ) on hydronic systems (see *Figure 25*). RPZs protect the potable water supply in case of back pressure or back siphonage. Install the backflow preventer on the cold water supply line before the pressure regulator valve. Install a low-water safety cutoff switch on the boiler's burner. The switch will turn off the burner if the water level falls below the boiler's heat exchangers. Ensure that all valves are installed correctly. Use accessories that are permitted by local code. Ensure that all controls are rated for the system.

Zone control valves (*Figure 26*) manage the flow of water in each zone of a zoned system. They are usually two-position, thermostatically controlled valves. They can be either two-way or three-way valves. Zone control valves can be operated by heat or with an electric motor. In a heat-operated valve, a resistance wire around the valve heats the valve's bimetallic element when the zone thermostat calls for heat. This causes the bimetallic element to expand and slowly open the valve. When the zone thermostat is satisfied, the bimetallic element cools, allowing the valve to slowly close. Electric-motor operated valves also operate to open and close the valve slowly, under control of

the zone thermostat. They may contain a switch to operate the circulating pump. Slow opening and closing of zone control valves is necessary to reduce expansion noise and prevent water hammer. The time required for a typical zone control valve to open and close ranges from about 15 to 30 seconds. Some zone control valves have a feature that enables the valve to be opened manually in the event of a power failure. This feature is also useful when troubleshooting heating problems in a zoned system. Some zone control valves have a flow indicator dial that aids in system balancing.

The aquastat (*Figure 27*) is a control that is used almost exclusively on boilers. This thermally operated switch is used to control hot-water boilers. An aquastat works basically the same way as a thermostat with the exception that it is designed

20   NCCER – *Plumbing Level Four*   02405-14

02405-14_F27.EPS

Figure 27 Aquastat.

to control water temperatures instead of air temperatures. Aquastats may have the sensing element directly or indirectly immersed in the water, such as the one in *Figure 27*. Most use a well that is first placed into the boiler, and then the aquastat probe is inserted into the well. This gives the sensing element excellent exposure to the water temperature, while allowing it to be replaced without draining the water.

> **CAUTION**
>
> Use a tee to connect the house supply line to the pipe that runs between the boiler and the expansion tank. Install a globe valve, pressure-reducing valve, and check valve on the supply line before the tee. Follow the manufacturer's instructions or the local applicable code, whichever is more strict. Install a globe valve on the expansion tank.

3.3.2 Installing Solar Heating System Controls

Controls play an important part in solar heating systems. They maintain the system's temperatures and pressures. One of the most common controls is a differential thermostat. This device activates the pump when the water in the collector reaches a preset temperature above that of the water in the tank. As the temperature in the tank rises, so does the temperature at which the pump turns on. The differential thermostat improves the system's efficiency. For example, imagine that the differential thermostat in a solar hot water system is set to ac-

tivate at a 15°F differential. When the temperature of the water in the collector is 15° hotter than the water in the tank, the pump will turn on.

> **NOTE**
>
> Refer to the local applicable code to identify whether plumbers or electricians are responsible for installing low-voltage wiring in solar heating systems.

Limit switches work much the same way as differential thermostats. They activate only when the water in the collector reaches a preset temperature. Using the example mentioned above, imagine that instead of turning on the pump when the water in the collector is 15° hotter than the water in the tank, the limit switch turns on the pump when the water reaches 140°F and shuts off when it reaches 125°F.

Limit switches control temperature and pressure more efficiently than differential thermostats. However, they can cause problems if they are not set properly. The upper and lower temperatures of a limit switch must be manually set. Limit switches cannot adjust to daily temperature fluctuations. Afternoon sunlight might warm the system above the switch's upper setting. The switch will then try to lower the temperature to the lower limit. As a result, the pump will be forced to run for a long time. The action will waste all the heat collected in the system. A new system must be carefully adjusted to keep pump cycling to a minimum and stored heat at a maximum.

Install a thermistor to control pump cycling. A thermistor adjusts a pump's speed as the water temperature rises and falls. The term *thermistor* is short for thermal resistor. A thermistor allows a pump to increase flow in proportion to a drop in temperature. As the temperature rises, the pump operates more slowly. As a result, the water temperature in the system remains steady throughout the day.

Install check valves to prevent reverse thermo-siphoning of hot water. Ensure that they are installed along the direction of flow. Install a T/P relief valve in the hot water loop, especially on a closed system, and install a T/P valve near the expansion tank.

3.4.0 Testing and Balancing the System

After you install the piping, boiler, heating units, and controls, test the system. Begin by charging the system with water through the filler inlet. Connect a garden hose to the system drain outlet. Run the hose to a laundry sink or other container. Then purge the entire system through the relief valve. Open the boiler's purge valve to drain the boiler. Do not operate the pump or boiler during the purge. Allow the system to purge all water and air before you shut off the filler inlet.

When the system has been completely purged through the relief valve, open the pressure regulator valve and the boiler's return line. Allow the boiler to fill. Then open the boiler's supply line. Allow a small amount of air to flow through the expansion tank. Then turn on the pump and fire the boiler. Ensure that all valves are working correctly. Set the thermostat and temperature controls.

> **WARNING!**
>
> Hot water under pressure can escape with explosive force. It can also cause dangerous scalding. At 125°F, it takes about 1½ to 2 minutes for hot water to cause scalding. At 155°F, it takes only about one second. Wear appropriate personal protective equipment. Review and follow all construction drawings.

Allow air to bleed out from the highest points in the system. This step may require several cycles of purging and venting. When the system operates quietly, the air has been completely bled from the system. Finally, install all covers. Balance all zones using the zone control valves.

3.5.0 Preventing Corrosion in Hydronic and Solar Heating Systems

Once a heating system has been installed, tested, and balanced, it will operate reliably for years. Ensure that the components of the heating system are protected against corrosion.

The most common type of deterioration in a hydronic or solar heating system is galvanic corrosion. Galvanic corrosion can occur when two different types of metal are joined. Contact between the two metals can create a small electrical current. This current causes metal ions to flow from one pipe to the other. If the pipe is filled with mineral-rich water, the process will be accelerated. *Figure 28* shows the galvanic scale. There are several ways to reduce or eliminate the threat of galvanic corrosion in a heating system:

- Use pipes that have chemically similar metals. Similar metals will reduce the strength of the electrical current between the two pipes.
- Join pipes using dielectric fittings.

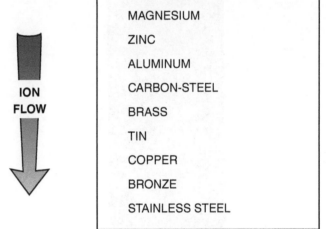

02405-14_F28.EPS

Figure 28 The galvanic scale.

- Install a sacrificial anode (*Figure 29*) made of magnesium. Magnesium has a high positive charge. It will attract the electrical current in the system's pipes. Sacrificial anodes must be replaced periodically.
- Apply an *anodic inhibitor* to the pipes. Anodic inhibitors are chemicals that prevent corrosion. Follow the manufacturer's directions when you apply an anodic inhibitor.

> **CAUTION**
>
> Antifreeze liquids are extremely toxic in any form. When used in closed systems they can cause corrosion if they chemically decompose. A collector will reach 400°F if a pump breaks down. Ethylene glycol will break down if it is heated to that temperature. The resulting chemicals—glycolic acid and oxalic acid—are highly corrosive. Periodic maintenance will ensure that the system is operating correctly and that there is no corrosion damage.

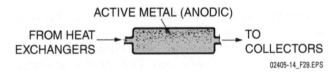

02405-14_F29.EPS

Figure 29 Magnesium sacrificial anode.

Removing Air from a Hydronic System

When installing pipes, pitch them in the direction of flow. This will allow the air to bleed out of the system faster. Inspect the pitch of existing systems to ensure that pipes have not sagged over time.

Pump placement can also affect the way air escapes from the system. Install the pump on the supply line upstream of the expansion tank. This allows the pump to add a small amount of pressure to the line. This pressure will help air dissolve into the water. If air remains in the system after startup, raise the static fill pressure. The increased pressure will cause the air to dissolve into the water. Once the system begins to operate, lower the pressure to normal. Normal pressure will ensure that the relief valve does not open.

Additional Resources

Quick & Basic Hydronic Controls: A Contractor's Easy Guide to Hydronic Controls, Wiring, and Wiring Diagrams. 2001. Carol Fey. Littleton, CO: P.I.G. Press.

Repairing Hot Water Heating Systems. 2012. Robert VanNorden. Seattle, WA: Amazon Digital Services, Inc.

3.0.0 Section Review

1. Pressurized expansion tanks often come pre-charged with _____.

   a. helium
   b. water
   c. nitrogen
   d. air

2. When installing a boiler in a hydronic system, the return line is fitted with _____.

   a. a purging valve
   b. a T/P relief valve
   c. high-temperature adhesive
   d. ¾-inch noncorroding straps

3. A device that controls temperature and pressure more efficiently than a differential thermostat is the _____.

   a. thermistor
   b. thermosiphon
   c. aquastat
   d. limit switch

4. The first step in testing a hydronic or solar heating system is to _____.

   a. charge the system with water through the filler inlet
   b. purge the entire system through the relief valve
   c. connect a garden hose to the system drain outlet
   d. open the boiler's purge valve to drain the boiler

5. The most common type of deterioration in a hydronic or solar heating system is _____.

   a. oxidation
   b. galvanic corrosion
   c. precipitation of salts
   d. backflow

SUMMARY

This module reviewed the components, operating principles, and installation of hydronic heating systems. Hydronic systems that use the sun to heat water are called solar heating systems. Hydronic and solar heating systems use water or steam to move heat from the heat source to the radiators or radiant loops.

This module explained the difference between one- and two-pipe hydronic systems. One-pipe hydronic systems feed water or steam through a single run of pipe. The pipe can be connected to the heating units in series or in parallel. One-pipe systems are usually found in smaller buildings. Two-pipe hydronic systems use separate supply and return lines. Two-pipe systems are suitable for larger buildings.

Solar heating systems can be installed as passive or active systems. Passive systems use convection to circulate hot water through the system. Active solar heating systems use pumps to circulate the heating water. They are more efficient than passive systems. The most common types of active systems are direct and indirect systems.

This module also discussed the installation of hydronic and solar heating systems. Ensure that the components are installed correctly. Always follow the manufacturer's instructions and your local code.

Hydronic and solar heating systems should provide years of trouble-free service. Like any other plumbing installation, heating systems enhance the quality of life by providing safe, sanitary living conditions. Plumbers are responsible for ensuring that these systems meet the needs of their customers.

1. Hydronic and solar heating systems can by classified as radiant systems or _____.

   a. gravity systems
   b. forced heat systems
   c. heat-sink systems
   d. blower-assisted systems

2. When one pound of steam condenses back to water, it gives off _____.

   a. 40 Btus
   b. 520 Btus
   c. 970 Btus
   d. 1,240 Btus

3. Steam heating systems are usually found in houses built _____.

   a. before World War I
   b. after 1950
   c. before World War II
   d. after 1985

4. To circulate potable water in a hydronic heating system, a pump should be made of stainless steel or _____.

   a. aluminum
   b. brass
   c. cast iron
   d. bronze

5. In a hydronic system, heat is emitted from a series of metal _____.

   a. louvers
   b. fins
   c. baffles
   d. dispersers

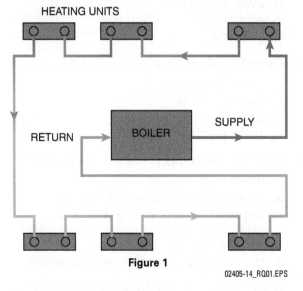

Figure 1

02405-14_RQ01.EPS

6. The one-pipe hydronic system illustrated in *Review Question Figure 1* has a _____.

   a. series arrangement
   b. primary-secondary arrangement
   c. parallel arrangement
   d. closed loop arrangement

7. A passive solar installation moves water through the system by means of _____.

   a. gravity
   b. conduction
   c. convection
   d. radiation

8. The simplest type of solar heating system is the _____.

   a. batch system
   b. direct system
   c. thermosiphon system
   d. indirect system

9. Water in a batch solar heating system does not move when there is no demand, making it vulnerable to _____.

   a. algae and mold growth
   b. evaporation loss
   c. stagnation
   d. freezing

10. Open loop systems are also called _____.

   a. direct systems
   b. batch systems
   c. indirect systems
   d. thermosiphon systems

11. Most heat exchangers are designed with a double wall to _____.

   a. prevent reverse thermosiphoning
   b. improve efficiency of heat transfer
   c. prevent contamination of potable water by toxic liquids
   d. increase flow rates through the system

12. When installing a hydronic system, the first step is to _____.

   a. identify locations for the zone control valves
   b. establish the location of the boiler
   c. plan the pipe runs between system components
   d. determine the most efficient flow path

13. In a pressurized expansion tank, system water is kept separate from the air by a(n) _____.

   a. bladder
   b. impermeable membrane
   c. bulkhead
   d. inflatable barrier

14. Check valves that allow hot water to flow only when the pump is operating are called _____.

   a. directional control valves
   b. diverter valves
   c. flow control valves
   d. safety valves

15. A chemical that can be applied to solar or hydronic system pipes to prevent corrosion is called a(n) _____.

   a. cathodic protector
   b. neutralizer coating
   c. antigalvanic solution
   d. anodic inhibitor

Trade Terms Quiz

Fill in the blank with the correct term that you learned from your study of this module.

1. In a one-pipe system, a tee that diverts the water flow by using an internal baffle to create a pressure drop is called a(n) _____.

2. A(n) _____ is a plumbing system that uses water or steam to heat a building.

3. A direct system in which water drains out of the collector when the pump shuts off is called a(n) _____.

4. A(n) _____ is a hydronic system in which water or steam is fed to the heating units and returned to the boiler through a single run of pipe.

5. A solar heating system that uses a pump and temperature controls to control the flow of hot water is called a(n) _____.

6. A(n) _____ is a two-pipe system in which the return water flows in the opposite direction from the flow in the supply pipe.

7. Another name for an indirect system is a(n) _____.

8. A(n) _____ is a solar heating system that uses a pump and controls to circulate hot water through the system.

9. In a hydronic system, a valve that allows the temperature in various parts of the building to be preset is called a(n) _____.

10. A(n) _____ is a two-pipe system in which the return water flows in the same direction as the flow in the supply pipe.

11. In a steam hydronic system, the place in the top of the boiler where steam collects is called a(n) _____.

12. A(n) _____ is a heating unit that consists of a network of hot water pipes in the floor or ceiling.

13. In a solar heating system, a control that adjusts a pump's speed according to changes in the water temperature is called a(n) _____.

14. A(n) _____ is a valve that purges air from the system into the expansion tank in a hydronic system.

15. In a hydronic system, a device used to heat or boil water is called a(n) _____.

16. _____ is a measure of how much heat a liquid can hold per pound, measured in British thermal units (Btus).

17. A one-pipe system in which each heating unit draws only the amount of water it requires from the supply line is called a(n) _____.

18. A(n) _____ is a one-pipe system in which all the water in the supply loop goes through each heating unit.

19. A solar heating system in which water is both heated and stored in the collector is called a(n) _____.

20. A(n) _____ is a hydronic system that uses the sun to heat water.

21. Another name for a direct system is a(n) _____.

22. A(n) _____ is a one-pipe system with a parallel arrangement.

23. A(n) _____ is a system that heats using heated air blown through ducts by fans.

24. In a thermosiphon system, a condition in which water flows backward through the system is called _____.

25. _____ is circulation caused by the sinking of dense cold water and the rising of hot water.

26. A passive system in which hot water is stored in a tank located higher than the solar collector is called a(n) _____.

27. _____ is the transfer of heat between bodies through space.

28. In a solar heating system, a control that activates a pump when water in the collector reaches a specified temperature above that of the water in the storage tank is called a(n) _____.

29. _____ is another name for a hydronic system.

30. An active system that uses a liquid other than water to distribute heat from the collector is called a(n) _____.

31. _____ is an alcohol-based liquid used in an indirect system to transfer heat from a collector to a heat exchanger.

32. A device used in an indirect system to transfer heat from one liquid to another in such a way that the two liquids do not come into contact with each other is called a(n) _____.

33. A(n) _____ is a device used in a solar heating system to receive heat from sunlight and transfer it to water or another liquid.

34. In a solar heating system, a switch that activates when the water in the collector reaches a preset temperature is called a(n) _____.

35. A(n) _____ is a heating unit that emits heat from water via a series of metal fins.

36. A hydronic system that uses separate supply and return lines is called a(n) _____.

37. A(n) _____ is a solar heating system that relies on convection to circulate water.

38. A chemical applied to a pipe to prevent galvanic corrosion is called a(n) _____.

39. _____ is the process whereby cold water flows back to the collector to be reheated in a hydronic system.

Trade Terms

Active system	Direct system	Open loop system	Reverse thermosiphoning
Airtrol valve	Diverter tee	Parallel	Series
Anodic inhibitor	Drain-down system	Passive system	Solar heating system
Batch system	Ethylene glycol	Primary-secondary	Specific heat
Boiler	Forced heat system	system	Steam dome
Closed loop system	Heat exchanger	Radiant loop	Thermistor
Collector	Hydronic system	Radiant system	Thermosiphon system
Convection	Indirect system	Radiation	Thermosiphoning
Differential thermostat	Limit switch	Radiator	Two-pipe system
Direct return system	One-pipe system	Reverse return system	Zone control valve

Trade Terms Introduced in This Module

Active system: A solar heating system that uses a pump and controls to circulate hot water through the system.

Airtrol valve: In a hydronic system, a valve that purges air from the system into the expansion tank.

Anodic inhibitor: A chemical applied to a pipe to prevent galvanic corrosion.

Batch system: A solar heating system in which water is both heated and stored in the collector. The system uses the principle of convection to circulate water.

Boiler: In a hydronic system, a device used to heat or boil water. Boilers differ from typical water heaters in that they have a greater capacity, operate at higher temperatures and pressures, and have a greater heat output.

Closed loop system: Another name for an indirect system.

Collector: A device used in a solar heating system to receive heat from sunlight and transfer it to water or another liquid.

Convection: Circulation caused by the sinking of dense cold water and the rising of hot water.

Differential thermostat: In a solar heating system, a control that activates a pump when water in the collector reaches a specified temperature above that of the water in the storage tank.

Direct return system: A two-pipe system in which the return water flows in the opposite direction from the flow in the supply pipe.

Direct system: A solar heating system that uses a pump and temperature controls to control the flow of hot water. It is also called an open loop system.

Diverter tee: In a one-pipe system, a tee that diverts the water flow by using an internal baffle to create a pressure drop.

Drain-down system: A direct system in which water drains out of the collector when the pump shuts off.

Ethylene glycol: An alcohol-based liquid used in an indirect system to transfer heat from a collector to a heat exchanger.

Forced heat system: A type of hydronic system that uses fans and ducts to circulate air heated by passing it over coils containing hot water.

Heat exchanger: A device used in an indirect system to transfer heat from one liquid to another in such a way that the two liquids do not come into contact with each other.

Hydronic system: A plumbing system that uses water or steam to heat a building. It is also called a radiant system.

Indirect system: An active system that uses a liquid other than water to distribute heat from the collector. It is also called a closed loop system.

Limit switch: In a solar heating system, a switch that activates when the water in the collector reaches a preset temperature.

One-pipe system: A hydronic system in which water or steam is fed to the heating units and returned to the boiler through a single run of pipe.

Open loop system: Another name for a direct system.

Parallel: A one-pipe system in which each heating unit draws only the amount of water it requires from the supply line.

Passive system: A solar heating system that relies on convection to circulate water.

Primary-secondary system: A one-pipe system with a parallel arrangement.

Radiant loop: A heating unit that consists of a network of hot water pipes in the floor or ceiling.

Radiant system: Another name for a hydronic system.

Radiation: The transfer of heat between bodies through space.

Radiator: A heating unit that emits heat from water via a series of metal fins.

Reverse return system: A two-pipe system in which the return water flows in the same direction as the flow in the supply pipe.

Reverse thermosiphoning: In a thermosiphon system, a condition in which water flows backward through the system.

Series: A one-pipe system in which all the water in the supply loop goes through each heating unit.

Solar heating system: A hydronic system that uses the sun to heat water.

Specific heat: A measure of how much heat a liquid can hold per pound, measured in British thermal units (Btus).

Steam dome: In a steam hydronic system, the place in the top of the boiler where steam collects.

Thermistor: In a solar heating system, a control that adjusts a pump's speed according to changes in the water temperature.

Thermosiphon system: A passive system in which hot water is stored in a tank located higher than the solar collector.

Thermosiphoning: In a thermosiphon system, the process whereby cold water flows back to the collector to be reheated.

Two-pipe system: A hydronic system that uses separate supply and return lines.

Zone control valve: In a hydronic system, a valve that allows the temperature in various parts of the building to be preset.

Additional Resources

This module presents thorough resources for task training. The following resource material is suggested for further study.

Hydronic Radiant Heating: A Practical Guide for the Nonengineer Installer. 1998. Dan Holohan. Bethpage, NY: Dan Holohan Associates.

Quick & Basic Hydronic Controls: A Contractor's Easy Guide to Hydronic Controls, Wiring, and Wiring Diagrams. 2001. Carol Fey. Littleton, CO: P.I.G. Press.

Repairing Hot Water Heating Systems. 2012. Robert VanNorden. Seattle, WA: Amazon Digital Services, Inc.

Solar Hot Water Fundamentals: Siting, Design, and Installation. 2011. Peter Skinner, et al. Albany, NY: E2G Solar.

Figure Credits

Answer	Section Reference	Objective
Section One		
1. b	1.1.0	1a
2. c	1.2.0	1b
3. b	1.3.0	1c
Section Two		
1. c	2.1.2	2a
2. a	2.2.3	2b
Section Three		
1. d	3.1.0	3a
2. a	3.2.0	3b
3. d	3.3.2	3c
4. a	3.4.0	3d
5. b	3.5.0	3e

NCCER CURRICULA — USER UPDATE

NCCER makes every effort to keep its textbooks up-to-date and free of technical errors. We appreciate your help in this process. If you find an error, a typographical mistake, or an inaccuracy in NCCER's curricula, please fill out this form (or a photocopy), or complete the online form at **www.nccer.org/olf**. Be sure to include the exact module ID number, page number, a detailed description, and your recommended correction. Your input will be brought to the attention of the Authoring Team. Thank you for your assistance.

Instructors – If you have an idea for improving this textbook, or have found that additional materials were necessary to teach this module effectively, please let us know so that we may present your suggestions to the Authoring Team.

NCCER Product Development and Revision
13614 Progress Blvd., Alachua, FL 32615

Email: curriculum@nccer.org
Online: www.nccer.org/olf

❏ Trainee Guide   ❏ Lesson Plans   ❏ Exam   ❏ PowerPoints   Other _____

Craft / Level: _____   Copyright Date: _____

Module ID Number / Title: _____

Section Number(s): _____

Description: _____

_____

_____

_____

_____

Recommended Correction: _____

_____

_____

_____

_____

Your Name: _____

Address: _____

_____

Email: _____   Phone: _____

02406-14

Codes

OVERVIEW

Codes are adopted to protect public health and safety, and to ensure that contractors perform their work according to recognized standards. Many jurisdictions develop their codes from templates called model codes. Professional organizations work closely with government agencies, administrators, and industry professionals to develop and revise model codes. Codes change over time to include new materials and technologies, and to reflect changing standards and practices. Model code organizations meet regularly to vote on code change proposals. Jurisdictions revise model codes to accommodate local requirements. Professional plumbers must keep up with changes to the local code and model codes.

Module Six

Trainees with successful module completions may be eligible for credentialing through the NCCER Registry. To learn more, go to **www.nccer.org** or contact us at **1.888.622.3720**. Our website has information on the latest product releases and training, as well as online versions of our *Cornerstone* magazine and Pearson's product catalog.

Your feedback is welcome. You may email your comments to **curriculum@nccer.org**, send general comments and inquiries to **info@nccer.org**, or fill in the User Update form at the back of this module.

This information is general in nature and intended for training purposes only. Actual performance of activities described in this manual requires compliance with all applicable operating, service, maintenance, and safety procedures under the direction of qualified personnel. References in this manual to patented or proprietary devices do not constitute a recommendation of their use.

Objectives

When you have completed this module, you will be able to do the following:

1. Describe the model and local plumbing codes and their purposes.
   a. Describe the ICC model plumbing code.
   b. Describe the IAPMO model plumbing code.
2. Explain how plumbing codes are developed and revised.
   a. Explain model code standards.
   b. Explain how model codes are revised and adopted.
   c. Explain typical code changes.

Performance Task

Under the supervision of your instructor, you should be able to do the following:

1. Use the local applicable plumbing code to find and cite references for the questions in *Appendix A*.

Trade Terms

Code
Comprehensive Consensus Codes®
International Plumbing Code® (IPC)

Model code
Uniform Plumbing Code® (UPC)

Industry-Recognized Credentials

If you're training through an NCCER-accredited sponsor, you may be eligible for credentials from NCCER's Registry. The ID number for this module is 02406-14. Note that this module may have been used in other NCCER curricula and may apply to other level completions. Contact NCCER's Registry at 888.622.3720 or go to **www.nccer.org** for more information.

Code Note

Codes vary among jurisdictions. Because of the variations in code, consult the applicable code whenever regulations are in question. Referring to an incorrect set of codes can cause as much trouble as failing to reference codes altogether. Obtain, review, and familiarize yourself with your local adopted code.

Contents

Topics to be presented in this module include:

Figures

1.0.0 MODEL AND LOCAL PLUMBING CODES AND THEIR PURPOSES

Objective

Describe the model and local plumbing codes and their purposes.

   a. Describe the ICC model plumbing code.
   b. Describe the IAPMO model plumbing code.

Performance Task

Use the local applicable plumbing code to find and cite references for the questions in *Appendix A*.

Trade Terms

Code: A legal document enacted to protect the public and property, establishing the minimum standards for materials, practices, and installations.

Comprehensive Consensus Codes®: A collection of ANSI-compliant codes and standards developed jointly by the International Association of Plumbing and Mechanical Officials (IAPMO) and the National Fire Protection Association (NFPA).

International Plumbing Code® (IPC): The model plumbing code of the International Code Council (ICC). It was first issued in 1995.

Model code: A set of comprehensive, general guidelines that establish and define acceptable plumbing practices and materials and list prohibited installations.

Uniform Plumbing Code® (UPC): The model code of the International Association of Plumbing and Mechanical Officials (IAPMO). It has been published since 1945.

A ll construction work is governed by codes. A code is a legal document adopted in a jurisdiction that establishes the minimum acceptable standards, rules, and regulations for all materials, practices, and installations used in buildings and building systems. Codes are adopted to protect the health and safety of the public and property, ensuring that contractors perform their work according to recognized standards. In this module, you will learn about the different types of codes that affect plumbing installations, materials, and processes. You will also learn how codes are modified.

There are thousands of local codes in existence in the United States, and the potential total number of city, county, and state jurisdictions that could adopt codes is in the tens of thousands. Obviously, not all of these codes have been created entirely from scratch. Many of them were developed from templates called model codes.

Model codes are comprehensive sets of general guidelines that establish and define acceptable plumbing practices and materials. They also list prohibited installations and clarify obvious plumbing hazards. Model codes do not have the force of law, but they serve as the basis for more detailed, legally binding local codes developed by individual jurisdictions, which can adopt and amend the model code in whole or in part. Because they offer general guidance, model codes tend to focus more on processes than on specifics.

Model codes are created by professional organizations with the expertise and resources to undertake such a large-scale task. These organizations work closely with government agencies, code administrators, and building industry professionals to develop, publish, and regularly revise the model codes. As well as plumbing codes, model code organizations issue general standards for the following:

- Residential and commercial building construction
- Fire prevention
- Mechanical and electrical installations
- Energy conservation
- Property maintenance
- Sewage disposal
- Zoning

Most model code organizations are open to membership by professionals from the construction trades, architects and designers, regulation enforcement specialists, and industry representatives. These organizations use the internet to communicate with members and other interested parties. Websites feature the latest news releases, provide detailed information about pending and approved code changes, and offer membership information. Many organizations also offer the option to purchase model codes and other publications online.

In general, professional code-development organizations publish new editions of their codes every three years. However, states and local jurisdictions may choose to adopt new model codes only for some construction standards while keeping established codes for others. Some states even develop their own codes without reference to a model code, and some use more than one model code. Because all of these issues pose challenges for plumbing professionals trying to keep up with new developments, a summary review of the two major model plumbing-code organizations in the United States will help clarify this complicated situation.

1.1.0 Describing the ICC Model Plumbing Code

The *International Plumbing Code® (IPC)* (*Figure 1*) is the model code published by the International Code Council, Inc. (ICC), located in Washington, DC. The purpose of the IPC is to establish "the minimum requirements for providing safe water to a building as well as a safe manner in which liquid-borne wastes are carried away from a building." The IPC focuses primarily on laying out the specifications for various plumbing installations, and also provides additional information that tells plumbers how to design and install systems that meet those specifications. For example, Chapter 4 states simply that water supply lines and fittings must be installed in such a way as to prevent backflow; Chapter 6 lays out the procedures to be followed to ensure this, including the requirements for selecting and locating air gaps and other types of backflow preventers, protecting potable water outlets, and ensuring safe connections to the water supply and waste systems.

02406-14_F01.EPS

Figure 1 Cover of the 2012 *International Plumbing Code®*.

1.1.1 The International Code Council

The ICC (**www.iccsafe.org**) was founded in 1994 by three model-code development organizations:

History of Codes

The predecessors of today's plumbing standards and practices can be traced to guidelines developed in ancient times to regulate safe building construction and public health in large urban settlements. Rudimentary regulations were incorporated into one of the earliest sets of codified laws. The *Code of Hammurabi*, written around 2100 BC for Hammurabi, King of Babylonia, declared that if a house collapsed, the builder should give his own house to the owner of the fallen house. If people died as a result of poor workmanship, the builder would lose his life, too. Modern building codes are not quite so drastic!

The plumbing standards and practices observed in the United States today are based on those originally developed in Europe beginning in the Renaissance in the sixteenth century. Beginning in 1519, for example, laws passed in various regions of France required that all houses should have indoor toilets or cesspools. In the mid-nineteenth century, the British Parliament passed the 1848 Public Health Act, the first law of its kind in scale and scope. One of the first sanitation laws passed after Louis Pasteur discovered that infectious diseases were spread by microscopic organisms called germs was the New York Metropolitan Health Law, enacted in 1866, which served as a model for the health laws of other large cities.

Early in the twentieth century, insurance companies in the United States called for the adoption of building codes to help reduce the excessive loss of lives due to fires in overcrowded inner cities. Over the next forty years, code enforcement officials banded together in several regional professional organizations to develop and implement such codes, establishing the system that exists today.

the Building Officials and Code Administrators International, Inc. (BOCA), the International Conference of Building Officials (ICBO), and the Southern Building Code Conference International, Inc. (SBCCI). In 2003, the ICBO, BOCA, and SBCCI consolidated their services, products, and operations under the single entity of ICC.

The ICC issued its first IPC in 1995. The ICC also publishes model codes, nicknamed I-codes, for building construction, fire prevention, green construction, residential buildings, mechanical installations, fuel gas, energy conservation, property maintenance, private sewage disposal, swimming pools and spas, wildlife-urban interfaces, and zoning.

The ICC's goal is to develop a consistent and uniform set of international standards for the United States, and eventually, the world. The ICC sees a variety of benefits arising from global uniform standards, such as the following:

- Construction professionals and code officials would be able to work with a single set of requirements throughout the world.
- Manufacturers would no longer have to design products for multiple standards.
- Professionals and manufacturers would be able to compete in worldwide markets.
- Education and certification programs would be standardized.
- Construction and enforcement standards would become consistent.
- Concerned parties would require only one forum to resolve code-enforcement and regulation issues.

1.1.2 Chapters in the 2012 ICC Model Plumbing Code

The 2012 IPC is divided into 14 chapters and six appendixes (see *Figure 2*). This section provides

Local Codes Online

Many states and municipalities make their codes available online. These codes may be in the form of downloadable documents, smartphone apps, searchable databases, or simply the text of the code displayed on a web page. When referring to online versions of a local code, make sure that you are using the most recent and up-to-date version. Older versions of codes can remain on the web for many years after they have been superseded.

brief summaries of each chapter and appendix in the 2012 IPC. Note that the chapters and appendixes may vary from edition to edition.

Chapter 1 is titled *Scope and Administration*. It covers the legal conditions that govern the application, enforcement, and administration of the IPC's requirements. It also lists the buildings and structures that the code covers. It is important for jurisdictions that adopt the IPC to follow these conditions, to ensure that the code has been used properly. To help ensure compliance, the chapter also lays out the duties and powers of the code official responsible for enforcing the code in a jurisdiction, and the proper procedures to follow in the event of a code violation.

Chapter 2 is titled *Definitions*. This is where the terms of the trade used in the code are defined. This is important because, as a legal document, the code uses terms precisely. This is to ensure that plumbers are able to understand what the code is referring to when a specific term is used, and thus prevent mistakes or misunderstandings. This is especially important for terms that may have a specific technical meaning in the code, but also have other meanings when used in ca-

A Brief History of BOCA and SBCCI

Two of the organizations that founded the International Code Council were the Building Officials and Code Administrators International, Inc. (BOCA) and the Southern Building Code Congress International, Inc. (SBCCI). Founded in 1915, BOCA established the *National Plumbing Code*® (NPC), as well as national codes for building construction, mechanical installations, fire prevention, property maintenance, private sewage disposal, and energy conservation. The NPC was adopted primarily by cities, counties, and states in the eastern and midwestern United States.

SBCCI was established in 1940. The first edition of its *Standard Plumbing Code*™ (SPC) appeared in 1955. The SPC was adopted by cities, counties, and states in the southern and southeastern United States. SBCCI also issued model codes for mechanical installations, excavation and grading, gas, fire prevention, housing, swimming pools, existing-building maintenance, building construction, and unsafe-building abatement.

With the advent of the ICC's *International Plumbing Code*®, both the NPC and the SPC have ceased publication.

TABLE OF CONTENTS

02406-14_F02.EPS

Figure 2 A page from the table of contents for the 2012 *International Plumbing Code®.*

sual conversation. Such terms are shown in italics throughout the rest of the code, to remind readers to consult the definition as required to avoid confusion.

Chapter 3 is titled *General Regulations*. You may also hear this chapter referred to as *Miscellaneous*, because it is where the ICC puts all the requirements that don't fit in the other chapters. It includes safety requirements, lists of approved and prohibited materials, guidelines for excavating and filling trenches, tests and inspections, and even how to provide toilet facilities for trade workers on a job site.

Chapter 4 is titled *Fixtures, Faucets and Fixture Fittings*. This is the chapter where the minimum number of plumbing fixtures in a building is specified. It also provides guidelines for fixture and fitting installation, quality, and usability.

Chapter 5 is titled *Water Heaters*. As its title suggests, the chapter covers the design, approval, and installation of water heaters and safety devices that are used on them. While the IPC does not specify water-heater size, it regulates other aspects such as temperature and pressure relief (T/P) valves, drip pans, maximum water temperatures, and proper techniques for installing and connecting water heaters.

Chapter 6 is titled *Water Supply and Distribution*. It covers the requirements for public and private potable water supply sources and lays out the requirements for water distribution system design to prevent backflow conditions.

Chapter 7 is titled *Sanitary Drainage*. This chapter provides guidelines for reliable and properly sized sanitary drainage piping system materials, design, installation, and connections. It also covers the proper fittings to be used in sanitary drainage systems and servicing requirements.

Chapter 8 is titled *Indirect/Special Waste*. Indirect connections to sanitary drainage systems are covered in this chapter, as well as the protection of fixtures used in special waste systems such as food preparation and health care facilities. Hazardous wastes are also covered in this chapter.

Chapter 9 is titled *Vents*. It covers the venting of drain, waste, and vent (DWV) systems and specifies their allowable pressure differentials.

Chapter 10 is titled *Traps, Interceptors and Separators*. Design and installation requirements for traps are covered, as well as the types of traps that are specifically prohibited. It also lays out the requirements for traps in fixtures that are infrequently used and for venting of separators and interceptors.

Chapter 11 is titled *Storm Drainage*. It covers the design and installation of stormwater systems for rainfall.

Chapter 12 is titled *Special Piping and Storage Systems*. The design, installation, storage, handling, and use of medical gas and vacuum systems are covered in this chapter, as well as oxygen-fuel gas systems that are used for cutting and welding.

Chapter 13 is titled *Gray Water Recycling Systems*. This chapter covers design and installation requirements for systems that are used to collect and dispose of gray water, which is often used for the flushing of water closets and urinals, as well as in subsurface irrigation systems.

Chapter 14 is titled *Referenced Standards*. It lists all of the standards that are referenced in the IPC and which must also be followed in order to be in compliance with the code. The standards are listed alphabetically by agency, and under that alphabetically or numerically as appropriate.

The six appendixes cover plumbing permit fee schedules, rainfall rates for the regions and major cities of the United States, information on vacuum drainage systems, temperature information for areas prone to freezing temperatures, methods for sizing water supply piping, and limitations for structural modifications in wood and steel framing.

1.1.3 Using the 2012 ICC Model Plumbing Code

The 2012 IPC is available in print in both looseleaf and softcover formats, as well as a password-protected PDF (portable document format) document. A version of the IPC with commentary at the end of each section is also available. This version is available in softcover and CD-ROM (compact disc read-only memory), as well as in a password-protected PDF document. The ICC has also developed smartphone apps that allow users to shop for, purchase, and browse codes and other ICC reference books. The app also provides access to the ICC website and many state and local codes.

The print version of the IPC can be searched using the detailed table of contents as well as an alphabetical index in the back of the book. The PDF is searchable by keyword, section number, and figure and table numbering. Clicking on an item in the table of contents will take the reader to the appropriate section in the text. The ICC also offers adhesive tabs that can be attached to section breaks in the printed versions of the code.

1.2.0 Describing the IAPMO Model Plumbing Code

The *Uniform Plumbing Code® (UPC)* (*Figure 3*) is the model code published by the International Association of Plumbing and Mechanical Officials (IAPMO) of Ontario, CA. According to IAPMO, the purpose of the UPC is to establish "minimum requirements and standards for the protection of the public health, safety and welfare." It is intended to apply to "the erection, installation, alteration, repair, relocation, replacement, addition to, use, or maintenance of plumbing systems within [the] jurisdiction." It covers piping for water supply and DWV systems, fixtures and traps, building drains and sewers, water heaters, and other systems such as medical gas and vacuum systems and liquid- and fuel-gas systems. According to IAPMO, the goal of the UPC is "to provide consumers with safe and sanitary plumbing systems while, at the same time, allowing latitude for innovation and new technologies."

1.2.1 The International Association of Plumbing and Mechanical Officials

The International Association of Plumbing and Mechanical Officials (**www.iapmo.org**) was established in 1926. The organization has published the *Uniform Plumbing Code®* (UPC) since 1945. IAPMO has formed a partnership with the National Fire Protection Association (NFPA) to produce the Comprehensive Consensus Codes® (C3), each of which has been developed using an open-consensus model accredited by the American National Standards Institute (ANSI). The C3 standards include:

- NFPA 5000, *Building Construction and Safety Code®*
- NFPA 70, *National Electrical Code®*
- NFPA 101, *Life Safety Code®*

Freeze Depths

Model codes can be used by jurisdictions anywhere in the country. Because of this fact, they do not provide detailed guidance for specific concerns that vary from region to region due to weather, climate, and humidity.

One issue for which model codes do not provide specific guidance is freeze depths. Wide variations in freeze depths occur throughout the country. Model codes simply state that plumbing should be protected from freezing. It is up to local officials to define these depths in the local codes.

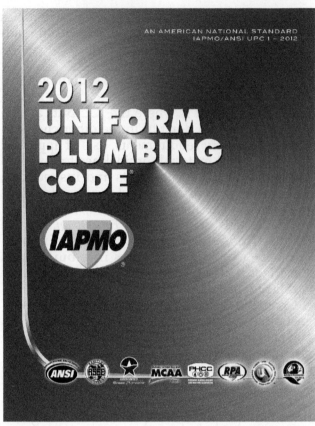

Figure 3 Cover of the 2012 *Uniform Plumbing Code®*.

- NFPA 1, *Fire Code*
- NFPA 54, *National Fuel Gas Code*
- NFPA 58, *Liquefied Petroleum Gas Code*
- NFPA 30, *Flammable and Combustible Liquids Code*
- NFPA 30A, *Code for Motor Fuel Dispensing Facilities and Repair Garages*
- IAPMO *Uniform Mechanical Code*™
- IAPMO *Uniform Plumbing Code®*
- NFPA 900, *Building Energy Code* (ASHRAE 90.1 and 90.2)

In addition to the UPC and the *Uniform Mechanical Code*™ included in the C3 set, IAPMO publishes two additional model codes: one that covers swimming pools, spas, and hot tubs, and another that covers solar energy and hydronic systems. IAPMO also publishes the plumbing and mechanical codes for the state of California and the plumbing codes for the states of Oregon and Idaho.

1.2.2 Chapters in the 2012 IAPMO Model Plumbing Code

The 2012 UPC is divided into 17 chapters and 12 appendixes (*Figure 4*). This section provides brief summaries of each chapter and appendix in the

TABLE OF CONTENTS

02406-14_F04.EPS

Figure 4 A page from the table of contents for the 2012 *Uniform Plumbing Code®*.

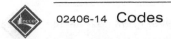

2012 UPC. Note that the chapters and appendixes may vary from edition to edition.

Chapter 1 is titled *Administration*. It covers the scope and purpose of the model code, and establishes the procedures for resolving conflicts between codes and submitting plans to the authority having jurisdiction (AHJ). The requirements for repairs, alterations, and maintenance of existing and new construction are also covered. Chapter 1 also addresses organizational and enforcement issues such as penalties for violations, issues related to permits, and requirements for inspections.

Chapter 2 is titled *Definitions*. As the title suggests, this section provides definitions for technical terms that are used throughout the code.

Chapter 3 is titled *General Regulations*. This chapter covers issues such as the minimum standards and alternates for materials, waste disposal, required connections, and the proper procedures to follow in the event of damage to the drainage system or the public sewer. Proper handling and discharge of industrial and detrimental wastes, proper and improper location, engineering practices, fittings, requirements for independent systems, standards for hanging and supporting pipe, and other specialized requirements are also covered in this chapter.

Chapter 4 is titled *Plumbing Fixtures and Fixture Fittings*. It covers general requirements for materials including fixtures, joints, and the proper design and installation of water supply and DWV system fixtures and components. It also covers requirements for installing backflow prevention valves and floor drains, and covers the minimum requirements for water temperature, number of fixtures, and other plumbing facilities.

Chapter 5 is titled *Water Heaters*. This chapter, as its title suggests, addresses all aspects of sizing, locating, and installing water heaters. Subjects covered include safety devices, insulation, venting and chimneys, and accessibility.

Chapter 6 is titled *Water Supply and Distribution*. It covers the requirements for potable and nonpotable water supply systems, including prohibited and unlawful connections, approved devices and assemblies, suitable materials for fittings and connections, approved valves, trenching, and water hammer. This chapter provides guidance on calculating water supply fixture units (WSFUs) and drinking-water treatment units, which is a measure used in determining the water softening needs of a system.

Chapter 7 is titled *Sanitary Drainage*. This chapter is divided into two parts: *Drainage Systems* and *Building Sewers*. *Drainage Systems* covers materials and sizing requirements for drainage piping

and fittings. It also covers aspects such as joints, cleanouts, drainage, and testing. *Building Sewers* addresses requirements related to public and private sewer systems, including materials, grading, support and protection, cleanouts, and disposal facilities.

Chapter 8 is titled *Indirect Wastes*. It defines indirect wastes and spells out the requirements for air gaps and air breaks, connections and traps, piping and receptors, condensate and cooking-water wastes, steam and hot-water drainage condensers and sumps, venting, clear-water wastes, and swimming pool wastewater.

Chapter 9 is titled *Vents*. General requirements for the various types of vents are covered, as are instances where vents are not required. The chapter covers interceptors, vent grades and horizontal drainage pipe, combination waste and vent systems, and engineered vent systems.

Chapter 10 is titled *Traps and Interceptors*. This chapter goes into detail about when and where various types of traps and interceptors are required, and how to size and install them.

Chapter 11 is titled *Storm Drainage*. It identifies the conditions under which storm drainage systems are required and their approved uses. The various types of storm drainage systems are discussed, as are the requirements for system components such as conductors, leaders, gutters, traps, and drains. The chapter also covers the testing of storm drainage systems.

Chapter 12 is titled *Fuel Gas Piping*. After introducing the scope and coverage of approved gas-piping systems, this chapter covers inspection requirements and the various authorities having jurisdiction over gas service and supply. Piping and the various types of approved fittings and valves are discussed in detail, as is testing. Gas supply requirements are also covered in this chapter.

Chapter 13 is titled *Health Care Facilities and Medical Gas and Vacuum Systems*. This chapter is divided into two parts: *Special Requirements for Health Care Facilities* and *Medical Gas and Vacuum Systems*. The first part thoroughly covers the installation requirements for medical gas and vacuum systems in health care facilities, including proper brazing procedures, sterilization, and venting. The second part of the chapter provides comprehensive details on system requirements and sizing; flow rates; accepted engineering practices; cleaning and joining piping; approved valves, labels, and alarms; and the testing and inspection process required for medical gas and vacuum systems.

Chapter 14 is titled *Referenced Standards*. As its name indicates, this chapter provides a list of all standards referenced in the code and which must also be followed in order to be in compliance with the code. The standards are provided in a table format by standard number, followed by title, application, and the sections of the code in which they are referenced.

Chapter 15 is titled *Firestop Protection*. This chapter covers the proper procedures for firestopping when installing water supply system and DWV systems. It addresses acceptable fire-resistance ratings, connections, couplings, sleeves, and penetrations; it also specifies the inspection procedure to be followed to ensure that the fire-stopping meets or exceeds the code's minimum requirements.

Chapter 16 is titled *Alternate Water Sources for Nonpotable Applications*. It covers the design, maintenance, testing, and inspection of gray water systems, and establishes minimum water quality requirements for these systems.

Chapter 17 is titled *Nonpotable Rainwater Catchment Systems*. This chapter discusses the process of installing rainwater catchment systems and how to connect them safely to potable or reclaimed water systems. It also covers backflow prevention and the coloring and marking of catchment systems.

The twelve appendixes cover the recommended rules for sizing water supply systems; combination waste and vent systems; sizing of alternate plumbing systems such as engineered plumbing systems, water heat exchangers, and vacuum systems; stormwater drainage systems;

parks for manufactured homes, mobile homes, and recreational vehicles; replenishment systems for firefighter breathing air; sizing of venting systems that serve appliances fitted with draft hoods and other special appliances; private sewage-disposal systems; installation standards; the design of combination indoor-outdoor combustion and ventilation systems; catchment systems for potable rainwater; and sustainable practices.

1.2.3 Using the 2012 IAPMO Model Plumbing Code

The 2012 UPC is available in print in loose-leaf and softcover formats with chapter tabs, on CD-ROM, and in e-book format. IAPMO also publishes an illustrated training manual, study guide, and guide to important code changes to accompany the UPC. These publications are available in print and e-book formats. IAPMO also offers printed sizing guides for natural gas, DWV, and water supply system piping.

The print version of the UPC can be searched using the detailed table of contents as well as an alphabetical index in the back of the book. The e-book is searchable by keyword, section number, and figure and table numbering. Clicking on an item in the table of contents will take the reader to the appropriate section in the text. The e-book version also provides a hierarchical table of contents that can be expanded or collapsed by clicking on icons to the left of the chapter or section title.

Additional Resources

International Plumbing Code®, Latest Edition. Falls Church, VA: International Code Council.

Uniform Plumbing Code®, Latest Edition. Ontario, CA: International Association of Plumbing and Mechanical Officials.

1.0.0 Section Review

1. In the IPC, the design, installation, storage, handling, and use of medical gas and vacuum systems are covered in the chapter titled _____.

    a. *Indirect/Special Waste*
    b. *Fixtures, Faucets and Fixture Fittings*
    c. *Health Care Facilities*
    d. *Special Piping and Storage Systems*

2. IAPMO publishes model codes for each of the following categories *except* _____.

    a. swimming pools, spas, and hot tubs
    b. fire
    c. plumbing
    d. solar energy and hydronic systems

2.0.0 PLUMBING CODE DEVELOPMENT AND REVISION

Objective

Explain how plumbing codes are developed and revised.

a. Explain model code standards.
b. Explain how model codes are revised and adopted.
c. Explain typical code changes.

Plumbing codes are intended to reflect the current state of the science, technology, and practice of the trade. However, codes must change over time to include new materials and technologies and to reflect changing standards and practices. Both model and local codes change, though not always at the same time. Changes to model codes can drive changes to local codes and vice versa. The code change process is a constant cycle (see *Figure 5*).

2.1.0 Explaining Model Code Standards

Several organizations oversee code standards for their specific industries. Because these standards organizations conduct detailed research and testing as part of the development of their standards, model code organizations such as the ICC and IAPMO can reference those standards in their model codes rather than duplicate that work. This section provides basic information about those organizations and the particular standards they set.

The Air-Conditioning, Heating, and Refrigeration Institute (AHRI) is a trade association with 70 established standards and guidelines. AHRI represents the manufacturers of more than 90 percent of US-produced central air conditioning, heating, and commercial refrigeration equipment.

The American National Standards Institute (ANSI) is a private, not-for-profit organization that administers and coordinates the US voluntary standardization and conformity assessment system. Its mission is to enhance global competitiveness by promoting and facilitating voluntary consensus standards.

The American Society of Mechanical Engineers (ASME) is a not-for-profit, educational, and technical organization serving 130,000 members in 158 countries. ASME sets many industrial and manufacturing standards to promote and enhance the technical competency and professional well-being of its members.

The American Society of Sanitary Engineering (ASSE) International is dedicated to continually improving the performance, reliability, and safety of plumbing systems through consensus product performance and professional qualification standards, active product seal control standards and credentialing, advocacy, and public awareness programs.

The American Welding Society (AWS) is a multifaceted, not-for-profit organization with over 66,000 members, including engineers, scientists, educators, researchers, welders, and inspectors. AWS supports welding education and technology development in order to advance the science, technology, and application of welding.

The American Water Works Association (AWWA) is an international, not-for-profit scientific and educational organization that promotes healthy drinking water and high-quality water supplies. AWWA has more than 50,000 members, including 4,000 utilities that supply water to 180 million people.

ASTM International, formerly the American Society for Testing and Materials International (ASTM), is a developer and provider of more than 12,000 voluntary consensus standards in a wide range of technical fields.

CSA Group, formerly the Canadian Standards Association (CSA), is a not-for-profit association serving business, industry, government, and consumers in Canada. CSA Group develops standards that address and enhance public safety and preserve the environment.

The Cast-Iron Soil Pipe Institute (CISPI) seeks to advance the manufacture, use, and distribution of cast-iron soil pipe and fittings. CISPI strives to improve industry products, achieve standardization of cast-iron soil pipe and fittings, and provide a continuous program of product testing, evaluation, and development.

Federal Specifications (FS) are standards recognized by the General Services Administration. FS standards are available from the Superintendent of Documents, US Government Printing Office.

The International Code Council (ICC) is a not-for-profit organization dedicated to developing a single set of comprehensive and coordinated national model construction codes.

The National Fire Protection Association (NFPA) promotes fire and electrical safety by publishing and advocating over 300 consensus codes and standards. NFPA's membership consists of more than 70,000 individual members from over 100 countries, as well as national trade and professional organizations.

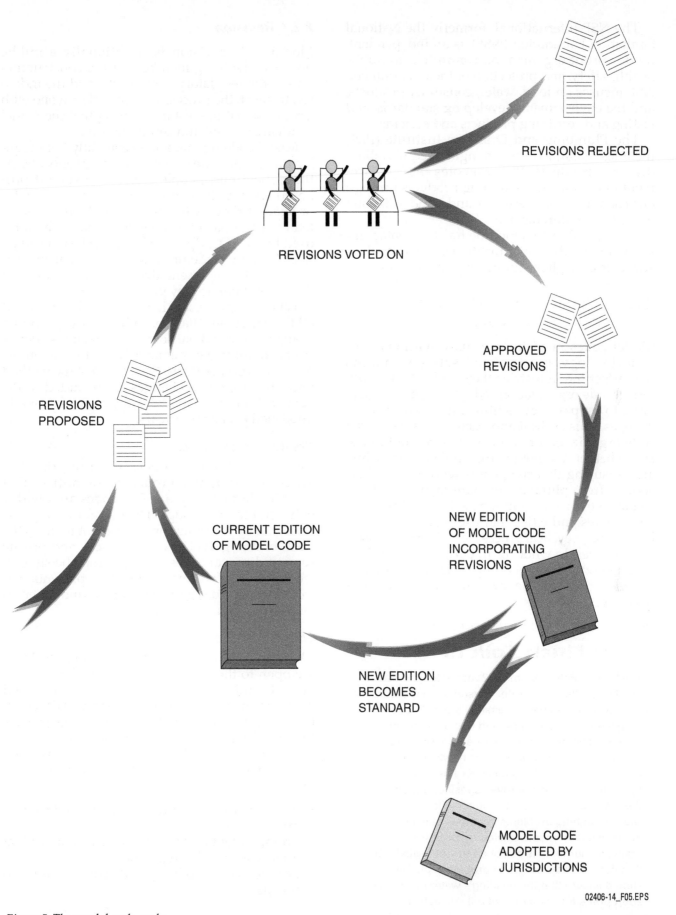

REVISIONS REJECTED

REVISIONS VOTED ON

APPROVED
REVISIONS

REVISIONS
PROPOSED

CURRENT EDITION
OF MODEL CODE

NEW EDITION
OF MODEL CODE
INCORPORATING
REVISIONS

NEW EDITION
BECOMES
STANDARD

MODEL CODE
ADOPTED BY
JURISDICTIONS

02406-14_F05.EPS

Figure 5 The model-code cycle.

The NSF International, formerly the National Sanitation Foundation (NSF), is an independent, not-for-profit organization committed to public health, safety, and protection of the environment. NSF focuses on food, water, consumer products, and the environment; developing standards; and testing and certifying products and systems.

The Plumbing and Drainage Institute (PDI) includes manufacturers of engineered plumbing drainage products. PDI promotes the advancement of engineered plumbing products through outreach, advocacy, research, and standardization of product requirements.

Appendix B in this module has the contact information for these organizations, including their addresses, telephone numbers, and websites.

2.2.0 Explaining the Revision and Adoption of Model Codes

Model codes need to keep pace with changes in technology, practice, and safety standards that affect the industries they cover. As a result, model code organizations adopt a regular update cycle to prepare new editions that reflect those changes. This cyclical approach allows the model code organization to review changes and to decide whether, and how, to incorporate them, while also ensuring that the people who rely on their codes—from plumbers to manufacturers—have a chance to provide input based on their experience. States and cities that use model codes follow a similar cyclical process for adopting revised model codes, in whole or in part, for use in their jurisdictions. This section provides a summary overview of how these revision and adoption processes typically work.

Low-Flush Toilets

The Energy Policy Act of 1992 mandated the use of low-flush toilets as a water conservation measure. The law required newly installed toilets to use 1.6 gallons per flush, less than half the amount they used before the law was passed. The latest version of the law now requires newly installed toilets to use just 1.28 gallons per flush. Although the technology has improved considerably since the law was introduced, low-flush toilets may cause plumbing problems in homes built before the new requirements existed because their drainage plumbing is designed to work with higher flow rates. To address these problems, plumbers should ensure that the building's water pressure is appropriate for the model of toilet installed.

2.2.1 Revision

Model code revision is traditionally a public process, involving members of the construction trades, the regulatory community, and the industry. In short, the process may include anyone with a professional interest in ensuring that the model codes are current and comprehensive.

Model code organizations generally hold regular meetings to hear code change proposals. At these meetings, people are free to submit proposed changes for consideration and to argue for or against them. Change proposals are submitted on a standard form. *Figure 6* shows the form used by the ICC; *Figure 7* shows the form used by IAPMO. Any trade professional, regulatory official, industry representative, or even a member of the general public may propose changes.

When all proposals have been submitted and all the arguments have been heard, the proposed changes are voted on. Some organizations allow a select committee of public officials to vote on the proposed code changes. Regardless of the method used, the approved changes are included in the next edition of the model code, which is usually published every three years.

The ICC Revision Process

Changes to the ICC codes, including the IPC, are reviewed by two groups of committees that meet in alternating years. The codes are divided between the two groups. The IPC, for example, is reviewed by the same committee that reviews the ICC's building, fuel gas, mechanical, and private sewage disposal codes. Changes may be submitted by code enforcement officials, representatives from industry, design professionals, and other interested parties.

The IAPMO Revision Process

Changes to the IAPMO codes, including the UPC, are open to the general public as well as to code officials, industry representatives, installers, and manufacturers. As discussed earlier, IAPMO follows an ANSI-accredited open-consensus process that encourages members and nonmembers alike to vote on proposed changes. IAPMO has identified three main goals for its code development process:

- Ensuring that public health, safety, and welfare are preserved effectively
- Prompt attention to technological changes that affect construction regulations
- Decisions reached by open discussion and consensus

ICC CODES - PUBLIC COMMENT FORM
FOR PUBLIC COMMENTS ON THE 2013 REPORT
OF THE PUBLIC HEARINGS ON THE 2012 EDITIONS OF:

Administrative Provisions© (ADM)
International Energy Conservation Code©
➢ Commercial Energy (CE)
➢ Residential Energy (RE)
International Existing Building Code© (EB)
International Fire Code© (F)
ICC Performance Code© (PC)
International Residential Code©
➢ Building (RB)
➢ Mechanical (RM)
➢ Plumbing (RP)
International Property Maintenance Code© (PM)
International Swimming Pool and Spa Code© (SP)
International Wildland-Urban Interface Code© (WUIC)

CLOSING DATE: <u>All Comments Must Be Received by:</u> July 15, 2013

1) Please type or print clearly: Public comments will be returned if they contain unreadable information.

Name:			Date:	
Jurisdiction/Company:				
Submitted on Behalf of:				
Address:				
City:		State:		Zip +4:
Phone:		Ext:		Fax:
e-mail:				

2) Copyright Release: In accordance with Council Policy #28 Code Development, all Code Change Proposals, Floor Modifications and Public Comments are required to include a copyright release. A copy of the copyright release form is included at the end of this form. Please follow the directions on the form. This form as well as an alternative release form can also be downloaded from the ICC website at <u>www.iccsafe.org</u>. If you have previously executed the copyright release, please check the box below:

☐ **2012-2014 Cycle copyright release on file**

3) Code Change Proposal Number:
Indicate the Code Change Proposal Number that is being addressed by this Public Comment: _____

4) Public Comment: The Final Action requested on this Code Change Proposal is: (Check Box)

| ☐ Approved as Submitted (AS): | ☐ Approved as Modified by this Public Comment (AMPC): | ☐ Approved as Modified by the Code Committee as Published in the ROH (AM): | ☐ Approved as Modified by Assembly Floor Action as Published in the ROH (AMF): | ☐ Disapproved (D): |

Attached Proposed Modifications and/or Reason Statements:

See Attached Individual Consideration Form

PLEASE USE SEPARATE FORM FOR EACH PUBLIC COMMENT
SUBMITTAL AS A DOCUMENT ATTACHED TO AN EMAIL IS PREFERRED
SEE BACK OF FORM FOR DIRECTIONS ON WHERE TO SEND PUBLIC COMMENTS

02406-14_F06.EPS

Figure 6 ICC public comment form for code changes.

FORM FOR PROPOSALS ON IAPMO UPC/UMC COMMITTEE DOCUMENTS

NOTE: All Proposals MUST be received by 5:00 PM PST on January 3, 2013.

For further information on the standards-making process, please contact the Codes and Standards Administration at 909-472-4110.

For technical assistance, please call IAPMO at 909-472-4111 or 909-230-5535.

FOR OFFICE USE ONLY

LOG # : _____

DATE REC'D: _____

PLEASE USE SEPARATE FORM FOR EACH PROPOSAL

IAPMO's green initiative is to lead the organization paper free by providing the Proposed Monographs, Annual Report on Proposals and Comments in digital Adobe PDF http://www.iapmo.org/Pages/RequestFormforROPROC.aspx

Note printed copies of Report on Proposals and Report on Comments will not be available at the hearings. All requested printed copies will be mailed 30 days prior to the hearing date. Please submit by the following dates listed by the code below on the request form.

Date _____ Name _____ Tel. No. _____

Company _____

Street Address _____ City _____ State ____ Zip. _____

Please Indicate Organization Represented (if any) _____

1. **IAPMO Document Title** _____ **IAPMO No. & Year** _____

    Section/Paragraph _____

2. **Proposal Recommends** (check one): ☐ New Text ☐ Revised Text ☐ Deleted Text

3. **Proposal** (Include proposed new wording, or identification of wording to be deleted): [Note: Proposed text should be in legislative format: i.e., use underscore to denote wording to be inserted (inserted wording) and strike-through to denote wording to be deleted (deleted wording).] Please note if you are referencing a standard or other publication, please provide two copies.

4. **Statement of Problem and Substantiation for Proposal**: [Note: State the problem that will be resolved by your recommendations: give the specific reason for your proposal including copies of tests, research papers, etc. If more then 200 words, it may be abstracted for publication.]

5. ☐ **This Proposal is original material**. [Note: Original material is considered to be the submitter's own idea based on or as a result of his/her own experience, thought or research and, to the best of his/her knowledge, is not copied from another source.)

    ☐ **This Proposal is not original material; its source (if known) is as follows:** _____

I hereby grant the IAPMO all and full rights in copyright, in this proposal, and I understand that I acquire no rights in any publication of IAPMO in which this proposal in this or another similar or analogous form is used.

Signature (Required): _____

PLEASE USE SEPARATE FORM FOR EACH PROPOSAL • IAPMO FAX • (909) 472-4198 or (877) 852-6337
Mail to: Code Development • IAPMO • 4755 E Philadelphia St • Ontario • CA • 91761-2816
Email to: codechange@iapmo.org

viii

02406-14_F07.EPS

Figure 7 IAPMO public comment form for code changes.

2.2.2 Adoption

Model codes are comprehensive in scope, but they can offer only the minimum acceptable standards and the broadest guidelines. Therefore, the jurisdictions that adopt them are free to make changes and add amendments to accommodate their state and local regulations. Changes to model codes are typically determined by an appointed advisory commission of regulatory experts, the members of which are determined by law. The commission reviews the code that is current in the jurisdiction. They also review the most recent edition of various model codes unless the jurisdiction has developed its own. The commission decides which model code to adopt and what, if any, changes or amendments are necessary. Adopted codes must be signed into law before they can take effect. Codes usually have to be submitted as an ordinance and be approved by the regulatory officials of the jurisdiction, as well as by officials all the way up to the state government.

Adopted codes—for example, building, mechanical, electrical, plumbing, and fire protection codes—are reviewed on a regular basis, usually every three to five years. Reviews ensure that the jurisdiction's codes remain up to date and comprehensive. Otherwise, the constant need to amend codes with variances and exceptions for each new technology and process would result in a patchwork code and bureaucratic headaches. A local code may not change the requirements that are specified in the model code on which it is based.

The review process also gives regulators an opportunity to adopt a different model code from the one they have used in the past. Regulators may choose to adopt a new model code to bring the jurisdiction into closer alignment with neighboring jurisdictions, or the model code from which a local code was adopted may no longer be updated.

Authority for code administration, inspection, and enforcement varies from jurisdiction to jurisdiction. The responsibility may lie with a building inspector, the health department, the city engineer, or a commissioner's office. Become familiar with how codes are regulated in your area so that you will know who to call when you have questions or concerns about code compliance.

2.3.0 Explaining Typical Code Changes

Changes in model and local codes frequently result from developments in plumbing materials, the growing demand for energy and water conservation, and the availability of alternative energy sources. New materials are approved for inclusion in model and local codes only after they are proven consistently safe and reliable. For example, local codes increasingly accept the use of new technologies in plumbing installations that may have been considered unsuitable for use in certain plumbing installations years ago. In addition to the piping material itself, codes also approve the new fastening and joining methods that these pipes require.

To illustrate some of the factors that can influence what is included in model codes, the 2012 UPC incorporated changes that were adopted to address the following developments:

- New industry guidance for using alternative water sources in nonpotable applications
- New methods of joining piping in water supply and DWV systems
- Changes in plumbing fixture requirements for greater ease of use by the elderly
- New research conducted by the American Society of Plumbing Engineers (ASPE) and other organizations into design methods that reflect the concept of "potty parity," or the standard ratio of women's restroom facilities to those for men

Forbidden Fittings

Some codes prohibit certain fittings, while others may permit them. Plumbers are expected to know what fittings may or may not be used in plumbing installations, but consumers may be unaware of plumbing code requirements and install prohibited fittings. You may come across such fittings from time to time. Learn from local plumbing experts the proper course of action to take if you discover something that is in violation of the local code.

One prohibited fitting is the saddle fitting, which is used to make a connection to a pre-installed pipe. Saddle fittings are often used in humidifiers and ice makers. These widely used and easily available fittings may or may not be acceptable in the local code. Homeowners run the greatest risk of violating a code that forbids saddle fittings because these fittings can be easily obtained at local hardware stores.

Codes increasingly stress energy conservation measures, such as insulation for flow restrictors on hot- and cold-water outlets, pipes, and water heaters. Solar hot water installations are a popular alternative source of hot water in residential and commercial installations. They are addressed in special model codes, such as IAPMO's *Uniform Solar Energy and Hydronics Code*™ or the ICC's *International Energy Conservation Code*®. Model solar-energy codes do not address specific technical issues, such as whether to use double-wall or single-wall heat exchangers, or whether it is more appropriate to use toxic or nontoxic circulating fluids in solar water heater installations.

Code changes reflect the most recent consensus among the construction trades, the regulatory community, and the industry. As technology, practice, and laws change, model codes and adopted codes will change as well. As a result, there will never be a perfect or ultimate code, just the latest code.

House Traps and Regional Code Differences

Differences between codes may arise from a variety of factors, such as geography, climate, urban population density, or even the age of a city and its infrastructure. A good example of the variations between local codes and model codes is the case of house traps. Both the IPC and the UPC model codes prohibit house traps, but provide an exception when the local authority requires them. This exception is necessary because many counties and municipalities in the northeastern part of the United States actually require them in every structure, especially in cities with older sections and high population density.

House-trap rules demonstrate the dramatic differences in codes that can be found in different regions and states, and how important it is to know the code governing your specific area.

Additional Resources

Code Check: A Field Guide to Building a Safe House. 2000. Redwood Kardon, Michael Casey, and Douglas Hansen. Newtown, CT: Taunton Press.

Code Check Plumbing: A Field Guide to the Plumbing Codes. 2006. Michael Casey, Douglas Hansen, and Redwood Kardon. Newtown, CT: Taunton Press.

The Engineering Resources Code Finder for Building and Construction. 2001. Dennis Phinney. Anaheim, CA: BNI Building News.

2.0.0 Section Review

1. The international organization dedicated to promoting healthy drinking water and high-quality water supplies is _____.

   a. AHRI
   b. ASME
   c. AWWA
   d. AWS

2. The open-consensus code revision process followed by IAPMO has been accredited by _____.

   a. NFPA
   b. ASTM International
   c. ANSI
   d. ASSE International

3. New materials are approved for inclusion in model and local codes _____.

   a. when the manufacturer demonstrates that they can be produced safely
   b. when they are considered affordable
   c. immediately upon their introduction
   d. when they are proven consistently safe and reliable

SUMMARY

The purpose of a code is to provide safe plumbing installations for the public. Professional organizations develop model codes that serve as general guidelines for plumbing installations, materials, and practices. Jurisdictions can adopt and amend model codes to meet their specific requirements. Codes are regularly revised to reflect changing technologies, practices, and safety standards. The plumber's work must be performed according to the standards and practices outlined in a code. The best way to understand a code is to take the time to read it thoroughly. Ask local officials and experienced plumbers to explain the similarities and differences between model, state, and local codes that apply to your area. Remember, codes are among the most important tools you will ever use on the job.

1. In addition to establishing and defining acceptable practices and materials, a model code also _____.

   a. lists prohibited installations
   b. regulates contractor/owner responsibilities
   c. outlines apprenticeship requirements
   d. establishes work hours and conditions

2. Establishing uniform global standards is a goal of _____.

   a. BOCA
   b. IAPMO
   c. ICC
   d. SBCCI

3. Requirements for systems that collect and dispose of water used in applications such as subsurface irrigation systems are outlined in the IPC chapter entitled _____.

   a. *Storm Drainage*
   b. *Gray Water Recycling Systems*
   c. *Water Supply and Distribution*
   d. *Indirect/Special Waste*

4. The *International Plumbing Code*® is available in either loose-leaf or softcover versions as well as a(n) _____.

   a. audiobook version
   b. searchable online version
   c. e-book version
   d. password-protected PDF version

5. The NFPA and IAPMO have partnered to produce a series of _____.

   a. Code Classification Collections®
   b. Comprehensive Consensus Codes®
   c. Construction Code Categories®
   d. Comprehensive Construction Codes®

6. The *Uniform Plumbing Code*® consists of 17 chapters and 12 _____.

   a. addendums
   b. accessory sections
   c. allied trades listings
   d. appendixes

7. The *Uniform Plumbing Code*® is available in either loose-leaf or softcover versions as well as a(n) _____.

   a. audiobook version
   b. searchable online version
   c. e-book version
   d. password-protected PDF version

8. The private, not-for-profit group that administers the voluntary standardization and conformity assessment system in the United States is _____.

   a. ANSI
   b. ASME
   c. AWS
   d. AWWA

9. Model code organizations prepare new editions based on _____.

   a. supply and demand
   b. government agency requirements
   c. a regular update cycle
   d. equipment and materials improvements

10. A jurisdiction that adopts a model code _____.

    a. can adapt and amend it to meet local requirements
    b. must implement it exactly as written
    c. may purchase a version customized to its needs
    d. is required to get permission before making changes

Trade Terms Quiz

Fill in the blank with the correct term that you learned from your study of this module.

1. The _____ is a collection of ANSI-compliant codes and standards developed jointly by the International Association of Plumbing and Mechanical Officials (IAPMO) and the National Fire Protection Association (NFPA).

2. A set of comprehensive, general guidelines that establish and define acceptable plumbing practices and materials and list prohibited installations is called a(n) _____.

3. The _____ is the model plumbing code of the International Code Council (ICC).

4. A legal document enacted to protect the public and property, establishing the minimum standards for materials, practices, and installations is called a(n) _____.

5. The _____ is the model code of the International Association of Plumbing and Mechanical Officials (IAPMO).

Trade Terms

Code
Comprehensive Consensus Codes®
International Plumbing Code® (IPC)

Model code
Uniform Plumbing Code® (UPC)

Trade Terms Introduced in This Module

Code: A legal document enacted to protect the public and property, establishing the minimum standards for materials, practices, and installations.

Comprehensive Consensus Codes®: A collection of ANSI-compliant codes and standards developed jointly by the International Association of Plumbing and Mechanical Officials (IAPMO) and the National Fire Protection Association (NFPA).

International Plumbing Code® (IPC): The model plumbing code of the International Code Council (ICC). It was first issued in 1995.

Model code: A set of comprehensive, general guidelines that establish and define acceptable plumbing practices and materials and list prohibited installations.

Uniform Plumbing Code® (UPC): The model code of the International Association of Plumbing and Mechanical Officials (IAPMO). It has been published since 1945.

Use a copy of your local applicable code to answer the following questions about requirements in your area for various types of plumbing installations. Be sure to cite the reference following your answer.

1. What are the duties and powers entrusted to your local code officials?

2. What work is specifically exempt from the requirement for a permit?

3. What are the requirements for connecting the drainage piping to offsets and bases of stacks?

4. How are vacuum system station receptacles to be installed?

5. What standard is referenced for the installation of nonflammable medical gas systems?

6. What are the minimum required air gaps for lavatories and fixtures with similarly sized openings?

7. How is the maximum water-consumption flow rate determined for a public lavatory?

8. What is the maximum allowable amount of lead content in water supply pipe and fittings?

9. What are the requirements for joints between copper and copper-alloy tubing?

10. What are the size requirements for a shower compartment?

Appendix B

MODEL CODE ORGANIZATIONS

Air-Conditioning, Heating, and Refrigeration Institute (AHRI)
2111 Wilson Boulevard, Suite 500
Arlington, VA 22201
(703) 524-8800
www.ari.org

American National Standards Institute (ANSI)
1899 L Street, NW, 11th Floor
Washington, DC 20036
(202) 293-8020
www.ansi.org

American Society of Mechanical Engineers (ASME)
Two Park Avenue
New York, NY 10016-5990
(800) 843-2763
www.asme.org

American Society of Sanitary Engineering (ASSE) International
18927 Hickory Creek Drive, Suite 220
Mokena, IL 60448
(708) 995-3019
www.asse-plumbing.org

American Water Works Association (AWWA)
6666 West Quincy Avenue
Denver, CO 80235
(303) 794-7711
www.awwa.org

American Welding Society (AWS)
8669 NW 36 Street, #130
Miami, FL 33166-6672
(305) 443-9353
www.aws.org

ASTM International
100 Barr Harbor Drive
West Conshohocken, PA 19428-2959
(610) 832-9585
www.astm.org

CSA Group
178 Rexdale Boulevard
Toronto, Ontario, Canada M9W 1R3
(416) 747-4000
www.csagroup.ca

Cast-Iron Soil Pipe Institute (CISPI)
3008 Preston Station Drive
Hixson, TN 37343
(423) 842-2122
www.cispi.org

Federal Specifications (FS)
General Services Administration
One Constitution Square
1275 First Street, NE
Washington, DC 20417
(800) 333-4636
www.gsa.gov

International Association of Plumbing and Mechanical Officials (IAPMO)
4755 E. Philadelphia Street
Ontario, CA 91761
(909) 472-4100
www.iapmo.org

International Code Council, Inc. (ICC)
500 New Jersey Avenue, NW, 6th Floor
Washington, DC 20001
(888) 422-7233
www.iccsafe.org

National Fire Protection Association (NFPA)
1 Batterymarch Park
Quincy, MA 02169-7471
(617) 770-3000
www.nfpa.org

NSF International
P.O. Box 130140
789 N. Dixboro Road
Ann Arbor, MI 48105
(734) 769-8010
www.nsf.org

Plumbing and Drainage Institute (PDI)
800 Turnpike Street, Suite 300
North Andover, MA 01845
(800) 589-8956
www.pdionline.org

Plumbing-Heating-Cooling Contractors Association (PHCC)
180 South Washington Street, Suite 100
Falls Church, VA 22046
(800) 533-7694
www.phccweb.org

Additional Resources

This module presents thorough resources for task training. The following resource material is suggested for further study.

Code Check: A Field Guide to Building a Safe House. 2000. Redwood Kardon, Michael Casey, and Douglas Hansen. Newtown, CT: Taunton Press.

Code Check Plumbing: A Field Guide to the Plumbing Codes. 2006. Michael Casey, Douglas Hansen, and Redwood Kardon. Newtown, CT: Taunton Press.

The Engineering Resources Code Finder for Building and Construction. 2001. Dennis Phinney. Anaheim, CA: BNI Building News.

International Plumbing Code®, Latest Edition. Falls Church, VA: International Code Council.

Uniform Plumbing Code®, Latest Edition. Ontario, CA: International Association of Plumbing and Mechanical Officials.

Figure Credits

Excerpted from the 2012 International Plumbing Code. Copyright 2011. Washington DC. International Code Council. Reproduced with permission. All rights reserved. **www.ICCSAFE. org**, Figures 1–2, Figure 6

Courtesy of IAPMO®, Figures 3–4, Figure 7

Answer	Section Reference	Objective
Section One		
1. d	1.1.2	1a
2. b	1.2.1	1b
Section Two		
1. c	2.1.0	2a
2. c	2.2.1	2b
3. d	2.3.0	2c

NCCER CURRICULA — USER UPDATE

NCCER makes every effort to keep its textbooks up-to-date and free of technical errors. We appreciate your help in this process. If you find an error, a typographical mistake, or an inaccuracy in NCCER's curricula, please fill out this form (or a photocopy), or complete the online form at **www.nccer.org/olf**. Be sure to include the exact module ID number, page number, a detailed description, and your recommended correction. Your input will be brought to the attention of the Authoring Team. Thank you for your assistance.

Instructors – If you have an idea for improving this textbook, or have found that additional materials were necessary to teach this module effectively, please let us know so that we may present your suggestions to the Authoring Team.

NCCER Product Development and Revision

13614 Progress Blvd., Alachua, FL 32615

Email: curriculum@nccer.org
Online: www.nccer.org/olf

❑ Trainee Guide ❑ Lesson Plans ❑ Exam ❑ PowerPoints Other _____

Craft / Level: _____ Copyright Date: _____

Module ID Number / Title: _____

Section Number(s): _____

Description: _____

_____

_____

_____

_____

Recommended Correction: _____

_____

_____

_____

_____

Your Name: _____

Address: _____

_____

Email: _____ Phone: _____

02408-14

Private Water Supply Well Systems

OVERVIEW

Private water supply systems provide potable water in rural areas. Most private systems in the United States use wells with pumps that lift water from aquifers. The four most common well types are bored, drilled, driven, and dug. Pumps commonly used with wells include shallow- and deep-well jet pumps, and submersible pumps. Codes require well water to be tested for turbidity, pH value, hardness, and biochemical oxygen demand (BOD).

Module Seven

Trainees with successful module completions may be eligible for credentialing through the NCCER Registry. To learn more, go to **www.nccer.org** or contact us at **1.888.622.3720**. Our website has information on the latest product releases and training, as well as online versions of our *Cornerstone* magazine and Pearson's product catalog.

Your feedback is welcome. You may email your comments to **curriculum@nccer.org**, send general comments and inquiries to **info@nccer.org**, or fill in the User Update form at the back of this module.

This information is general in nature and intended for training purposes only. Actual performance of activities described in this manual requires compliance with all applicable operating, service, maintenance, and safety procedures under the direction of qualified personnel. References in this manual to patented or proprietary devices do not constitute a recommendation of their use.

Objectives

When you have completed this module, you will be able to do the following:

1. Explain how to drill wells.
   a. Explain how to locate wells.
   b. Explain how to size wells.
   c. Explain how to construct wells.
   d. Explain how to clean wells.
2. Explain the operation of various types of pumps and pump components.
   a. Explain the operation of a shallow-well jet pump.
   b. Explain the operation of a deep-well jet pump.
   c. Explain the operation of a submersible pump.
3. Explain how to select and install water supply and storage components.
   a. Explain how to select and install water supply lines.
   b. Explain how to select and install water storage tanks.

Performance Task

Under the supervision of your instructor, you should be able to do the following:

1. Assemble and disassemble given components of private water supply well systems.

Trade Terms

Aquifer	Drilled well	pH value	Turbidity
Artesian aquifer	Driven well	Pitless adapter	Two-pipe system
Biochemical oxygen demand (BOD)	Drop pipe	Pressure control valve	Venturi
	Dug well	Seven-minute peak demand period	Water table
Bored well	Foot valve		Water table aquifer
Chlorinate	Hydrologic cycle	Shallow well	Well
Cistern	Low-water cutoff device	Shallow-well jet pump	Well casing
Deep well	Packer-type ejector	Submersible pump	Well-casing adapter
Deep-well jet pump	Perched water table	Tail pipe assembly	Well head

Industry-Recognized Credentials

If you're training through an NCCER-accredited sponsor, you may be eligible for credentials from NCCER's Registry. The ID number for this module is 02408-13. Note that this module may have been used in other NCCER curricula and may apply to other level completions. Contact NCCER's Registry at 888.622.3720 or go to **www.nccer.org** for more information.

Code Note

Codes vary among jurisdictions. Because of the variations in code, consult the applicable code whenever regulations are in question. Referring to an incorrect set of codes can cause as much trouble as failing to reference codes altogether. Obtain, review, and familiarize yourself with your local adopted code.

Contents

Topics to be presented in this module include:

Figures and Tables

SECTION ONE

1.0.0 DRILLING WELLS

Objective

Explain how to drill wells.
  a. Explain how to locate wells.
  b. Explain how to size wells.
  c. Explain how to construct wells.
  d. Explain how to clean wells.

Trade Terms

Aquifer: An underground reservoir containing water-saturated soil, sand, and rock.

Artesian aquifer: A deep aquifer that does not lose water through runoff. Also called a confined aquifer.

Bored well: A type of well drilled by an auger and lined with a well casing.

Chlorinate: To disinfect a well with a chlorine compound such as calcium hypochlorite or liquid bleach.

Cistern: A water storage tank used to supply some private water supply well systems with potable water.

Deep well: A well that is drilled deeper than 25 feet.

Drilled well: A type of well created by chisels or rotary drills.

Driven well: A type of well drilled by a pointed well head driven by a hammer.

Dug well: A type of well excavated by shovel and backhoe and lined with brick, stone, or concrete.

Hydrologic cycle: The process through which water cycles through the environment as both a liquid and a vapor.

Perched water table: A small water table that is cut off from a larger reservoir.

Seven-minute peak demand period: An estimate of the greatest possible water demand in a private water supply well system.

Shallow well: A well that is no more than 25 feet deep.

Water table: The top of an aquifer, below which rock, sand, and soil are completely saturated with groundwater.

Water table aquifer: An aquifer in which water flows toward bodies of water. Also called an unconfined aquifer.

Well: A hole in the ground that connects with an aquifer. Wells are equipped with pumps to lift water from the aquifer into water storage tanks.

Well casing: A pipe section inserted into a well that maintains the shape of the well and prevents contamination.

Well head: A pointed rod used to create a driven well by being forced into the ground with a hammer.

Many residences and businesses in rural areas and in other remote locations do not have access to municipal water supplies. These people rely on private sources for their potable (drinkable) water. Some private water supply systems get water from springs or from storage tanks called cisterns. However, most private systems in the United States get water from wells. A well is a hole in the ground that connects with an aquifer, or underground reservoir. Wells use pumps to lift water from the aquifer into pipes that connect to a home or business. A reliable private water supply well system begins with a good well. A good well is a reliable source of clean, potable water. It should also be a sound investment for the customer. Customers have a right to expect years of reliable service from a private water supply well system.

Plumbers should know how wells work. In many areas, plumbers install the pump, supply piping, and storage tanks. In some states, certified well drillers perform these tasks. Elsewhere, the duties are split between the plumber and the well driller. Local codes specify the plumber's responsibilities.

As mentioned earlier, wells draw water from aquifers. Aquifers are filled with water naturally as part of the hydrologic cycle, the process by which water circulates through the environment. This process is illustrated in *Figure 1*.

At the beginning of the process, water vapor evaporates, or rises into the air from the ground, bodies of water, and plants. Once in the air, water condenses into clouds and returns to Earth in the form of rain, snow, or hail. This process is called precipitation. Back on the ground, the water percolates, or seeps into the soil, where it becomes part of an aquifer. Surface runoff and underground flow fill lakes, rivers, and oceans. Then the process begins again.

The sand, soil, and rocks in an aquifer are completely saturated with water. The upper limit of

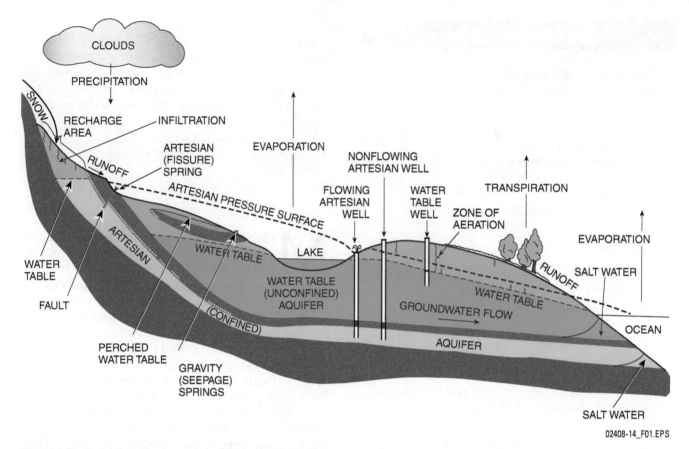

Figure 1 Hydrologic cycle.

an aquifer is called the water table. The water table can be close to or far below the land surface. Whatever its depth, the water table usually follows the contours of the land. The two types of aquifers are water table aquifers and artesian aquifers.

Water table aquifers are refilled, or recharged, primarily by water percolating down from Earth's surface. Artesian aquifers are recharged only by precipitation that falls in the recharge area (usually a higher-elevation formation of rock and rock outcrops). These aquifers are restricted between impermeable layers of soil and rock, and are therefore under pressure. This force, called artesian pressure, slopes downward from the recharge area, and the artesian-pressure surface is the level to which the water in the aquifer would rise if allowed. Drilling a well into an artesian aquifer with the outlet positioned below the pressure surface will produce a flowing well.

Wells can be drilled into either water table aquifers or artesian aquifers. Artesian aquifers do not lose groundwater through runoff. They are often called confined aquifers for this reason. Water table aquifers are not as deep as artesian aquifers. The water in water table aquifers flows toward bodies of water. These aquifers are also called unconfined aquifers.

A perched water table is a smaller water table that is cut off from a larger, nearby aquifer by an impermeable layer of rock. If a perched water table is tapped by many wells, its water supply may eventually be exhausted, and the wells will run dry.

1.1.0 Locating Wells

Wells must be correctly located over the water supply. An incorrectly located well will cause a variety of problems for the customer. In most cases, plumbers are not directly involved in the well-drilling process. However, customers often ask plumbers to recommend a reputable well driller. Take the time to become familiar with dependable drillers in your area.

Private Water Supply Systems

The US Environmental Protection Agency (EPA) estimates that approximately 42 million people in the United States—16 percent of the population—obtain their potable water from private water supply systems.

A driller should be able to determine which type of well would work best for each installation. The driller should review drilling logs to learn about other water wells in the area. Drilling logs list the following information for each well:

- The depth at which water was first found when the well was drilled
- The well's total depth
- The different types of rock, sand, and soil encountered during drilling
- The water quality

Finally, the driller should obtain all required permits. In most jurisdictions, well permits are issued by the health department or a similar agency.

Work closely with the customer, driller, and code officials to find a good location for a well. Be sure to review state and local regulations that apply to wells. They will list what types of wells are approved for household and commercial use. Regulations will also specify the minimum allowable distances between a well and property lines, septic tanks, buried fuel tanks, and livestock fields. Most codes require at least a 100-foot distance between a well and a leach field, for example. Consider the distance from the well to the structure being serviced, as it will influence the amount of pipe and the number of fittings needed. Locating a well on high ground will ensure that surface water drains away from, not into, the well opening.

There are always risks when locating a well. The well might not be able to replenish itself fast enough to meet demand; worse, the water table could eventually fall below the well, causing the well to go dry. Study the drilling logs carefully and talk to drillers who know the area. If you are concerned that the well may not meet the customer's needs, talk to the customer and the driller. Taking the time to locate the well correctly will save everyone time and money in the long run.

1.2.0 Sizing Wells

Sizing the well and selecting a pump can be difficult. One popular way to determine residential water needs is to calculate the seven-minute peak demand period. This is an estimate of the heaviest possible water demand, usually presented in a table. Local plumbing officials can provide a table for your area.

Some fixtures are less likely than others to operate at their maximum output. For example, a kitchen sink is not likely to be run at its peak output for seven minutes at a time. Three gallons per minute (gpm) is a reasonable compromise (see *Table 1*). Because demand may exceed the sample minimum, use a flow rate higher than the minimum acceptable rate for the system.

1.3.0 Constructing Wells

There are four common types of wells. These four wells are classified by how they are created:

- Bored wells
- Drilled wells
- Driven wells
- Dug wells

Bored wells (*Figure 2*) are drilled out by an auger where the ground is relatively soft. Once the hole has been completely drilled, a pipe section called a well casing is inserted. The well casing holds the shape of the well and prevents contamination of the water. The diameter of a bored well can range from 2 to 30 inches. In many areas, bored wells are less common than other types of wells.

Drilled wells are constructed where the ground is too hard to be dug or driven and/or where the water table is deep underground. Chisels or rotary drills are used to construct drilled wells (*Figure 3*). The diameter of a drilled well depends on the size of the chisel or drill, but is usually smaller than bored or dug wells. Generally speaking, drilled wells are currently the most common type of well.

Driven wells are used where the soil is relatively free of rocks. A pointed well head forced into the ground with a hammer creates driven wells (see *Figure 4*). Well heads are usually between 1½ and 2 inches in diameter.

Dug wells are excavated with shovels and backhoes and are lined with brick, stone, concrete, tile, or metal (*Figure 5*). Their diameters depend on the excavating tools used to dig them.

Wells that are 25 feet or less in depth are called shallow wells. Wells that are deeper than 25 feet are called deep wells. Bored, dug, and driven wells are usually shallow wells, while drilled wells are most often deep wells. Regardless of how the well is created, it is correct to use the term *drilling a well*.

Due to the enormous variety of geological conditions in different regions, the types and methods of well construction differ from region to region. Some types of wells may be common in some regions and rarely used, or even prohibited, in others. Be sure to check your local code when making decisions on what type of well to construct.

Table 1 Sample Table for Estimating the Seven-Minute Peak Demand Period

Outlets	Flow Rate (gpm)	Total Usage (gal.)	Bathrooms in House			
			1	1½	2 to 2½	3 to 4
Shower or bathtub	5	35	35	35	53	70
Lavatory	4	2	2	4	6	8
Toilet	4	5	5	10	15	20
Kitchen sink	5	3	3	3	3	3
Automatic washer	5	35	—	18	18	18
Dishwasher	2	14	—	—	3	3
Normal seven-minute peak demand (gal.)			45	70	98	122
Minimum-size pump required to meet peak demand without supplemental supply			7 gpm (420 gph)	10 gpm (600 gph)	14 gpm (840 gph)	17 gpm (1,020 gph)

Notes: Values given are average and do not include extremes, Peak demand can occur several times during morning and evening hours.
Farm, irrigation, and sprinkler requirements are not shown. These values must be added to the peak demand figures if usage will occur during normal demand periods.

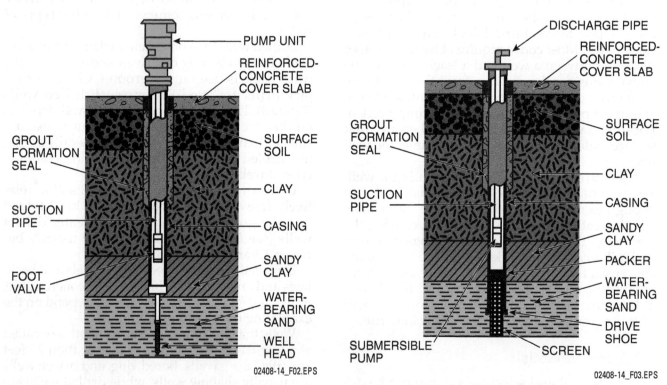

Figure 2 Cross section of a typical bored well.

02408-14_F02.EPS

Figure 3 Cross section of a typical drilled well.

02408-14_F03.EPS

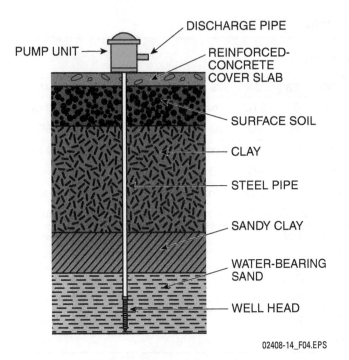

Figure 4 Cross section of a typical driven well.

02408-14_F04.EPS

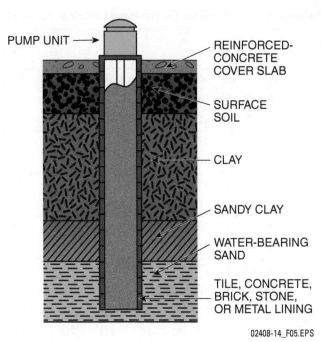

Figure 5 Cross section of a typical dug well.

02408-14_F05.EPS

> **WARNING!**
>
> A flooded well is a serious health hazard. Flooded wells may not be safe for months after a flood. Floodwater contains bacteria, wastewater, and chemicals. Repeated testing is usually required to ensure the safety of the water. The EPA does not recommend disinfecting or using flooded dug wells.

1.4.0 Cleaning Wells

After all the sand and dirt left in the well pit from the drilling operation have been removed, chlorinate, or disinfect, the well. Add a chlorine compound, such as calcium hypochlorite or liquid household bleach, to the well water. Dilute the chlorine with water before adding it to the well. For calcium hypochlorite, use a 70 percent

solution with water. For liquid bleach, use a 5.25 percent solution with water. Your local code will provide guidelines on how much disinfectant to use (see *Table 2*).

Health officials are responsible for testing the water quality of private water supplies. Most testing for contaminants occurs after installation, annually, or when a problem is suspected.

> **WARNING!**
>
> Disinfectants do not protect against pesticides, heavy metals, and other nonbiological contaminants. Consult local health officials if you suspect that a well contains nonbiological contaminants. Home water-treatment units, also called point-of-use or point-of-entry units, may be required to eliminate such contaminants.

Arsenic Levels

Arsenic is a natural chemical that is released by rock, mineral, and soil erosion. Arsenic is also a by-product of industrial processes. Scientific studies have linked continued exposure to arsenic in drinking water to certain types of cancer. In 2006, the limit on the amount of arsenic in drinking water was lowered to 10 parts per billion (ppb).

Table 2 Sample Table for Determining Chlorination Levels in Wells

Water Depth (ft)		2	3	4	5	6	8	10	12	16	20	24	28	32	36	42	48
											Well Diameter (in)						
5	A	1T	1T	1T	1T	1T	1T	2T	3T	5T	6T	3 oz	4 oz	5 oz	7 oz	9 oz	12 oz
	B	1C	1C	1C	1C	1C	1C	1C	1C	2C	4C	1Q	2Q	3Q	3Q	4Q	5Q
10	A	1T	1T	1T	1T	1T	2T	3T	5T	8T	4 oz	6 oz	8 oz	10 oz	13 oz	1½ lbs	1½ lbs
	B	1C	1C	1C	1C	1C	1C	2C	2C	1C	2Q	3Q	4Q	4Q	6Q	8Q	2½ G
15	A	1T	1T	1T	1T	2T	3T	5T	8T	4 oz	6 oz	9 oz	12 oz	1 lb	1½ lbs	1½ lbs	2 lbs
	B	1C	1C	1C	1C	1C	2C	3C	4C	2Q	2½ Q	4Q	5Q	6Q	2G	3G	4G
20	A	1T	1T	1T	2T	3T	4T	6T	3 oz	5 oz	8 oz	—	—	—	—	—	—
	B	1C	1C	1C	1C	1C	2C	4C	1Q	2½ Q	3½ Q	—	—	—	—	—	—
30	A	1T	1T	2T	3T	4T	6T	3 oz	4 oz	8 oz	12 oz	—	—	—	—	—	—
	B	1C	1C	1C	1C	2C	4C	1½ Q	2Q	4Q	5Q	—	—	—	—	—	—
40	A	1T	1T	2T	4T	6T	8T	4 oz	6 oz	10 oz	1 lb	—	—	—	—	—	—
	B	1C	1C	1C	2C	2C	1Q	2Q	2½ Q	4½ Q	7Q	—	—	—	—	—	—
60	A	1T	2T	3T	5T	8T	4 oz	6 oz	9 oz	—	—	—	—	—	—	—	—
	B	1C	1C	2C	3C	4C	2Q	3Q	4Q	—	—	—	—	—	—	—	—
80	A	1T	3T	4T	7T	9T	5 oz	8 oz	12 oz	—	—	—	—	—	—	—	—
	B	1C	1C	2C	4C	1Q	2Q	3½ Q	5Q	—	—	—	—	—	—	—	—
100	A	2T	3T	5T	8T	4 oz	7 oz	10 oz	1 lb	—	—	—	—	—	—	—	—
	B	1C	2C	3C	1Q	1½ Q	2½ Q	4Q	6Q	—	—	—	—	—	—	—	—
150	A	3T	5T	8T	4 oz	6 oz	10 oz	1 lb	1½ lbs	—	—	—	—	—	—	—	—
	B	2C	2C	4C	2Q	2½ Q	4Q	6Q	2½ G	—	—	—	—	—	—	—	—

Notes:
A = Calcium hypochlorite solution
B = Liquid household-bleach solution
T = Tablespoon; oz = ounces (by weight); C = cup; lb = pound; Q = quart; G = gallon

Water Quality Testing

Health officials test well water when private water supply installations are complete. However, tests for nitrate and coliform bacteria should be performed annually. Tests for radon and pesticides should be performed more frequently if a problem is suspected.

Private laboratories will test for nitrate and bacteria for a small fee. Tests for other types of contaminants could cost hundreds or thousands of dollars. Contact your local water-quality or certification officer for a list of testing laboratories in your area.

Laboratories usually mail test results within a few weeks. If the results are positive for contamination, contact the public health department for guidance and instructions.

Additional Resources

Finding Water: A Guide to the Construction and Maintenance of Private Water Supplies. Rick Brassington. Chichester, New York: John Wiley & Sons.

International Plumbing Code®, Latest Edition. Falls Church, VA: International Code Council.

Wells and Septic Systems. Second Edition. 1992. Max Alth, Charlotte Alth, and S. Blackwell Duncan. Blue Ridge Summit, PA: Tab Books.

1.0.0 Section Review

1. Each of the following types of information can be found in a drilling log *except* _____.

   a. the well's total depth
   b. the quality of the water in the well
   c. the type of pump used in the well
   d. the types of rock, sand, and soil encountered during drilling

2. When estimating fixture output, use a flow rate that is _____.

   a. At least 75 percent of the flow rate of the fixture with the greatest seven-minute peak demand
   b. the same as the minimum acceptable rate for the system
   c. lower than the minimum acceptable rate for the system
   d. higher than the minimum acceptable rate for the system

3. The type of well that can be used where the ground is relatively soft is the _____.

   a. bored well
   b. drilled well
   c. driven well
   d. dug well

4. Disinfect wells using a(n) _____.

   a. muriatic acid solution
   b. chlorine compound
   c. mixture of lime and water
   d. ultraviolet purifier

2.0.0 OPERATION OF PUMPS AND WELLS

Objective

Explain the operation of various types of pumps and pump components.

   a. Explain the operation of a shallow-well jet pump.
   b. Explain the operation of a deep-well jet pump.
   c. Explain the operation of a submersible pump.

Trade Terms

Deep-well jet pump: A pump used in a deep well that uses a nozzle and venturi to accelerate the water flow.

Drop pipe: Another name for the discharge pipe that connects the well to the water storage tank.

Foot valve: A valve used to prime water-lubricated pumps in wells deeper than 32 feet.

Low-water cutoff device: A device that deactivates a pump when the water level in a well falls to a preset low point. See also *tail pipe assembly*.

Packer-type ejector: For jet pumps, an alternative discharge method to the two-pipe system in which the space between the water supply pipe and the well casing acts as a pressure pipe.

Pitless adapter: A device that permits a discharge pipe to exit a well and connect to a water storage tank. Pitless adapters protect well water from contamination and freezing.

Pressure control valve: A valve that diverts a portion of a pump's discharge through the ejector, thereby maintaining adequate pressure in the ejector.

Shallow-well jet pump: A pump used in a shallow well that uses a nozzle and venturi to accelerate the water flow.

Submersible pump: A pump consisting of a multistage centrifugal pump and a submersible electric motor. Submersible pumps are usually installed in the bottom of deep wells.

Tail pipe assembly: A device that matches a pump's discharge rate to a well's recovery rate. See also *low-water cutoff device*.

Two-pipe system: For jet pumps, an alternative to the packer-type ejector discharge method. One pipe draws well water through an ejector, and another pipe feeds the pressurized water to the jet.

Venturi: A device that accelerates water and lowers its pressure by passing it through a narrow pipe opening. It is also called a diffuser.

Well-casing adapter: A well cover that protects the well casing from air leaks.

After a well has been drilled and cleaned, it must be fitted with a suitable pump, along with the components that allow the pump to operate. In fact, pump selection may be the most critical decision of the entire project. In some jurisdictions, only licensed installers are allowed to install pumps. If you are allowed to install pumps, be sure to follow the manufacturer's specifications and the local codes. You learned how to install and troubleshoot sewage, sump, and recirculation pumps earlier in this curriculum. Well pumps have many mechanical similarities to these pumps, but well installations have their own requirements. Regardless, you can use the same troubleshooting techniques you learned earlier for well pumps.

Consider the following factors when selecting a well pump (see *Figure 6*):

- The total vertical height that the water must be lifted in the well
- The total vertical and horizontal distance that the water must be pumped to the structure
- The rate at which the water is pumped

Remember that 1 cubic foot of pure water weighs 62.4 pounds. Every vertical foot that water has to be lifted, first in the well and then to the structure, adds to the weight that the pump must overcome. Long vertical and horizontal pipe runs require more pipe and fittings, resulting in friction loss. A high pumping rate will also add to friction loss. Select a pump that will be able to overcome these obstacles and still provide water at the correct pressure.

Ensure that the pump is set at the proper water depth within the well. If the pump is set at the bottom of the well, sediment could clog the pump and burn out the motor. If the pump is placed too high in the well, it may not be able to pump enough water. Refer to the manufacturer's specifications for proper depth placement.

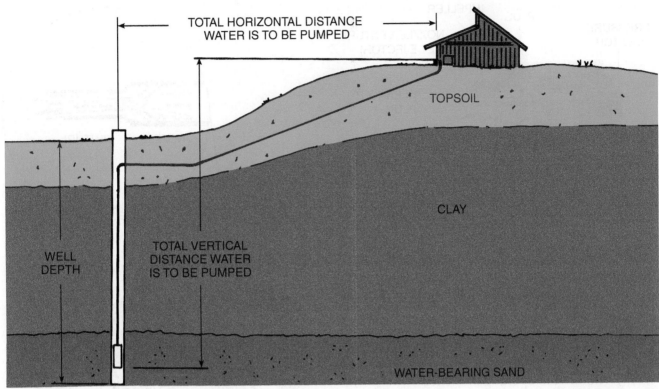

TOTAL HORIZONTAL DISTANCE
WATER IS TO BE PUMPED

TOPSOIL

CLAY

WELL
DEPTH

TOTAL VERTICAL
DISTANCE WATER
IS TO BE PUMPED

WATER-BEARING SAND

02408-14_F06.EPS

Figure 6 Factors to consider when selecting an appropriate pump.

Three types of pumps are commonly used in wells:

- Shallow-well jet pumps
- Deep-well jet pumps
- Submersible pumps

Each type of pump is reviewed in the following sections. Take time to read the information closely. The safe and efficient operation of the well depends largely on selecting the proper pump.

2.1.0 Understanding the Operation of Shallow-Well Jet Pumps

As the name suggests, shallow-well jet pumps (*Figure 7*) are used in wells that are no more than 25 feet deep. The pump uses an impeller to draw water through a jet assembly, which consists of a nozzle and a venturi. A venturi, also called a diffuser, is a narrowing of the pipe that accelerates water flowing through it. The accelerated water lowers the pressure in the pump body so that normal air pressure pushes the well water up the suction pipe and into the structure (*Figure 8*). Note that at high altitudes, the lower air pressure will reduce the depth at which shallow-well jet pumps can function.

Follow these steps when installing shallow-well jet pumps (refer to *Figure 8*):

Step 1 Attach the foot valve to the end of the pipe. This is a self-priming valve for water-lubricated pumps in wells deeper than 32 feet, which is the maximum height that a single pump can pump a column of water.

Step 2 Lower the assembly into the well, adding lengths of water line pipe as the assembly descends.

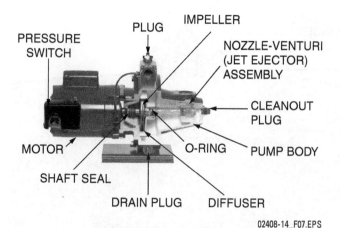

Figure 7 Cut-away of a shallow-well jet pump.

02408-14_F07.EPS

Step 3 When the assembly is properly positioned, place a well seal on top of the well. The seal will protect the well from contamination.

Step 4 Place a tee at the top of the vertical suction pipe, and plug the top opening. Attach the suction side of the pump to the horizontal opening of the tee.

Step 5 Connect the water storage tank to the discharge side of the pump.

Step 6 Install a shutoff valve and a union between the pump and the tank.

2.2.0 Understanding the Operation of Deep-Well Jet Pumps

Jet pumps used in deep wells work the same way as those used in shallow wells. However, they differ slightly in the arrangement of the jet assembly. In a deep-well jet pump, the nozzle and venturi are located in the well, usually about 5 feet below the low waterline. Some water is diverted from the discharge line back into the jet assembly, allowing the jet cycle to continue (*Figure 9*).

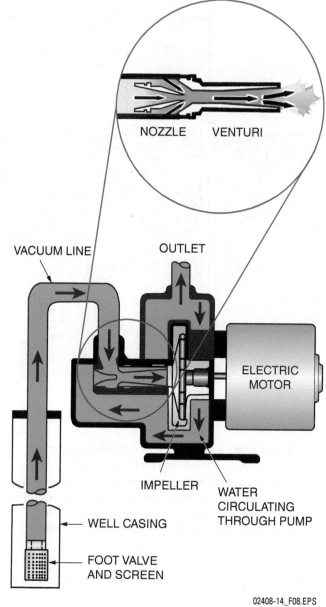

02408-14_F08.EPS

Figure 8 Water flow in a shallow-well jet pump.

Hydraulic Fracturing

Most of the wells that plumbers help install are designed to remove water from the ground. Wells can also be used the other way, to inject water deep into the ground to cause rocks to fracture. This process is called hydraulic fracturing, or fracking. In hydraulic fracturing, water mixed with sand and trace amounts of chemicals are injected under high pressure into a well to force underground rock to split apart, causing thin cracks called fissures. When the pressure is removed, natural gas, liquid petroleum, and even uranium can then flow through the fissures into the well, where they can then be pumped out and processed for use by consumers and industry.

Hydraulic fracturing allows previously inaccessible deposits of gas, petroleum, and minerals to be tapped more easily than other mining or drilling efforts, which results in lower fuel costs for consumers. In recent years, concerns over the potential contamination of aquifers from the chemicals in the pressurized water have led to increased scrutiny by regulators to ensure that hydraulic fracturing does not accidentally contaminate potable water supplies.

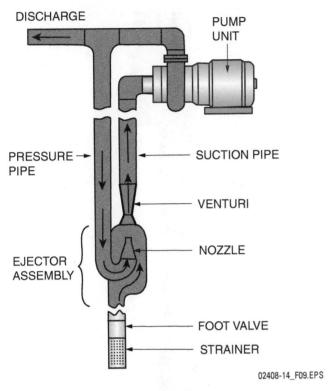

DISCHARGE

PUMP UNIT

PRESSURE PIPE

SUCTION PIPE

VENTURI

NOZZLE

EJECTOR ASSEMBLY

FOOT VALVE

STRAINER

02408-14_F09.EPS

Figure 9 Water flow in a deep-well jet pump.

The installation procedure for deep-well jet pumps is similar to that for shallow-well jet pumps. Many deep-well jet pumps use a two-pipe system to pump water from a well. Refer to *Figure 9*. One pipe (the pressure pipe) feeds pressurized water to the jet assembly, and the other (the suction pipe) draws up the well water. An ejector on the bottom of the pipe assembly allows well water to enter the suction pipe. Install the ejector after installing the foot valve. (Do not confuse the two-pipe system with the hydronic heating system of the same name. This other type of two-pipe system is discussed in the *Hydronic and Solar Heating Systems* module.)

If the well is too narrow for a two-pipe installation, install a packer-type ejector (*Figure 10*). In a packer-type ejector, the space between the water supply pipe and the well casing doubles as the pressure pipe. Though packer-type ejectors take

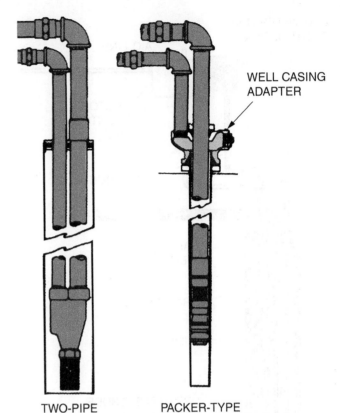

WELL CASING ADAPTER

TWO-PIPE   PACKER-TYPE

02408-14_F10.EPS

Figure 10 Comparison of a two-pipe system and a packer-type ejector in a deep-well jet pump installation.

up less space, their steel casings rust through over time. This makes two-pipe systems less costly to install and easier to replace.

To protect the pump if the well runs dry, install a low-water cutoff device on the pump (*Figure 11*). Low-water cutoff devices turn off the pump when the water level in the well falls to a predetermined low point. The device also turns the pump on when the water level recovers. Another option on two-pipe systems is to install a tail pipe assembly below the jet assembly and above the foot valve (*Figure 12*). The length of the tail pipe assembly adjusts the pumping rate to balance with the well's recovery rate. If the recovery slows, the pumping rate also slows.

The United States Geological Survey

The Water Resources Division of the United States Geological Survey (USGS) is responsible for providing up-to-date hydrologic information and resources. USGS maintains a database of more than 3.5 million water quality analyses. It also operates measuring instruments that continuously record and transmit water quality data.

USGS also maintains the Ground Water Site Inventory, with more than 850,000 records of wells, springs, test holes, tunnels, drains, and excavations throughout the country. This database is available through the USGS website at **waterdata.usgs.gov/nwis/gw**.

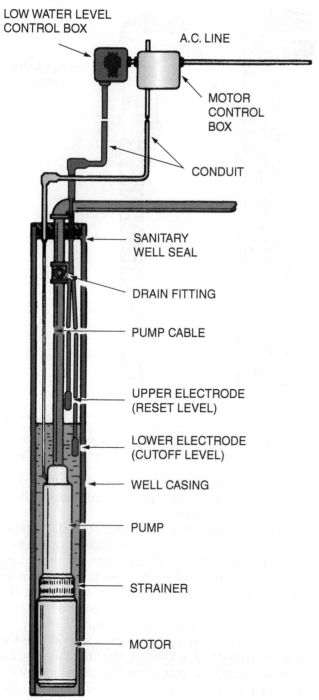

Figure 11 Typical low-water cutoff device.

LOW WATER LEVEL CONTROL BOX

A.C. LINE

MOTOR CONTROL BOX

CONDUIT

SANITARY WELL SEAL

DRAIN FITTING

PUMP CABLE

UPPER ELECTRODE (RESET LEVEL)

LOWER ELECTRODE (CUTOFF LEVEL)

WELL CASING

PUMP

STRAINER

MOTOR

02408-14_F11.EPS

Install pressure control valves on all deep-well jet pumps. These valves divert some of the pump's output down the ejector. This ensures that there is enough pressure to operate the ejector, even during periods of minimum discharge. Finally, install a well-casing adapter to seal off the top of the well casing from air leaks.

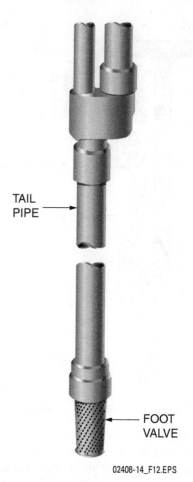

TAIL PIPE

FOOT VALVE

02408-14_F12.EPS

Figure 12 Tail pipe assembly.

2.3.0 Understanding the Operation of Submersible Pumps

Submersible pumps are mostly used in deep wells and are more efficient and easier to install than jet pumps. Submersible pumps are installed at the bottom of the well, below the water level. Submersible pumps have two major components: a multistage centrifugal pump and a submersible electric motor (*Figure 13*).

Well water enters the pump through the intake screen near the bottom of the pump. The impellers push the water up the pump body through the discharge pipe at the top of the pump. A check valve at the top of the pump prevents water in the discharge pipe from running back into the well when the pump shuts off. Note that in *Figure 13*, the motor-shaft extension at the top of the motor fits into the splined coupling at the bottom of the pump.

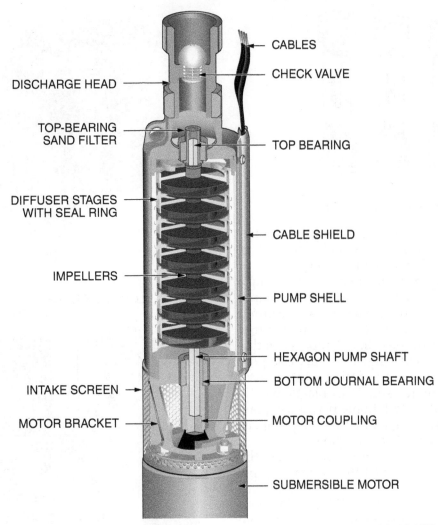

DISCHARGE HEAD

TOP-BEARING
SAND FILTER

DIFFUSER STAGES
WITH SEAL RING

IMPELLERS

INTAKE SCREEN

MOTOR BRACKET

CABLES

CHECK VALVE

TOP BEARING

CABLE SHIELD

PUMP SHELL

HEXAGON PUMP SHAFT

BOTTOM JOURNAL BEARING

MOTOR COUPLING

SUBMERSIBLE MOTOR

02408-14_F13.EPS

Figure 13 Components of a typical submersible pump.

WARNING!
In some jurisdictions, electricians must install and test pump electrical systems. Review your local codes and standards before installing and testing pump electrical systems. Never try to do electrical work by yourself if you have not had the proper training.

WARNING!
Always turn off the pump motor before attempting to repair or replace the pump. When wet, pump motors can cause injury and death from electric shock. Dry the pump motor thoroughly before testing. Wear appropriate personal protective equipment. Do not attempt to repair a pump or pump motor without proper training.

Test the pump before installing it, following the manufacturer's instructions. Use a collar clamp to support the pump and pipe assembly. Remember that submersible pumps are water lubricated, so submerge the pump in water before testing; otherwise, the pump will be damaged. Test the check valve according to the manufacturer's instructions.

When the pump passes all tests, it is ready to be installed. Splice the pump's electrical cable using a waterproof mechanical splice kit (*Figure 14*). Always be sure to use proper wiring techniques, and consult with local electrical officials if you have any questions about the electrical system.

02408-14_F14.EPS

Figure 14 Waterproof mechanical splice kits.

Tape the electrical line to the water pipe at intervals of no more than 10 feet. Install a secondary lightning arrester if required for safety.

Lower the pump assembly into the well using a hoist, and use the collar clamp to keep the pump assembly from falling out of control. As the pump descends, couplings on the end of each length of pipe will stop against the collar clamp. When this occurs, attach the next length of pipe. Use pipe dope or Teflon® tape on the fitting joints to prevent leaks. In many jurisdictions, plastic pipe is permitted for well installations. If you use plastic pipe, be careful when extracting the pump, as careless handling will break the pipe.

Install a pitless adapter in the well below the frost line. Pitless adapters let the discharge pipe, also called the drop pipe, leave the well and connect to the water storage tank (*Figure 15*). They are called pitless adapters because they eliminate the need to dig a pit to access the water line. Otherwise, contaminated water could enter the well through the pit. Because pitless adapters are installed below the frost line, they also protect the discharge line from freezing.

Connect the electrical cable to the building's electrical system only after installing the water supply line. Ensure that the building's electrical system has been de-energized before connecting the cable. The electrical supply line can be buried in the same trench as the water supply piping, but it must be at a separate level. Refer to your local code to determine if licensed electricians must perform this task. If you are allowed to perform the task, follow the local electrical code.

Most residential pump motors use 220 to 230 volts, so ensure that the electrical wire can handle the voltage for the length to be run. Select wire that is suitable for underground use. Bed the wire in soil or sand, which will protect the wire from cuts by sharp rocks. The wire should be laid loose

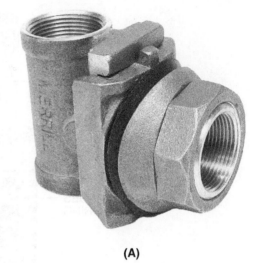

(A)

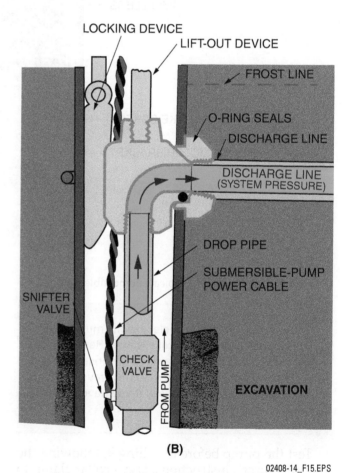

(B)

02408-14_F15.EPS

Figure 15 Pitless adapter installed in a well.

or snaked; otherwise, settling and freeze-thaw cycles may stress the wire. In most cases, bury the wire about 18 inches below the surface of the soil. Consult local electrical officials for requirements in your area.

Additional Resources

Efficient Building Design, Volume 3, *Water and Plumbing*. 1999. Ifte Choudhury and J. Trost. Upper Saddle River, NJ: Prentice-Hall, Inc.

International Plumbing Code®, Latest Edition. Falls Church, VA: International Code Council.

Audel Technical Trades Series, *Water Well Pumps and Systems Mini-Ref*. 2012, Roger D. Woodson. New York: John Wiley & Sons, Inc.

2.0.0 Section Review

1. The maximum well depth for using a shallow-well jet pump is _____.

   a. 15 feet
   b. 25 feet
   c. 35 feet
   d. 45 feet

2. In a deep-well jet pump, the nozzle and venturi are located in the well at a typical depth of around _____.

   a. 5 feet above the low waterline
   b. 5 feet above the high waterline
   c. 5 feet below the low waterline
   d. 5 feet below the high waterline

3. In a submersible pump, water in the discharge pipe is prevented from running back into the well when the pump shuts off by the installation of a _____.

   a. vacuum breaker
   b. butterfly valve
   c. check valve
   d. backwater valve

3.0.0 SELECTION AND INSTALLATION OF WATER SUPPLY AND STORAGE SYSTEM COMPONENTS

Objective

Explain how to select and install water supply and storage components.

   a. Explain how to select and install water supply lines.
   b. Explain how to select and install water storage tanks.

Performance Task

Assemble and disassemble given components of private water supply well systems.

Trade Terms

Biochemical oxygen demand (BOD): A measure of life-sustaining oxygen content in water.

pH value: A measure of the hydrogen ions in water. The term is an abbreviation for "potential hydrogen."

Turbidity: A measure of suspended solids in well water.

I n addition to pumps and their components, wells require water supply lines and storage tanks in order to operate. Like pumps, supply lines and storage tanks must be selected based on the needs of the particular installation. The local applicable code will specify the requirements for lines and tanks, including approved materials and sizing requirements. The following sections cover the general considerations for selecting and installing water supply lines and storage tanks. Always refer to the local applicable code for the selection and installation requirements that are applicable to your jurisdiction and to the needs of individual well installations.

3.1.0 Selecting and Installing Water Supply Lines

Your local plumbing code will specify approved piping materials in your area. The most popular pipe materials are steel, copper, and plastic. Copper usually lasts more than twice as long as galvanized steel pipe, as steel pipe corrodes in acidic soils. However, copper pipe corrodes in soil with high carbon dioxide or hydrogen sulfide content. Plastic water supply pipe lasts the longest. Consult your local code and plumbing officials.

Ensure that the pipe is properly sized for the installation. When selecting water supply pipe, consider the following factors:

- Cost
- Life expectancy
- Likelihood of corrosion and freezing
- System pressure
- Local plumbing codes
- Ease of installation

With experience, you will learn how to select the proper pipe for the types of installations you will encounter.

Bed water supply lines in soft, clean soil or sand, ensuring that there are no large rocks or boulders in the material used to backfill the pipe trench. Falling rocks and boulders can damage water pipes during an improper backfilling procedure. Install dielectric fittings on the supply lines to protect them from galvanic corrosion. Talk to local plumbing officials to determine the risk of galvanic corrosion in your area.

Most codes require that a water sample be drawn from the well and tested. Water tests are usually done after the well has been purged but before the supply line has been connected to the building. Local water-quality officials perform the tests. Water is tested for four characteristics:

- **Turbidity** – The presence of suspended solids in the water. Turbidity is less of a concern in groundwater than in surface water.
- **pH value** – The amount of hydrogen ions in the water. The term pH is an abbreviation for "potential hydrogen."
- *Hardness* – The presence of mineral salts in the water.
- **Biochemical oxygen demand (BOD)** – The amount of oxygen in the water available to support waterborne organisms.

Take the time to learn about the water quality of aquifers in your area. Your knowledge will help ensure that water supply line installations go smoothly.

3.2.0 Selecting and Installing Water Storage Tanks

Water storage tanks serve three main functions:

- They store water for intermittent use.
- They prevent the short cycling of the pump.
- They provide water under constant pressure throughout the building.

Pressurized air-bladder storage tanks (*Figure 16*) are widely used in private water supply well systems. In this type of tank, water is circulated at the proper system pressure by air pressure. The tank is equipped with an internal air bladder, and the pump fills the tank to a preset pressure. When the customer operates a fixture, air pressure in the tank pushes down on the bladder, allowing water to flow through the system at a constant pressure. The pump then refills the tank when it reaches a preset lower air pressure (*Figure 17*).

The main function of the pressure tank is to reduce pump cycling. Therefore, size the water tank according to the desired running time of the pump. Refer to the tank manufacturer's specifications to determine the proper-size tank. For example, some manufacturers recommend a running time of one minute for a submersible pump that has a flow rate of 10 gpm and an operating pressure range of 30 to 50 pounds per square inch (psi).

Locate the water storage tank before connecting the supply lines to the building. Install a pressure switch and a pressure gauge as close as possible to the tank itself. The pressure switch controls the amount of water and the range of pressure inside the tank, and turns the pump on and off based on whether a preset pressure differential has been

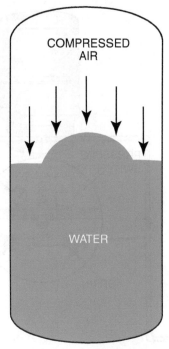
02408-14_F16.EPS

Figure 16 Typical pressurized air-bladder storage tank.

reached. The pressure gauge shows the pressure in both the tank and the piping system. Many tanks also require the installation of a drain and a relief valve. The drain is used to empty the tank when necessary. If the tank exceeds the maximum preset pressure because the pressure switch fails, the relief valve can drain the tank and relieve the pressure. *Figure 18* displays the typical installation of a pressure switch, pressure gauge, relief valve, and drain on a pressurized air-bladder storage tank.

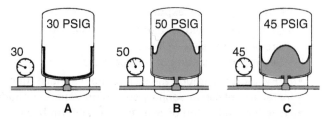

A THIS IS A FACTORY-INSTALLED PRECHARGED AIR CUSHION. THE PUMP IS OFF.

B WHEN THE PUMP STARTS, WATER ENTERS THE RESERVOIR. AT 50 PSIG, THE SYSTEM IS FILLED. THE PUMP SHUTS OFF.

C WHEN WATER IS DEMANDED, PRESSURE IN THE AIR CHAMBER FORCES WATER INTO THE SYSTEM. THE PUMP STAYS OFF.

02408-14_F17.EPS

Figure 17 Operation of a typical pressurized air-bladder storage tank.

Clean Pipes

Ensure that the water supply piping is as clean as possible before installation. Sand, gravel, or mud trapped in the pipes will flow through the piping system, lodge in the valves and faucets, and damage the washers and O-rings, causing the faucets to leak. Store pipe in a clean, dry area, and cap all pipe ends at the end of each workday. Be sure to flush the line as thoroughly as possible when installation is complete to remove any sand, mud, or gravel from the system.

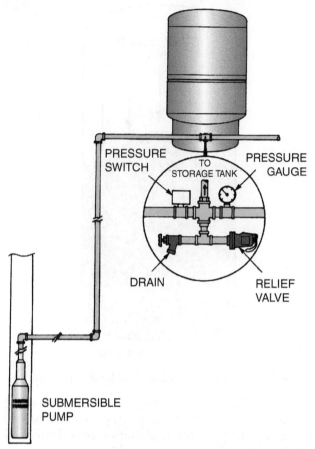

PRESSURE SWITCH

TO STORAGE TANK

PRESSURE GAUGE

DRAIN

RELIEF VALVE

SUBMERSIBLE PUMP

02408-14_F18.EPS

Figure 18 Typical installation of a pressure switch, pressure gauge, relief valve, and drain on a pressurized air-bladder storage tank.

Friction

The greater the number of fittings in the system (valves, tees, elbows, and so on), the greater the amount of friction, which causes a pressure drop in the system.

After installing the pressure tank and controls, purge the entire water system by flushing the potable water lines thoroughly with water. This will wash out the chlorine used to disinfect the well. It will also remove sediment, joint compound, and other foreign matter still in the lines. Following the purge, perform final checks on the tank's pressure-switch settings. At this point, a water-quality authority should test the water.

Following the water quality test, the installation of the private water supply well system is complete. From that point on, the installation of stacks, branches, and fixtures is just like in any other building. Install DWV and water supply systems using the techniques you have already learned. Refer to your local code for other conditions that apply to private water supply well systems.

Protecting Private Water Supplies

The EPA recommends the following steps for protecting private water supply well systems:

- Periodically inspect exposed parts of the well for problems such as:
  - Cracked, corroded, or damaged well casing
  - Broken or missing well cap
  - Settling and cracking of surface seals
- Slope the area around the well to drain surface runoff away from the well.
- Install a well cap or sanitary seal to prevent unauthorized use of or entry into the well.
- Have the well tested once a year for coliform bacteria, nitrates, and other contaminants of concern.
- Keep accurate records of any well maintenance, such as disinfection or sediment removal, requiring the use of chemicals in the well.
- Hire a certified well driller for any new well construction, modification, or abandonment and closure.
- Avoid mixing or using pesticides, fertilizers, herbicides, degreasers, fuels, and other pollutants near the well.
- Do not dispose of wastes in dry or abandoned wells.
- Do not cut off the well casing below the land surface.
- Pump and inspect septic systems as often as recommended by your local health department.
- Never dispose of hazardous materials in a septic system.

Additional Resources

Efficient Building Design, Volume 3, Water and Plumbing. 1999. Ifte Choudhury and J. Trost. Upper Saddle River, NJ: Prentice-Hall, Inc.

Finding Water: A Guide to the Construction and Maintenance of Private Water Supplies. Rick Brassington. Chichester, New York: John Wiley & Sons.

International Plumbing Code®, Latest Edition. Falls Church, VA: International Code Council.

Audel Technical Trades Series, *Water Well Pumps and Systems Mini-Ref.* 2012, Roger D. Woodson. New York: John Wiley & Sons, Inc.

3.0.0 Section Review

1. Each of the following is a popular material for water supply piping in a private water supply well system *except* _____.

   a. copper
   b. steel
   c. plastic
   d. cast iron

2. The main function of the pressure tank is to _____.

   a. store more water in a smaller space
   b. reduce pump cycling
   c. use less energy to heat the water
   d. equalize service pressure across all fixtures

SUMMARY

Private water supply well systems provide potable water to homes and businesses that do not have access to a municipal system. Wells get water from underground aquifers. Wells can be bored, drilled, driven, or dug. Each type of well has its own requirements. Wells must be carefully located. Well drillers use drilling logs to select likely sites. Wells are often sized according to the seven-minute peak demand period. Once a well has been drilled, it must be cleaned and disinfected.

Pumps provide well water to the system. The most common types of well pumps are shallow-well jet pumps, deep-well jet pumps, and submersible pumps. Galvanized steel, copper, and plastic pipe can be used to connect the pump to the building. Water storage tanks may store well water in or near the building. Pressurized air-bladder storage tanks are a widely used type of tank. When the pump, supply line, and storage tank are all installed, a local water-quality expert will test the private water supply well system. Always refer to your local code for installation requirements in your area.

1. People depend on wells to provide reliable water that is _____.
   a. diluted
   b. artesian
   c. filtered
   d. potable

2. For any given installation, the person who decides which type of well works best is the _____.
   a. plumber
   b. driller
   c. customer
   d. code official

3. Driven wells are typically used where the soil is relatively free of _____.
   a. rocks
   b. roots
   c. metals
   d. air pockets

4. The excavating tools used to drill a dug well will affect its _____.
   a. liner
   b. well head
   c. diameter
   d. well casing

5. Testing the water quality of a well is the responsibility of the _____.
   a. health officials
   b. code officials
   c. plumbers
   d. customers

6. A well that is no more than 25 feet deep uses a _____.
   a. venturi-well jet pump
   b. shallow-well jet pump
   c. diffuser-well jet pump
   d. pump-well jet pump

7. One disadvantage to using packer-type ejectors is that _____.
   a. they take up a lot of space
   b. they are difficult to install
   c. their steel casings eventually rust
   d. their use is prohibited in many jurisdictions

8. A low-water cutoff device is used to protect the pump if the well _____.
   a. runs dry
   b. overflows
   c. becomes turbid
   d. becomes contaminated

9. A submersible pump has two major components: a submersible electric motor and a _____.
   a. multistage impeller pump
   b. multistage centrifugal pump
   c. motor shaft pump
   d. water-lubricated pump

10. Most residential pump motors use _____.
    a. 180 to 190 volts
    b. 200 to 210 volts
    c. 220 to 230 volts
    d. 240 to 250 volts

11. The type of water supply piping that lasts the longest when used in private water supply well systems is _____.
    a. galvanized steel pipe
    b. plastic pipe
    c. copper pipe
    d. lead pipe

12. Local water-quality officials test water for four characteristics: hardness, biochemical oxygen demand, pH value, and _____.
    a. hydrogen ions
    b. mineral salts
    c. turbidity
    d. galvanic corrosion

13. Air pressure is used to maintain water pressure at the proper system pressure in a pressurized _____.

   a. air-bladder storage tank
   b. high-pressure storage tank
   c. submersible-pump tank
   d. supply-line tank

14. Draining the tank and relieving the pressure is the purpose of a _____.

   a. gauge valve
   b. drain valve
   c. pressure valve
   d. relief valve

15. After installing the pressure tank and controls, the next step is to _____.

   a. perform final checks on the settings
   b. test the water for impurities
   c. purge the entire water system
   d. install a vent system

Trade Terms Quiz

Fill in the blank with the correct term that you learned from your study of this module.

1. An aquifer in which water flows toward bodies of water is called a(n) _____.

2. A(n) _____ is another name for the discharge pipe that connects the well to the water storage tank.

3. A pointed rod used to create a driven well by being forced into the ground with a hammer is called a(n) _____.

4. _____ is a measure of suspended solids in well water.

5. A type of well drilled by an auger and lined with a well casing is called a(n) _____.

6. A(n) _____ is a valve that diverts a portion of a pump's discharge through the ejector, thereby maintaining adequate pressure in the ejector.

7. A pump consisting of a multistage centrifugal pump and a submersible electric motor, usually installed in the bottom of deep wells, is called a(n) _____.

8. The _____ is the top of an aquifer, below which rock, sand, and soil are completely saturated with groundwater.

9. A pump used in a deep well that uses a nozzle and venturi to accelerate the water flow is called a(n) _____.

10. A(n) _____ is an alternative to the packer-type ejector discharge method for jet pumps.

11. A device that permits a discharge pipe to exit a well and connect to a water storage tank is called a(n) _____.

12. A(n) _____ is a deep aquifer that does not lose water through runoff.

13. A device that accelerates water and lowers its pressure by passing it through a narrow pipe opening is called a(n) _____.

14. A(n) _____ is a well that is no more than 25 feet deep.

15. A water storage tank used to supply some private water supply well systems with potable water is called a(n) _____.

16. A(n) _____ is a well that is drilled deeper than 25 feet.

17. The measure of life-sustaining oxygen content in water is called _____.

18. A(n) _____ is a hole in the ground that connects with an aquifer.

19. A pump used in a shallow well that uses a nozzle and venturi to accelerate the water flow is called a(n) _____.

20. A(n) _____ is a device that deactivates a pump when the water level in a well falls to a preset low point.

21. A pipe section inserted into a well that maintains the shape of the well and prevents contamination is called a(n) _____.

22. A(n) _____ is a device that matches a pump's discharge rate to a well's recovery rate.

23. A type of well created by chisels or rotary drills is called a(n) _____.

24. A(n) _____ is a type of well excavated by shovel and backhoe and lined with brick, stone, or concrete.

25. The estimate of the greatest possible water demand in a private water supply well system is called a(n) _____.

26. A(n) _____ is an underground reservoir containing water-saturated soil, sand, and rock.

27. A type of well drilled by a pointed well head driven by a hammer is called a(n) _____.

28. _____ is the measure of the hydrogen ions in water.

29. A(n) _____ is an alternative discharge method to the two-pipe system for jet pumps in which the space between the water supply pipe and the well casing acts as a pressure pipe.

30. A well cover that protects the well casing from air leaks is called a(n) _____.

31. To _____ is to disinfect a well with a chlorine compound such as calcium hypochlorite or liquid bleach.

32. A valve used to prime water-lubricated pumps in wells deeper than 32 feet is a(n) _____.

33. A(n) _____ is a small water table that is cut off from a larger reservoir.

34. The process through which water cycles through the environment as both a liquid and a vapor is called _____.

Trade Terms

Aquifer	Drilled well	pH value	Turbidity
Artesian aquifer	Driven well	Pitless adapter	Two-pipe system
Biochemical oxygen demand (BOD)	Drop pipe	Pressure control valve	Venturi
	Dug well	Seven-minute peak demand period	Water table
Bored well	Foot valve		Water table aquifer
Chlorinate	Hydrologic cycle	Shallow well	Well
Cistern	Low-water cutoff device	Shallow-well jet pump	Well casing
Deep well	Packer-type ejector	Submersible pump	Well-casing adapter
Deep-well jet pump	Perched water table	Tail pipe assembly	Well head

Trade Terms Introduced in This Module

Aquifer: An underground reservoir containing water-saturated soil, sand, and rock.

Artesian aquifer: A deep aquifer that does not lose water through runoff. Also called a confined aquifer.

Biochemical oxygen demand (BOD): A measure of life-sustaining oxygen content in water.

Bored well: A type of well drilled by an auger and lined with a well casing.

Chlorinate: To disinfect a well with a chlorine compound such as calcium hypochlorite or liquid bleach.

Cistern: A water storage tank used to supply some private water supply well systems with potable water.

Deep well: A well that is drilled deeper than 25 feet.

Deep-well jet pump: A pump used in a deep well that uses a nozzle and venturi to accelerate the water flow.

Drilled well: A type of well created by chisels or rotary drills.

Driven well: A type of well drilled by a pointed well head driven by a hammer.

Drop pipe: Another name for the discharge pipe that connects the well to the water storage tank.

Dug well: A type of well excavated by shovel and backhoe and lined with brick, stone, or concrete.

Foot valve: A valve used to prime water-lubricated pumps in wells deeper than 32 feet.

Hydrologic cycle: The process through which water cycles through the environment as both a liquid and a vapor.

Low-water cutoff device: A device that deactivates a pump when the water level in a well falls to a preset low point. See also *tail pipe assembly*.

Packer-type ejector: For jet pumps, an alternative discharge method to the two-pipe system in which the space between the water supply pipe and the well casing acts as a pressure pipe.

Perched water table: A small water table that is cut off from a larger reservoir.

pH value: A measure of the hydrogen ions in water. The term is an abbreviation for "potential hydrogen."

Pitless adapter: A device that permits a discharge pipe to exit a well and connect to a water storage tank. Pitless adapters protect well water from contamination and freezing.

Pressure control valve: A valve that diverts a portion of a pump's discharge through the ejector, thereby maintaining adequate pressure in the ejector.

Seven-minute peak demand period: An estimate of the greatest possible water demand in a private water supply well system.

Shallow well: A well that is no more than 25 feet deep.

Shallow-well jet pump: A pump used in a shallow well that uses a nozzle and venturi to accelerate the water flow.

Submersible pump: A pump consisting of a multistage centrifugal pump and a submersible electric motor. Submersible pumps are usually installed in the bottom of deep wells.

Tail pipe assembly: A device that matches a pump's discharge rate to a well's recovery rate. See also *low-water cutoff device*.

Turbidity: A measure of suspended solids in well water.

Two-pipe system: For jet pumps, an alternative to the packer-type ejector discharge method. One pipe draws well water through an ejector, and another pipe feeds the pressurized water to the jet.

Venturi: A device that accelerates water and lowers its pressure by passing it through a narrow pipe opening. It is also called a diffuser.

Water table: The top of an aquifer, below which rock, sand, and soil are completely saturated with groundwater.

Water table aquifer: An aquifer in which water flows toward bodies of water. Also called an unconfined aquifer.

Well: A hole in the ground that connects with an aquifer. Wells are equipped with pumps to lift water from the aquifer into water storage tanks.

Well casing: A pipe section inserted into a well that maintains the shape of the well and prevents contamination.

Well-casing adapter: A well cover that protects the well casing from air leaks.

Well head: A pointed rod used to create a driven well by being forced into the ground with a hammer.

Additional Resources

This module presents thorough resources for task training. The following resource material is suggested for further study.

Efficient Building Design, Volume 3, *Water and Plumbing*. 1999. Ifte Choudhury and J. Trost. Upper Saddle River, NJ: Prentice-Hall, Inc.

Finding Water: A Guide to the Construction and Maintenance of Private Water Supplies. Rick Brassington. Chichester, New York: John Wiley & Sons.

International Plumbing Code®, Latest Edition. Falls Church, VA: International Code Council.

Audel Technical Trades Series, *Water Well Pumps and Systems Mini-Ref*. 2012, Roger D. Woodson. New York: John Wiley & Sons, Inc.

Wells and Septic Systems. Second Edition. 1992. Max Alth, Charlotte Alth, and S. Blackwell Duncan. Blue Ridge Summit, PA: Tab Books.

Figure Credits

Section Review Answer Key

Answer	Section Reference	Objective
Section One		
1. c	1.1.0	1a
2. d	1.2.0	1b
3. a	1.3.0	1c
4. b	1.4.0	1d
Section Two		
1. b	2.1.0	2a
2. c	2.2.0	2b
3. c	2.3.0	2c
Section Three		
1. d	3.1.0	3a
2. b	3.2.0	3b

NCCER CURRICULA — USER UPDATE

NCCER makes every effort to keep its textbooks up-to-date and free of technical errors. We appreciate your help in this process. If you find an error, a typographical mistake, or an inaccuracy in NCCER's curricula, please fill out this form (or a photocopy), or complete the online form at **www.nccer.org/olf**. Be sure to include the exact module ID number, page number, a detailed description, and your recommended correction. Your input will be brought to the attention of the Authoring Team. Thank you for your assistance.

Instructors – If you have an idea for improving this textbook, or have found that additional materials were necessary to teach this module effectively, please let us know so that we may present your suggestions to the Authoring Team.

NCCER Product Development and Revision
13614 Progress Blvd., Alachua, FL 32615

Email: curriculum@nccer.org
Online: www.nccer.org/olf

❑ Trainee Guide   ❑ Lesson Plans   ❑ Exam   ❑ PowerPoints   Other _____

Craft / Level: _____ Copyright Date: _____

Module ID Number / Title: _____

Section Number(s): _____

Description: _____

_____

_____

_____

_____

Recommended Correction: _____

_____

_____

_____

_____

Your Name: _____

Address: _____

_____

Email: _____ Phone: _____

02409-14

Private Waste Disposal Systems

OVERVIEW

Rural areas are often too far away from a city to connect to the municipal waste system. Farms and parks in such areas require private waste disposal systems. The three categories of private waste disposal systems are soil absorption systems, organic systems, and closed systems. The three types of soil absorption system are septic systems, aeration systems, and pressure distribution systems. Septic systems work by breaking down solid wastes and allowing liquid wastes, or effluent, to leach into the ground.

Module Eight

Trainees with successful module completions may be eligible for credentialing through the NCCER Registry. To learn more, go to **www.nccer.org** or contact us at **1.888.622.3720**. Our website has information on the latest product releases and training, as well as online versions of our *Cornerstone* magazine and Pearson's product catalog.

Your feedback is welcome. You may email your comments to **curriculum@nccer.org**, send general comments and inquiries to **info@nccer.org**, or fill in the User Update form at the back of this module.

This information is general in nature and intended for training purposes only. Actual performance of activities described in this manual requires compliance with all applicable operating, service, maintenance, and safety procedures under the direction of qualified personnel. References in this manual to patented or proprietary devices do not constitute a recommendation of their use.

Objectives

When you have completed this module, you will be able to do the following:

1. Describe the types of private waste disposal systems.
   a. Describe soil absorption systems.
   b. Describe organic systems.
   c. Describe closed systems.
2. Explain how to locate and size private waste disposal systems.
   a. Explain how to locate and size septic tanks.
   b. Explain how to locate and size leach fields.
3. Explain how to install private waste disposal systems.
   a. Explain how to install septic tanks.
   b. Explain how to install leach field distribution piping.
4. Explain how to clean and service private waste disposal systems.

Performance Tasks

This is a knowledge-based module; there are no performance tasks.

Trade Terms

Aeration system
Aeration tank
Aerobic bacteria
Anaerobic bacteria
Cesspool
Closed system
Distribution box
Drywell
Effluent
Humus

Leach
Leach field
Organic system
Pressure distribution system
Septic system
Septic tank
Soil absorption system
Soil mottle
Soil profile description

Industry-Recognized Credentials

If you're training through an NCCER-accredited sponsor, you may be eligible for credentials from NCCER's Registry. The ID number for this module is 02409-14. Note that this module may have been used in other NCCER curricula and may apply to other level completions. Contact NCCER's Registry at 888.622.3720 or go to **www.nccer.org** for more information.

Code Note

Codes vary among jurisdictions. Because of the variations in code, consult the applicable code whenever regulations are in question. Referring to an incorrect set of codes can cause as much trouble as failing to reference codes altogether. Obtain, review, and familiarize yourself with your local adopted code.

Contents

Topics to be presented in this module include:

Figures and Tables

1.0.0 TYPES OF PRIVATE WASTE DISPOSAL SYSTEMS

Objective

Describe the types of private waste disposal systems.

a. Describe soil absorption systems.
b. Describe organic systems.
c. Describe closed systems.

Trade Terms

Aeration system: A type of soil absorption system that uses an aeration tank to clarify effluent.

Aeration tank: A waste storage and separation tank used in an aeration system. In an aeration tank, anaerobic bacteria decompose sludge, and then aerobic bacteria clarify the effluent.

Aerobic bacteria: Bacteria that need oxygen to live. They are used in aeration tanks to clarify effluent.

Anaerobic bacteria: Bacteria that do not need oxygen to live. They are used in septic tanks and aeration tanks to decompose solid wastes.

Cesspool: A lined, covered pit for storing solid and liquid wastes. Cesspools allow effluent to seep into the soil.

Closed system: A private waste disposal system that stores wastes until they can be disposed of.

Distribution box: A watertight container that directs effluent to different parts of a leach field.

Drywell: A covered pit filled with loose sand and gravel that collects roof and basement drainage and allows it to percolate into the surrounding soil.

Effluent: Liquid waste that separates from solid waste. In soil absorption systems and cesspools, effluent is allowed to percolate into the soil.

Humus: In an organic system, the dried solid waste that results from decomposition. Humus can be used as fertilizer after a time.

Leach: To percolate into soil and be absorbed.

Leach field: In a soil absorption system, an area of soil that is designed to accept effluent from distribution pipes. Leach fields are also sometimes called absorption fields or drain fields.

Organic system: A private waste disposal system that allows waste to decompose through exposure to aerobic bacteria. The resulting dried waste is called humus.

Pressure distribution system: A soil absorption system in which effluent is pumped through the distribution pipes under pressure.

Septic system: A soil absorption system that uses a septic tank to separate out sludge.

Septic tank: A tank used in a soil absorption system to settle out sludge and scum and to decompose them using anaerobic bacteria.

Soil absorption system: A private waste disposal system that uses tanks to separate out effluent, then allows it to drain into a leach field.

Municipal waste systems serve urban homes, offices, and businesses. Rural areas, such as farms and parks, are often too far away from a municipal system to be connected. Just as they rely on private water supply systems, people who live and work in such areas also use private waste disposal systems. These systems are installations that treat and dispose of the sewage wastes from a single building. In this module, you will learn about the different types of private waste disposal systems and the way those systems treat and dispose of wastes.

Like other plumbing installations, private waste disposal systems must meet local code standards. The system must protect the health of the users and the safety of the environment. Plumbers working in rural and semirural areas must be familiar with the local code requirements that affect private waste disposal systems. They must also know how to choose the right system for the customer's needs, as well as how to install the different types of systems. In this module, you will learn how to lay out, install, and maintain the common types of private waste disposal systems.

The three main categories of private waste disposal systems are:

- Soil absorption systems
- Organic systems
- Closed systems

Soil absorption systems store solid wastes and allow wastewater to drain into the ground. Organic systems allow wastes to decompose through contact with air. Closed systems store wastes until they can be disposed of permanently. Each of these systems is described in detail in the following sections.

Local codes, the environment, and the customer's needs will determine the most appropriate system. Take time to learn about each system. Ask experienced plumbers and code officials for their advice on which systems work best under different conditions. The more you learn, the easier it will be for you to select the system that meets each customer's needs.

1.1.0 Identifying Soil Absorption Systems

Soil absorption systems are the most common type of private waste disposal system. These systems separate solid waste from liquid waste. The liquid waste, called effluent, is then allowed to seep into the ground, where it is absorbed. There are three types of soil absorption systems:

- Septic systems
- Aeration systems
- Pressure distribution systems

Soil absorption systems can be installed only where the soil is capable of absorbing effluent. Each location will have its own requirements.

1.1.1 Septic Systems

Septic systems are the most common type of soil absorption system. Septic systems consist of a septic tank, a distribution box, a leach field (also sometimes called an absorption field or a drain field), and the piping that connects these parts (see *Figures 1* and *2*). Septic systems work by breaking down solid wastes and allowing liquid wastes to percolate, or leach, into the ground. This is the most common system used today.

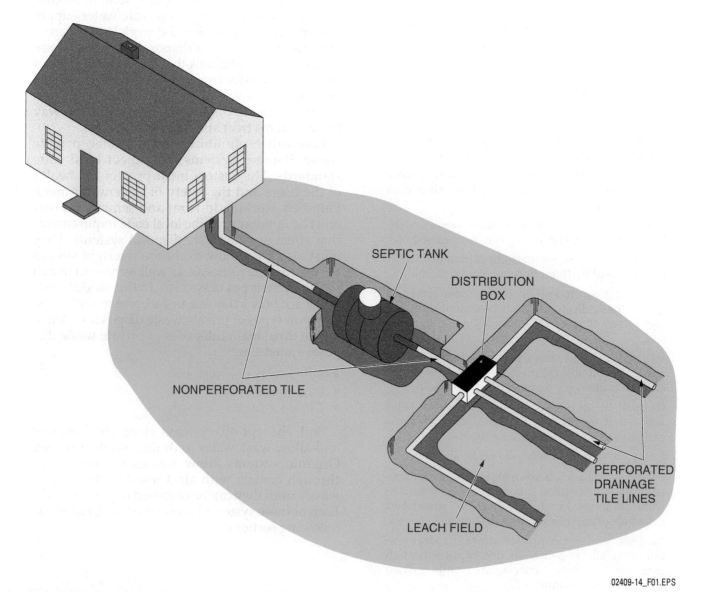

SEPTIC TANK

DISTRIBUTION BOX

NONPERFORATED TILE

PERFORATED DRAINAGE TILE LINES

LEACH FIELD

02409-14_F01.EPS

Figure 1 Typical layout for a septic system (1 of 2).

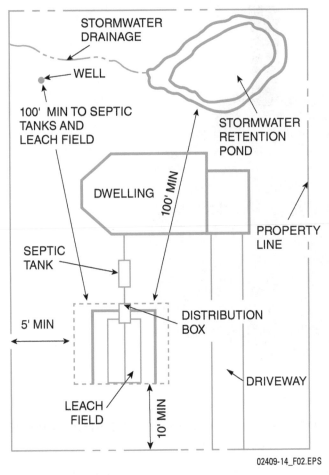

In septic systems, wastewater is first discharged into a septic tank directly from the building's waste system. A septic tank is a container that allows liquid wastes to separate from solid wastes. Wastewater remains in the tank for at least a day, allowing large solids to settle out as sludge. Anaerobic bacteria, or bacteria that do not need oxygen to live, decompose much of the sludge. Grease, oils, and floating particles rise to the top. Methane and carbon dioxide produced by the decomposition are vented from the effluent side of the tank. Septic tanks can be made from a variety of materials, including fiberglass, concrete (*Figure 3*), and plastic (*Figure 4*). On-site septic tanks can also be built with concrete blocks and tar-based sealers.

From the septic tank, the effluent flows through a sealed pipe to a distribution box. A distribution box (*Figure 5*) is a watertight container with one inlet and many outlets. It directs the effluent to different parts of a leach field or distributes it among several leach fields. A distribution box can also be used to direct effluent flow to a second or adjacent field. Most distribution boxes have lids for inspection and cleaning. Often, the outlets have knockout plugs, which allow the installer to use only the number of outlets required. Like septic tanks, distribution boxes can be made from different types of materials.

Figure 2 Typical layout for a septic system (2 of 2).

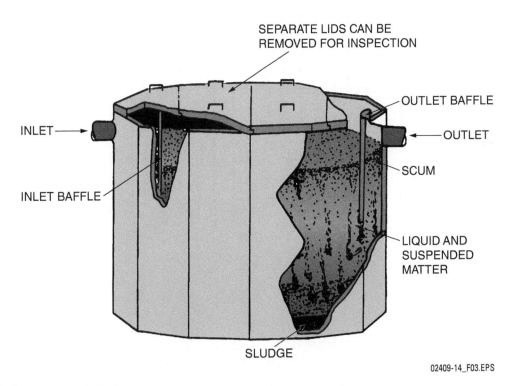

Figure 3 Typical concrete septic tank.

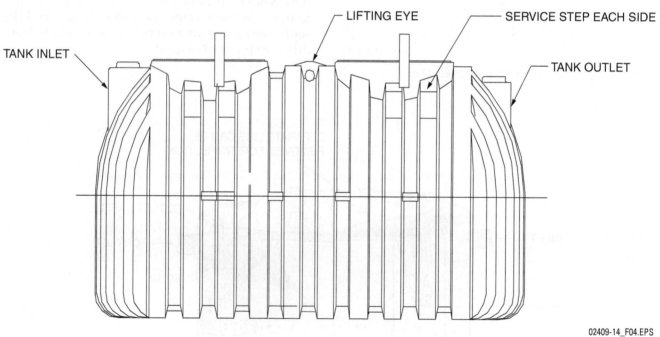

LIFTING EYE — SERVICE STEP EACH SIDE

TANK INLET

TANK OUTLET

02409-14_F04.EPS

Figure 4 Typical plastic septic tank.

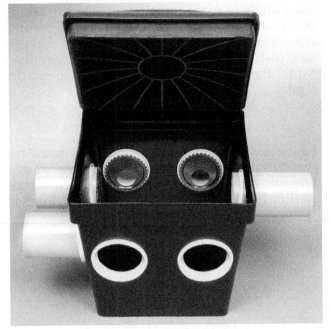

02409-14_F05.EPS

Figure 5 Plastic distribution box.

From the distribution box, effluent drains through distribution pipes. Distribution pipes have holes or gaps that allow effluent to drain out of them. Distribution pipes are laid underground in a leach field.

1.1.2 Aeration Systems

The second type of soil absorption system is an aeration system. Aeration systems are similar to septic systems. The major difference is that they use an underground aeration tank instead of a septic tank. Aeration tanks clarify wastewater more completely than septic tanks do. However, they are more mechanically complex. Because of this, some codes prohibit aeration systems. Consult your local code before installing an aeration system.

Aeration tanks are designed with two chambers (*Figure 6*) or three (*Figure 7*). In a two-chamber aeration tank, solid waste first enters the tank, where sludge settles out and is decomposed by anaerobic bacteria. The liquid waste then flows into a second chamber, into which air is injected. The air contains aerobic bacteria, bacteria that need oxygen to live. Aerobic bacteria consume the organic matter in the wastewater. Then the clarified wastewater flows out to the distribution box and into the leach field.

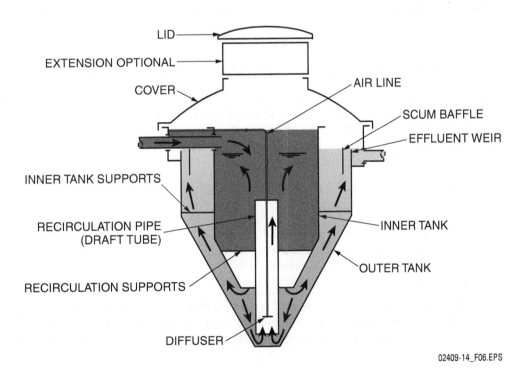

02409-14_F06.EPS

Figure 6 Cross section of a two-chamber aeration tank.

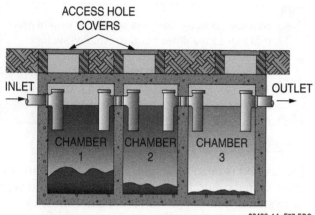

Figure 7 Cross section of a three-chamber aeration tank.

The wastewater that comes out of aeration tanks is mostly clear and odorless. Because of this, aeration systems often use smaller leach fields than septic systems. Aeration systems are often used in places where codes limit the types of wastes that can be disposed of into the ground. Some jurisdictions allow wastewater to be used for irrigation and in flush toilets. Check with your local code to determine how best to handle aeration system wastewater in your area.

1.1.3 Pressure Distribution Systems

The third type of soil absorption system is the pressure distribution system. Like other soil absorption systems, pressure distribution systems separate out effluent and allow it to drain into the ground. The main difference is that they use pumps to move the effluent. The effluent is pumped down a main to a manifold, where it is channeled into distribution pipes (*Figure 8*). Pressure distribution systems are more efficient than septic and aeration systems and require smaller pipe.

1.2.0 Identifying Organic Systems

Septic and aeration systems use large amounts of water to drain, separate, and decompose wastes. Organic systems (*Figure 9*) break down wastes without using water. Bathroom and kitchen

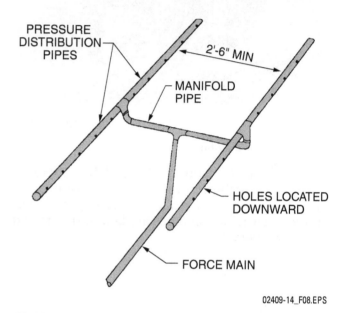

Figure 8 Typical design of a pressure distribution system.

The *International Private Sewage Disposal Code*®

The International Code Council (ICC) publishes the *International Private Sewage Disposal Code*® (IPSDC) as part of a series of model codes that are collectively nicknamed the I-codes. Other I-codes cover plumbing, building construction, fire prevention, residential buildings, mechanical installations, fuel gas, energy conservation, property maintenance, electrical installations, and zoning. Many jurisdictions are adopting I-codes as their current codes become obsolete. The IPSDC covers the following:

- Regulations governing the installation of a private waste disposal system
- Methods for evaluating a site
- Guidelines for layout and design
- Materials permitted for use in a private waste disposal system
- Requirements for each type of private waste disposal system
- Installation and maintenance of system components
- Inspection of installed systems

If your local private waste disposal system code is based on the IPSDC, take the time to familiarize yourself with its requirements. Ask a local code expert to explain differences between the model code and your local code. Develop a good working knowledge of your local private waste disposal system codes. The quality of your work will improve as a result.

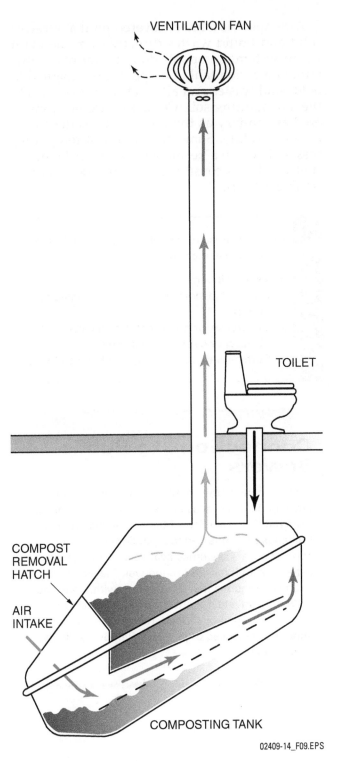

VENTILATION FAN

TOILET

COMPOST REMOVAL HATCH

AIR INTAKE

COMPOSTING TANK

02409-14_F09.EPS

Figure 9 Typical residential installation for an organic waste disposal system.

wastes fall into a sealed composting tank. In the tank are peat and topsoil that contain bacteria. The bacteria decompose the waste. Air flows through the tank to assist in the decomposition. As the waste decomposes, it emits carbon dioxide and water vapor. These are vented out of the tank. When the cycle is complete, the dried waste that remains is called humus. Humus can eventually be added to soil to fertilize it.

Organic systems are small, self-contained units. They can be located in basements or crawl spaces. They are often used in vacation houses. Some jurisdictions do not permit organic systems. Refer to your local code before installing an organic system.

1.3.0 Identifying Closed Systems

If you have worked on-site at a construction job, you have probably used a portable toilet (*Figure 10*). If you look at how a portable toilet is designed, you will see that the entire waste cycle is self-contained. A portable toilet is the simplest form of a closed waste disposal system. Closed systems are used where either the terrain or the soil is not suitable for waste drainage. The process is illustrated in *Figure 11*. Waste settles in a tank filled with water. The water is pumped through a reservoir, where it is mixed with chemicals that deodorize and purify it. This clean water is then used to flush new waste. Closed systems require little or no plumbing. However, they must be cleaned and recharged regularly.

Licensed Septic Tank Installers

Some jurisdictions require plumbers to have a separate license to install septic tanks. If your job will involve septic system installations, find out what your local requirements are. Do not attempt to install a system if you are not authorized by your local code. If you need a license, ask your local code official how and where you can get one.

02409-14_F10.EPS

Figure 10 Portable toilet.

1.3.1 Drywells and Cesspools

Some private systems may include drywells. These are covered pits that accept roof and basement drainage. Drywells are filled with loose sand and gravel. They collect drainage wastewater and allow it to leach into the surrounding soil. Do not allow drywells to drain into private waste disposal systems.

A cesspool is a lined, covered pit that receives solid and liquid wastes directly from the main waste system. Cesspools do not require a septic tank or aeration tank. Cesspools are designed to hold solid wastes and allow effluent to seep into the surrounding soil. Cesspools are not considered as sanitary as the other systems described in this module. Some codes permit temporary cesspools while a permanent system is being installed. Always consult your local code before installing a cesspool.

> **WARNING!**
>
> Do not dispose of the following materials in a private waste disposal system:
>
> - Ashes, cinders, or rags
> - Flammable, poisonous, or explosive liquids or gases
> - Oil, grease, or other material that can obstruct or damage the system
> - Surface water, rainwater, or other clear water

Connections to Public Sewers

In many parts of the country, new suburbs and developments are reaching areas that were once far from municipal services. Most codes require that buildings be connected to a public sewer system as soon as such a system is available. Codes usually set a time limit for the changeover. The IPSDC, for example, requires that it take place within one year.

When a private system is abandoned, all pipes must be plugged or capped. The remaining wastes should be disposed of properly. Remove or fill the septic tanks.

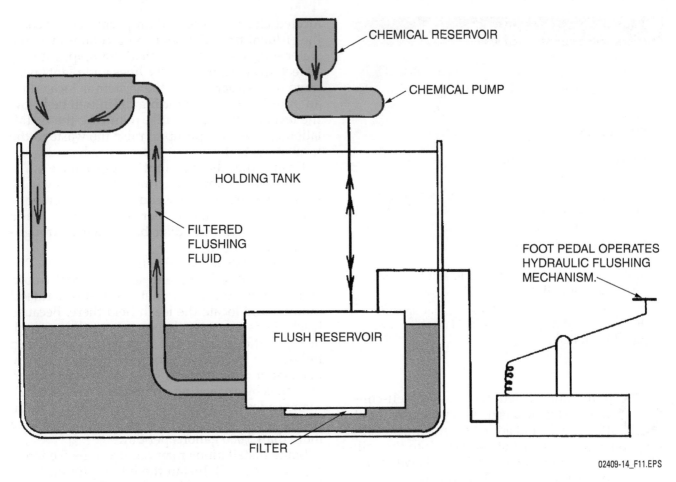

CHEMICAL RESERVOIR

CHEMICAL PUMP

HOLDING TANK

FILTERED FLUSHING FLUID

FOOT PEDAL OPERATES HYDRAULIC FLUSHING MECHANISM.

FLUSH RESERVOIR

FILTER

02409-14_F11.EPS

Figure 11 Closed waste-disposal system.

Additional Resources

Advanced Onsite Wastewater Systems Technologies. 2006. Anish R. Jantrania and Mark A. Gross. Boca Raton, FL: CRC Press.

International Private Sewage Disposal Code®, Latest Edition. Falls Church, VA: International Code Council.

Wells and Septic Systems, Second Edition. 1992. Max Alth, Charlotte Alth, and S. Blackwell Duncan. Blue Ridge Summit, PA: Tab Books.

1.0.0 Section Review

1. The type of soil absorption system that uses pumps to move the effluent is called a(n) _____.

   a. septic system
   b. pressure distribution system
   c. aeration system
   d. closed system

2. Covered pits that do not require a septic tank or aeration tank are called _____.

   a. distribution boxes
   b. organic systems
   c. cesspools
   d. drywells

2.0.0 LOCATING AND SIZING PRIVATE WASTE DISPOSAL SYSTEMS

Objective

Explain how to locate and size private waste disposal systems.
  a. Explain how to locate and size septic tanks.
  b. Explain how to locate and size leach fields.

Trade Terms

Soil mottle: Spots and streaks in soil that indicate water saturation.

Soil profile description: A report that specifies the depths and thicknesses of soil layers in a proposed leach field.

Organic and closed systems are self-contained. This means that they work without affecting the outside environment. Soil absorption systems distribute effluent into the soil. Therefore, they must be safely located away from houses, property lines, wells, and trees. Tanks and leach fields must be big enough to handle the wastes. In this section, you will learn how to locate and size a private soil absorption system. Note that these are general guidelines only. Refer to your local code to determine whether effluent can safely be discharged on the property. Consult local health officials for specific details.

A permit is required before a private system can be installed. Contact your local code official to find out how to apply for permits. Applications must include the following information:

- Property survey
- Two copies of the system's design drawings
- A site plan
- Soil-quality data
- Permit fees

Code officials usually inspect the system while it is being installed and again when it is completed. Follow the local code when installing a private waste disposal system. Doing so will help ensure that the system passes inspection.

Most codes specify the minimum distances from specific structures that septic tanks can be installed. Refer to *Figure 2*. The local applicable code will also specify the minimum distance from wells that leach fields can be located. If the field is located closer to a well than permitted by code, the effluent from the field could contaminate the water supply. Install leach field drainpipes at least 10 feet away from property lines and plants with large root structures. If the system is located on an uphill slope, the building drain will be below the level of the septic tank, requiring the installation of a sewage pump to raise the waste to the level of the tank.

Ensure that the distribution pipes have sufficient grade to allow the effluent to seep slowly by the force of gravity. Most codes suggest a grade of between 2 and 4 inches per 100 feet. This grade will allow effluent to distribute equally throughout the leach field.

Do not locate leach fields in swampy areas. The soil cannot absorb effluent, so the effluent would contaminate the area. Instead, build a raised mound, and locate the leach field there. Because leach fields are near the surface of the ground, heavy traffic areas are not suitable for use as leach fields. The leach field ground must have a slow percolation rate; clay soil is ideal. Prevent erosion on slopes by excavating and filling (see *Figure 12*). Many codes require two separate fields for drain, waste, and vent (DWV) systems that produce more than 5,000 gallons per day.

Some installations may require a grease interceptor (*Figure 13*). Install the interceptor on a different line from the septic tank. The two waste lines can be connected downstream from the septic tank (*Figure 14*).

2.1.0 Locating and Sizing Septic Tanks

Local codes specify the minimum distances that a septic tank can be located from other objects (see *Table 1*). Size the septic tank based on the number of bedrooms in the building (see *Table 2*). Tank capacity should not be less than 750 gallons. Many codes allow up to four septic tanks to be connected in series in place of a larger single tank.

2.2.0 Locating and Sizing Leach Fields

The requirements for locating and sizing leach fields are specified in the local applicable code. Be sure to follow these requirements closely, as failure to do so can result in contamination and health issues. Codes require evaluation and testing of soil in the area selected for the leach field to establish the percolation rate. The percolation rate is then used to determine the size of the leach field. The following sections describe these procedures in more detail. Refer to the local applicable code for guidelines that are specific to your area.

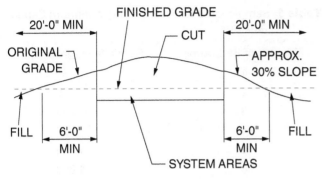

EXCAVATION OF COMPLETE HILLTOP

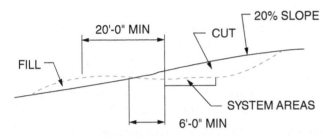

A SURFACE WATER DIVERSION
MAY BE NEEDED AT ONE OF
THESE POINTS IF LONG
SLOPES ARE PRESENT.

EXCAVATION INTO HILLSIDE

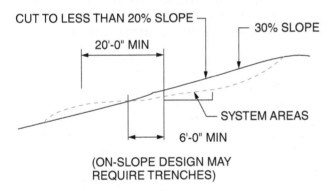

(ON-SLOPE DESIGN MAY
REQUIRE TRENCHES)

REGRADE OF HILLSIDE

02409-14_F12.EPS

Figure 12 Prevent erosion in hillside leach fields by excavating and filling.

2.2.1 Soil Evaluation and Percolation Testing

Before installing a leach field, determine the characteristics of the soil in the area. Bore at least three observation holes. Codes require each hole to be at least 3 feet deep and wide enough for a person to observe the characteristics of the soil. Holes can be dug by backhoe, auger, or hand.

Write a soil profile description for each bore hole. This is a report that indicates the depths and thickness of the different soil layers encountered. Also note the different types of soil mottle,

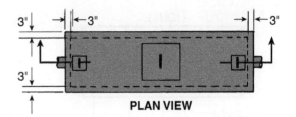

PLAN VIEW

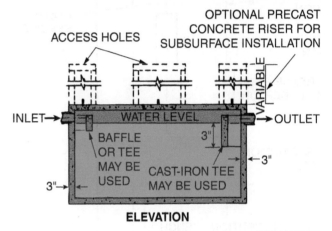

ELEVATION

02409-14_F13.EPS

Figure 13 Typical site-built grease interceptor.

or spots and streaks that indicate saturated soil, in each bore hole. In some cases, you may have to determine the depth of the bedrock. This information will allow the code official to determine whether the area is suitable to serve as a leach field.

When the soil test indicates that the area can be used as a leach field, the next step is to determine the rate at which water percolates into the soil. This will help you determine the size of the leach field. Refer to your local code for specific guidelines on how to conduct a percolation test.

Local health departments usually perform soil tests. You should know how these tests are performed, however. Use the following steps to test the soil's average percolation rate in a leach field:

Step 1 Dig at least six evenly spaced holes in the area of the proposed leach field. Dig the holes at least 3 feet deep and 4 to 12 inches in diameter.

Step 2 Roughen the sides of the holes. This will allow them to absorb water.

Step 3 Remove all loose dirt from each hole. Then add 2 inches of sand or gravel to each hole.

Step 4 Using a hose, fill each hole with about 1 foot of water. Add water to the hole as necessary to maintain this level for at least 4 hours. This is called the swelling period. Some codes require a swelling period of between 16 and 30 hours.

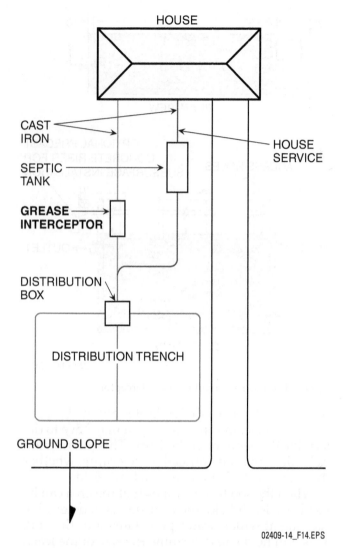

HOUSE

CAST IRON

SEPTIC TANK

GREASE INTERCEPTOR

DISTRIBUTION BOX

DISTRIBUTION TRENCH

GROUND SLOPE

HOUSE SERVICE

02409-14_F14.EPS

Figure 14 Typical grease interceptor installation in a septic system.

Table 1 Sample Table for Determining Minimum Separation of Septic Tanks from Other Objects

Element	Distance (feet)
Building	5
Cistern	25
Foundation wall	5
Lake, high-water mark	25
Lot line	2
Pond	25
Reservoir	25
Spring	50
Stream or watercourse	25
Swimming pool	15
Water service	5
Well	25

Table 2 Sample Table for Determining Minimum Septic Tank Size

Number of Bedrooms	Septic Tank (gallons)
1	750
2	750
3	1,000
4	1,200
5	1,425
6	1,650
7	1,875
8	2,100

Step 5 At the end of the swelling period, allow each hole to retain 6 inches of water. Then measure the drop in water in each hole over 30 minutes. Multiply the result by two. This will be the rate of drop, in inches per hour, for each hole.

Step 6 Add the rate of drop for each hole. Divide the sum by the total number of holes (in this example, by 6). This will be the average rate of drop.

Step 7 Convert the average rate of drop per hour to the average percolation rate in minutes. Divide the average rate of drop per hour by 60. The result is the soil's average percolation rate in minutes per inch.

Next, convert the average percolation rate into the allowable rate of effluent application. This is a measure of how much effluent can be discharged into the leach field. It is measured in gallons per square foot (see *Figure 15*). Consult your local code or contact your local code official. The sample chart in *Figure 15* uses the following common formula:

$$\text{Rate of application} = \frac{5}{\div \sqrt{(\text{average percolation rate})}}$$

Inspections

Local code officials are required to inspect all private waste disposal systems as they are being built and upon completion. Do not backfill over any part of the installation until the inspector has approved the entire installation. Any modifications, additions, or alterations must also be inspected and approved.

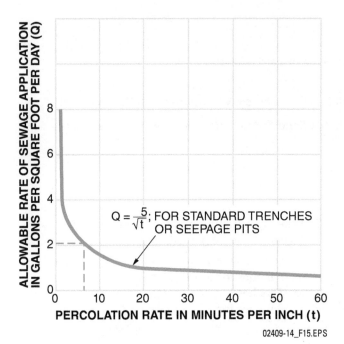

$$Q = \frac{5}{\sqrt{t}}; \text{ FOR STANDARD TRENCHES OR SEEPAGE PITS}$$

02409-14_F15.EPS

Figure 15 Sample chart for determining the rate of sewage application.

In other words, if the average percolation rate is 6 minutes per inch, the allowable rate of discharge is 2.04 gallons per square foot per day, or $5 \div \sqrt{6}$.

Percolation rates affect the size of the leach field. Slower percolation rates require larger fields. If the soil is extremely sandy, local codes may require a raised leach field. Raised leach fields are made of large piles of sod. Sod provides an adequate percolation rate for effluent.

If water seeps into the holes during the percolation test, the water table is too high to permit a septic system. If the percolation rate is less than 1 inch per hour, the soil is too dense or too saturated for use as a leach field. If the holes will not hold water because the soil is too sandy, shorten the test. Hold the water levels at 6 inches for 50 minutes. Then measure the drop in water over a period of 10 minutes.

2.2.2 Sizing the Leach Field

Once the percolation test has been completed, size the leach field using the test information. Your local code will provide information on how to determine the leach field size based on the percolation rate (see *Table 3*). Codes may provide different tables for sizing leach fields for residential and nonresidential installations.

Table 3 Sample Table for Calculating Size of Leach Field Using Percolation Rate Data

Percolation Class	Percolation Rate (minutes required for water to fall 1 inch)	Seepage Trenches or Pits (square feet per bedroom)	Seepage Beds (square feet per bedroom)
1	0 to less than 10	165	205
2	10 to less than 30	250	315
3	30 to less than 45	300	375
4	45 to 60	330	415

George E. Waring, Jr., Sanitation Pioneer

Colonel George E. Waring, Jr., was an educator and administrator who helped Americans live healthier, cleaner lives. He served as New York City's street-cleaning commissioner from 1895 until his death three years later. During that time, his department became famous for its army of street sweepers. They dressed in white uniforms and were known as the White Wings. In three years, the city went from one of the nation's dirtiest cities to one of its cleanest.

Waring was a retired Civil War officer as well as a sanitation engineer. Before he became a New York commissioner, he made his mark in the plumbing field. In 1876, he wrote a book called *The Sanitary Drainage of Houses and Towns*. It became a model for many cities throughout the country. It was only the first of many plumbing books that he wrote. His books made it easy for the average reader to understand the importance of sanitary plumbing. Historians credit him with helping to usher in the era of safe and sanitary plumbing.

Additional Resources

Anatomy of a Disaster. 2008. Washington, DC: Chemical Safety Board (video).
www.csb.gov/videos/anatomy-of-a-disaster

International Private Sewage Disposal Code®, Latest Edition. Falls Church, VA: International Code Council.

Wells and Septic Systems, Second Edition. 1992. Max Alth, Charlotte Alth, and S. Blackwell Duncan. Blue Ridge Summit, PA: Tab Books.

2.0.0 Section Review

1. Size septic tanks based on the number of a building's _____.
   a. bedrooms
   b. occupants
   c. water closets
   d. floors

2. When conducting a percolation test, the water levels should be held at _____.
   a. 6 inches for 30 minutes
   b. 6 inches for 50 minutes
   c. 3 inches for 30 minutes
   d. 3 inches for 50 minutes

3.0.0 Installing Private Waste Disposal Systems

Objective

Explain how to install private waste disposal systems.
  a. Explain how to install septic tanks.
  b. Explain how to install leach field distribution piping.

Most private waste disposal systems use a tank to store solid wastes and a leach field to drain liquid wastes. The size and arrangement of each system is unique. However, the steps required to install a private waste system are fairly standard no matter how large or small the system. You have already learned how to join pipes and connect them to building drains. In this section, you will learn how to install septic tanks, distribution pipe, and leach fields.

When doing so, ensure that all components are the proper size and that they are made of approved materials. Your local code will provide sizing guidelines. It will also list materials that can be used safely in a private waste disposal system. All pipes from the building that drain to the entrance to the leach field should be sloped downward to allow the waste to flow by gravity. The standard slope is ¼ inch per foot of pipe.

3.1.0 Installing Septic Tanks

Refer to the local applicable code for the minimum distance that septic tanks may be located away from buildings. Install the tank so that its top is at least 1 foot below ground level. In colder climates, the top should be 2 or 3 feet below ground. Refer to your local code for the appropriate depth. Provide access to the tank's manholes to allow for inspection, cleaning, and repair. Ensure that the tank has an inspection port above either the inlet baffle or the outlet baffle.

Pipes leading from the building drain should be watertight. Common pipe materials include cast-iron soil pipe, plastic pipe, and sealed clay pipe. Consult your local code for approved pipe materials. Use approved watertight pipe to connect the septic tank to the distribution box. Most codes specify 4-inch pipe for the connection between septic tanks and distribution boxes.

The decomposition process creates three layers of waste in a septic tank. Solids form a layer of sludge on the bottom of the tank. Greasy scum floats to the top of the tank. Between these two layers is the effluent that drains into the leach field. Locate inlets and outlets to allow the wastes to separate while allowing effluent to drain.

On the inlet, install a sanitary tee or ell that extends at least 15 inches into the effluent (*Figure 16*). This will deposit new sewage below the scum layer. Another option is to direct the flow to a baffle (*Figure 17*). Effluent entering the tank through the inlet will hit the baffle and follow it deep into the tank. Outlets should extend 18 inches into the tank. Install a sanitary tee on the outlet. Leave the top of the tee open to act as a vent (see *Figure 18*). On prefabricated tanks, ensure that the existing inlets and outlets meet local code standards.

> **WARNING!**
>
> Do not attempt to join different types of plastic pipe with solvent cement. If the bond fails, sewage waste can leak and contaminate the groundwater and soil.

3.2.0 Installing Leach Field Distribution Piping

Distribution pipe is similar to regular DWV pipe, but it has drain holes (see *Figure 19*). Some codes require covering the pipe with a nylon screen to prevent silt from entering the pipe. Lay the pipe

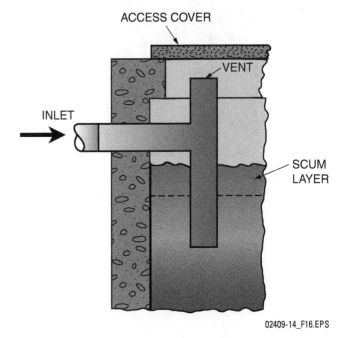

02409-14_F16.EPS

Figure 16 Septic tank inlet extending below the scum layer.

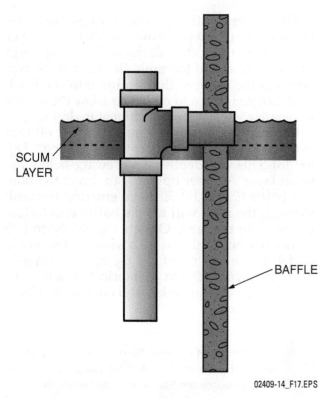

Figure 17 Septic tank inlet with inlet baffle.

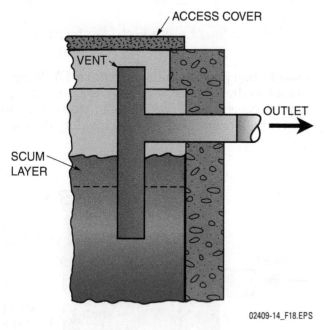

Figure 18 Septic tank outlet with sanitary tee vent.

Figure 19 Perforated plastic distribution pipe.

with the perforations facing down. Join plastic perforated pipes using solvent weld or friction fit. Rigid-wall plastic pipe may be used instead of specially made distribution pipe. Clay pipe is no longer installed for use as leach field distribution piping, but you may encounter it on service calls. When repairing clay pipe, do not seal the butt joints. This will allow the effluent to leach through the butts (*Figure 20*).

Flexible plastic distribution pipe bundled in synthetic aggregate (*Figure 21*) has become one of the most widely used methods of wastewater drainage into leach fields in the United States and Canada. Synthetic aggregate systems (often referred to by the trade name EZflow®) use aggregate material that is typically manufactured from recycled plastic. They can be installed in areas that have a smaller footprint than trenches or bed systems constructed solely using perforated pipe laid in natural aggregate material. Synthetic aggregate has been shown to improve leach field efficiency and capacity because it eliminates the use of stones that can clog drain perforations or the soil's surface, both of which can significantly reduce the infiltration of effluent. Synthetic aggregate systems are available in standard lengths and can typically be installed in less time, and with less site disruption, than a leach field constructed using mineral aggregate.

Install distribution pipe in trenches lined with washed gravel, crushed stone, or slag ranging in size from ½ inch to 2½ inches (see *Figure 22*). This material should be at least 6 inches deep below the pipe, extend from side to side of the trench, and cover the pipe. Your local code provides allowable dimensions for leach field trenches (see *Table 4*).

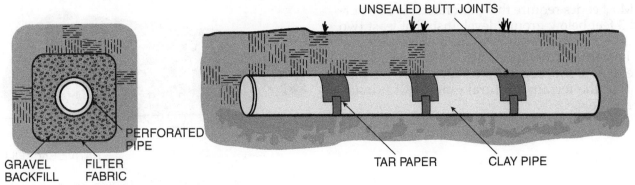

GRAVEL BACKFILL · FILTER FABRIC · PERFORATED PIPE · UNSEALED BUTT JOINTS · TAR PAPER · CLAY PIPE

02409-14_F20.EPS

Figure 20 Typical installation of clay pipe in a leach field.

02409-14_F21.EPS

Figure 21 Distribution piping in bundled synthetic aggregate.

Protect distribution pipe from being crushed under the weight of soil or aboveground traffic by installing heavy-duty nylon covers over the pipe once it has been laid (*Figure 23*). Also called chambers or crush protectors, these covers are designed with holes to allow effluent to flow into the surrounding leach field. They are available in a variety of standard lengths and can be joined together with snap fittings that allow for limited deflection. End caps are fitted with one or more knockout ports through which distribution piping can be fed. Covers eliminate the need to use gravel and other fine natural aggregate materials in leach field trenches.

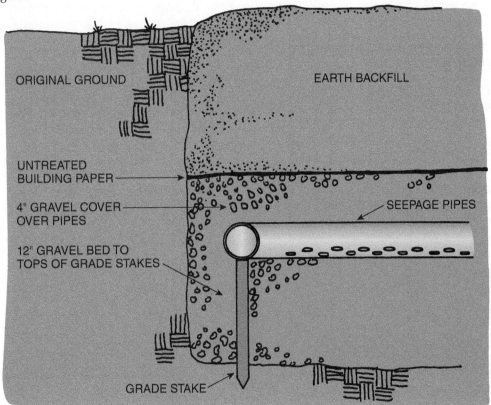

ORIGINAL GROUND · EARTH BACKFILL · UNTREATED BUILDING PAPER · 4" GRAVEL COVER OVER PIPES · SEEPAGE PIPES · 12" GRAVEL BED TO TOPS OF GRADE STAKES · GRADE STAKE

02409-14_F22.EPS

Figure 22 Typical trench for a leach-field distribution pipe.

Most codes require that perforated pipe be buried 3 feet below ground level. Install at least two lines of pipe per leach field. Each line should not be longer than 100 feet. Separate lines by at least 6 feet. A 10-foot separation is ideal. Lay the pipe so that the terrain's natural grade will provide steady flow.

Table 4 Sample Table for Determining Leach Field Trench Sizes

Width of Trench at Bottom (in)	Recommended Trench Depth (in)	Effective Absorption Area (in/sq ft)
18	18–30	1.5
24	18–30	2.0
30	18–30	2.5
36	24–36	3.0

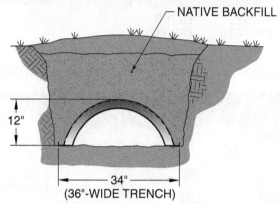

NATIVE BACKFILL

12"

34"
(36"-WIDE TRENCH)

02409-14_F23.EPS

Figure 23 Nylon covers protect distribution piping from being crushed.

Additional Resources

Advanced Onsite Wastewater Systems Technologies. 2006. Anish R. Jantrania and Mark A. Gross. Boca Raton, FL: CRC Press.

International Private Sewage Disposal Code®, Latest Edition. Falls Church, VA: International Code Council.

Septic Systems Handbook, Second Edition. 1991. O. Benjamin Kaplan. Boca Raton, FL: CRC Press.

3.0.0 Section Review

1. In most codes, the size of pipe specified for the connection between septic tanks and distribution boxes is _____.

    a. 2 inches
    b. 3 inches
    c. 4 inches
    d. 5 inches

2. Flexible plastic distribution pipe that has been bundled with loose polystyrene aggregate material bound in mesh is often called _____.

    a. SharkBite®
    b. EZflow®
    c. Big Hawg®
    d. FlowRite®

4.0.0 CLEANING AND SERVICING PRIVATE WASTE DISPOSAL SYSTEMS

Objective

Explain how to clean and service private waste disposal systems.

You have learned that both septic and aeration tanks use anaerobic bacteria to break down sludge. These bacteria can reduce the volume of sludge by only about 40 percent. Over time, tanks will gradually fill with sludge. As a rule of thumb, sludge should be pumped out and disposed of when it fills about one-third of the tank's volume. In most tanks, this will take two or three years. Codes specify when to clean tanks, depending on the tank's capacity and the amount of waste generated. For manufactured tanks, consult the instructions for cleaning requirements.

If sludge is allowed to overfill the tank, the solids will not be able to settle out. That means that the effluent will contain large amounts of solid waste. If these wastes are allowed to percolate into the leach field, they will clog the soil. If enough solid wastes contaminate the leach field, the field will fail. If a field fails, sewage could back up into the structure or contaminate groundwater.

Antibacterial soaps and other cleaning substances that are designed to kill bacteria should never be disposed of in private waste disposal systems. They will also kill the anaerobic bacteria that break down the waste in the tank. The undigested waste will be unable to flow out of the tank into the field, causing the tank to overflow. Furthermore, the entire drain field will have to be dug up and replaced to remove the antibacterial contamination.

Stress to the customer the importance of regular tank cleaning. Schedule the first cleaning when you finish installing the system. Then, at every cleaning, schedule another appointment. Remind customers a few days before each appointment. Failure to perform regular cleaning could be both harmful and expensive.

Clean the tank according to local code requirements. Professional contractors usually service tanks using pumper trucks. Before attempting to clean a tank, allow the tank to ventilate for several days. Do this by opening the manhole covers. Install signs and barriers around the tank to prevent people from falling into it.

After ventilating the tank, begin the cleaning process by breaking up the scum layer. Then mix the sludge with the liquid in the tank. To mix the sludge, pump out some liquid from the tank. Then pump it back into the bottom of the tank. Repeat the cycle until the sludge and liquid are completely mixed. Then, pump out the contents of the tank completely. Always pump out the tanks through the manhole. Do not attempt to pump the tank through the inspection ports. Doing so could damage the baffles.

Remember to take all necessary safety precautions when cleaning tanks. Do not enter a septic or aeration tank to clean it or perform maintenance. Provide adequate ventilation when working on a tank. Use electric lights only. Never use open flames around a septic or aeration tank. Wear appropriate personal protective equipment.

When the cleaning is complete, inspect the inlet and outlet baffles. If they have deteriorated or are missing, replace them. Secure the manhole covers to ensure a tight seal. Finally, dispose of the septic tank waste according to your local code.

> **WARNING!**
>
> Septic tanks contain gases, such as methane, carbon dioxide, and radon, which are toxic and flammable. Septic tanks are confined spaces. Confined-space safety procedures must be followed, because lack of oxygen in the tank can suffocate unprotected workers.
>
> Take proper precautions when working in or near septic tanks. Wear appropriate personal protective equipment. Do not attempt to work near a septic tank without prior emergency-rescue training. Never use an open flame near a septic tank. If someone falls into an unventilated tank, do not enter the tank without proper respiratory equipment. Install a fan over the manhole to provide fresh air, and call for emergency assistance.

> **WARNING!**
>
> Septic tank sludge is a biohazard. Dispose of sludge according to approved disposal procedures. Check with local health officials for the procedures in your area. Never dump septic tank sludge into a city sewage system without first verifying that this method is permitted by code.

Cold Weather Installations

Codes require that private waste disposal systems be installed only when the weather conditions are suitable. Do not try to test and install a leach field when the ground is frozen. If there is snow on the ground, remove it from the area of the leach field. Always protect the backfill material from freezing. To prevent freezing, use loose, unfrozen soil for the first foot of backfill.

Additional Resources

Anatomy of a Disaster. 2008. Washington, DC: Chemical Safety Board (video). **www.csb.gov/videos/anatomy-of-a-disaster**

International Private Sewage Disposal Code®, Latest Edition. Falls Church, VA: International Code Council.

Wells and Septic Systems, Second Edition. 1992. Max Alth, Charlotte Alth, and S. Blackwell Duncan. Blue Ridge Summit, PA: Tab Books.

4.0.0 Section Review

1. Anaerobic bacteria can reduce the volume of sludge only by approximately _____.

   a. 40 percent
   b. 30 percent
   c. 20 percent
   d. 10 percent

SUMMARY

Private waste disposal systems serve rural buildings that are not connected to a municipal system. This module discussed the three main types of private waste disposal systems: soil absorption, organic, and closed. Soil absorption systems are the most common. They separate solid wastes from liquid effluent and allow the effluent to leach into the soil. Septic, aeration, and pressure distribution are the three types of soil absorption systems.

Soil absorption systems must be located so that they do not contaminate the environment. Local codes specify the appropriate distance from wells and other structures. Size septic tanks and leach fields using guidelines in the local code. The soil must be tested to determine whether it is suitable for leaching.

The module also addressed the importance of installing septic or aeration tanks, leach fields,

and distribution piping according to the local code. Septic tanks must provide access to the inlet and outlet sides of the tank. Determine the size of the leach field based on the permeability of the soil. Install distribution pipe with the drain holes down. Lay the pipes in trenches dug at least 6 feet apart. Fill the trenches with washed gravel, crushed stone, or slag, and then backfill them.

Septic tanks and aeration tanks must be cleaned regularly. Make customers aware of the importance of regular tank cleaning. Clean tanks by ventilating them, mixing their contents, and then pumping them out. Observe proper safety precautions when working on or near septic tanks. The contents and gases are hazardous.

Ensure that private waste disposal systems are well installed and maintained. Your professional reputation and the health and satisfaction of the users require it.

1. A private waste disposal system that allows wastes to decompose through contact with air is called a(n) _____.
   a. decomposition system
   b. anaerobic system
   c. aeration system
   d. organic system

2. In a septic system, liquid effluent from the septic tank is directed into the leach field by a _____.
   a. distribution box
   b. gate valve
   c. diverter manifold
   d. wye connector

3. Decomposition of organic matter in a septic tank produces the gases carbon dioxide and _____.
   a. hydrogen
   b. propane
   c. methane
   d. nitrous oxide

4. In some jurisdictions, codes allow wastewater from aeration systems to be used for _____.
   a. washing vehicles
   b. irrigation
   c. concrete mixing
   d. food processing

5. In a portable toilet, wastes are deodorized and purified by _____.
   a. anaerobic bacteria
   b. filtration and aeration
   c. chemicals
   d. aerobic bacteria

6. Local codes may require two separate leach fields if the daily output of a DWV system is greater than _____.
   a. 10,000 gallons
   b. 5,000 gallons
   c. 3,000 gallons
   d. 1,000 gallons

7. The size of a leach field is determined by the soil's _____.
   a. percolation rate
   b. chemical composition
   c. water-retention index
   d. average particle size

8. Observation holes bored to permit examination of soil characteristics must have a depth of at least _____.
   a. 24 inches
   b. 36 inches
   c. 48 inches
   d. 60 inches

9. The preliminary step in percolation testing that involves holding a 1-foot water level in holes for at least four hours is referred to as the _____.
   a. saturation phase
   b. settlement period
   c. swelling period
   d. sealing phase

10. Soil is too dense or saturated for use as a leach field if it has a per-hour percolation rate of less than _____.
    a. 1 inch
    b. 2 inches
    c. 3 inches
    d. 4 inches

11. The standard slope used to permit gravity flow of waste through a pipe is _____.
    a. 1 inch to the foot
    b. ¾ inch to the foot
    c. ½ inch to the foot
    d. ¼ inch to the foot

12. The minimum distance between a building and a septic tank, according to most codes, is _____.
    a. 3 feet
    b. 5 feet
    c. 6 feet
    d. 10 feet

13. The bottom layer of solid in a septic tank is referred to as the _____.

    a. settlement layer
    b. decomposition layer
    c. debris layer
    d. sludge layer

14. A material no longer installed as distribution pipe in leach fields is _____.

    a. clay pipe
    b. perforated plastic pipe
    c. rigid-wall plastic pipe
    d. cast-iron pipe

15. To properly clean a septic tank, pumping should always be done through _____.

    a. an inspection port
    b. the distribution box
    c. the manhole
    d. a tank vent

Trade Terms Quiz

Fill in the blank with the correct term that you learned from your study of this module.

1. To percolate into soil and be absorbed is to _____.

2. A(n) _____ is a soil absorption system that uses a septic tank to separate out sludge.

3. A lined, covered pit for storing solid and liquid wastes that allows effluent to seep into the soil is called a(n) _____.

4. _____ are bacteria that do not need oxygen to live.

5. In a soil absorption system, an area of soil that is designed to accept effluent from distribution pipes is called a(n) _____.

6. A(n) _____ is a soil absorption system in which effluent is pumped through the distribution pipes under pressure.

7. Bacteria that need oxygen to live are called _____.

8. A report that specifies the depths and thicknesses of soil layers in a proposed leach field is called a(n) _____.

9. A(n) _____ is a private waste disposal system that uses tanks to separate out effluent, then allows it to drain into a leach field.

10. A watertight container that directs effluent to different parts of a leach field is called a(n) _____.

11. A(n) _____ is a tank used in a soil absorption system to settle out sludge and scum and to decompose them using anaerobic bacteria.

12. Liquid waste that separates from solid waste and that is allowed to percolate into the soil in soil absorption systems and cesspools is called _____.

13. A(n) _____ is a waste storage and separation tank used in an aeration system.

14. A private waste disposal system that allows waste to decompose through exposure to aerobic bacteria is called a(n) _____.

15. A(n) _____ is a type of soil absorption system that uses an aeration tank to clarify effluent.

16. A covered pit filled with loose sand and gravel that collects roof and basement drainage and allows it to percolate into the surrounding soil is called a(n) _____.

17. Spots and streaks in soil that indicate water saturation are called _____.

18. A(n) _____ is a private waste disposal system that stores wastes until they can be disposed of.

19. The dried solid waste in an organic system that results from decomposition is called _____.

Trade Terms

Aeration system
Aeration tank
Aerobic bacteria
Anaerobic bacteria
Cesspool
Closed system
Distribution box

Drywell
Effluent
Humus
Leach
Leach field
Organic system
Pressure distribution system

Septic system
Septic tank
Soil absorption system
Soil mottle
Soil profile description

Trade Terms Introduced in This Module

Aeration system: A type of soil absorption system that uses an aeration tank to clarify effluent.

Aeration tank: A waste storage and separation tank used in an aeration system. In an aeration tank, anaerobic bacteria decompose sludge and then aerobic bacteria clarify the effluent.

Aerobic bacteria: Bacteria that need oxygen to live. They are used in aeration tanks to clarify effluent.

Anaerobic bacteria: Bacteria that do not need oxygen to live. They are used in septic tanks and aeration tanks to decompose solid wastes.

Cesspool: A lined, covered pit for storing solid and liquid wastes. Cesspools allow effluent to seep into the soil.

Closed system: A private waste disposal system that stores wastes until they can be disposed of.

Distribution box: A watertight container that directs effluent to different parts of a leach field.

Drywell: A covered pit filled with loose sand and gravel that collects roof and basement drainage and allows it to percolate into the surrounding soil.

Effluent: Liquid waste that separates from solid waste. In soil absorption systems and cesspools, effluent is allowed to percolate into the soil.

Humus: In an organic system, the dried solid waste that results from decomposition. Humus can be used as fertilizer after a time.

Leach: To percolate into soil and be absorbed.

Leach field: In a soil absorption system, an area of soil that is designed to accept effluent from distribution pipes. Leach fields are also sometimes called absorption fields or drain fields.

Organic system: A private waste disposal system that allows waste to decompose through exposure to aerobic bacteria. The resulting dried waste is called humus.

Pressure distribution system: A soil absorption system in which effluent is pumped through the distribution pipes under pressure.

Septic system: A soil absorption system that uses a septic tank to separate out sludge.

Septic tank: A tank used in a soil absorption system to settle out sludge and scum and to decompose them using anaerobic bacteria.

Soil absorption system: A private waste disposal system that uses tanks to separate out effluent, then allows it to drain into a leach field.

Soil mottle: Spots and streaks in soil that indicate water saturation.

Soil profile description: A report that specifies the depths and thicknesses of soil layers in a proposed leach field.

Additional Resources

This module presents thorough resources for task training. The following resource material is suggested for further study.

Advanced Onsite Wastewater Systems Technologies. 2006. Anish R. Jantrania and Mark A. Gross. Boca Raton, FL: CRC Press.

Anatomy of a Disaster. 2008. Washington, DC: Chemical Safety Board (video). **www.csb.gov/videos/anatomy-of-a-disaster**

International Private Sewage Disposal Code®, Latest Edition. Falls Church, VA: International Code Council.

Septic Systems Handbook, Second Edition. 1991. O. Benjamin Kaplan. Boca Raton, FL: CRC Press.

Wells and Septic Systems, Second Edition. 1992. Max Alth, Charlotte Alth, and S. Blackwell Duncan. Blue Ridge Summit, PA: Tab Books.

Figure Credits

Courtesy of Infiltrator Systems, Inc., Module Opener, Figure 21, Figure 23

Courtesy of Den Hartog Industries, Inc., Figure 4

Courtesy of Tuf-Tite, Inc., Figure 5

Consolidated Treatment Systems, Inc., Figure 6

Excerpted from the 2012 International Plumbing Code. Copyright 2011. Washington DC. International Code Council. Reproduced with permission. All rights reserved. **www.ICCSAFE.org**, Figure 8, Figure 12, Tables 1–3

Courtesy of Satellite Industries, Figure 10

Courtesy of Advanced Drainage Systems, Inc., Figure 19

Section Review Answer Key

Answer	Section Reference	Objective
Section One		
1. b	1.1.3	1a
2. c	1.3.1	1c
Section Two		
1. a	2.1.0	2a
2. b	2.2.1	2b
Section Three		
1. c	3.1.0	3a
2. b	3.2.0	3b
Section Four		
1. a	4.0.0	4

NCCER CURRICULA — USER UPDATE

NCCER makes every effort to keep its textbooks up-to-date and free of technical errors. We appreciate your help in this process. If you find an error, a typographical mistake, or an inaccuracy in NCCER's curricula, please fill out this form (or a photocopy), or complete the online form at **www.nccer.org/olf**. Be sure to include the exact module ID number, page number, a detailed description, and your recommended correction. Your input will be brought to the attention of the Authoring Team. Thank you for your assistance.

Instructors – If you have an idea for improving this textbook, or have found that additional materials were necessary to teach this module effectively, please let us know so that we may present your suggestions to the Authoring Team.

NCCER Product Development and Revision

13614 Progress Blvd., Alachua, FL 32615

Email: curriculum@nccer.org
Online: www.nccer.org/olf

❏ Trainee Guide ❏ Lesson Plans ❏ Exam ❏ PowerPoints Other _____

Craft / Level: _____ Copyright Date: _____

Module ID Number / Title: _____

Section Number(s): _____

Description: _____

_____

_____

_____

Recommended Correction: _____

_____

_____

_____

Your Name: _____

Address: _____

_____

Email: _____ Phone: _____

02410-14

Swimming Pools and Hot Tubs

OVERVIEW

Swimming pools, hot tubs, and spas are popular installations. Swimming pools use a recirculation system to clean and circulate the pool water. Pools also use pumps, filters, gutters, drains, and chlorinators. When installing hot tubs or spas, follow local code and the manufacturer's instructions.

Module Nine

From *Plumbing, Level Four, Trainee Guide*, Fourth Edition. NCCER.
Copyright © 2014 by NCCER. Published by Pearson Education. All rights reserved.

Objectives

When you have completed this module, you will be able to do the following:

1. Explain how to size and install swimming pool systems and components.
   a. Explain how to size a swimming pool.
   b. Explain how to install water supply and recirculation systems.
   c. Explain how to install gutters and drains.
   d. Explain how to maintain water quality.
2. Identify hot tub and spa systems and their components.
   a. Identify hot tub and spa components.
   b. Explain how to install hot tubs and spas.

Performance Tasks

Under the supervision of your instructor, you should be able to do the following:

1. Calculate the volume of a pool.
2. Identify the components of piping for a spa.

Trade Terms

Backwater valve
Bathing load
Capacity
Floater
Header pipe
Hydrojet
Hydrostatic pressure

Sidewall drain
Skimmer
Turnover rate
Vacuum fitting
Water supply fitting

Industry-Recognized Credentials

If you're training through an NCCER-accredited sponsor, you may be eligible for credentials from NCCER's Registry. The ID number for this module is 02410-14. Note that this module may have been used in other NCCER curricula and may apply to other level completions. Contact NCCER's Registry at 888.622.3720 or go to **www.nccer.org** for more information.

Code Note

Codes vary among jurisdictions. Because of the variations in code, consult the applicable code whenever regulations are in question. Referring to an incorrect set of codes can cause as much trouble as failing to reference codes altogether. Obtain, review, and familiarize yourself with your local adopted code.

Contents

Topics to be presented in this module include:

Figures and Table

SECTION ONE

1.0.0 SWIMMING POOL SYSTEMS AND COMPONENTS

Objective

Explain how to size and install swimming pool systems and components.

a. Explain how to size a swimming pool.
b. Explain how to install water supply and recirculation systems.
c. Explain how to install gutters and drains.
d. Explain how to maintain water quality.

Performance Task

Calculate the volume of a pool.

Trade Terms

Backwater valve: A valve that prevents the backflow of sewage into a pool drain.

Bathing load: The maximum number of people allowed in a pool per hour, based on capacity and turnover rate.

Capacity: The volume of a pool in gallons.

Floater: A device that releases chlorine through vents while floating in a pool.

Header pipe: A pipe used to relieve hydrostatic pressure. The pipe runs underneath the pool to a pump connection on the pool deck.

Hydrostatic pressure: The force applied to the bottom of a pool, spa, or hot tub by high groundwater levels beneath it.

Sidewall drain: A drain installed on a swimming pool wall below the waterline.

Skimmer: A device that traps pool debris in a removable basket.

Turnover rate: The time required for all the water in a pool to cycle through the filter.

Vacuum fitting: A pool drain that serves as a connection for cleaning hoses.

Water supply fitting: A fixture that allows water to flow from a recirculation system back into a swimming pool.

Pools designed for public use adhere to very strict codes and regulations. Officials from the Public Board of Health inspect them regularly. Because of these restrictions, plumbers usually do not install public pools. The focus of this module, therefore, is on private swimming pools.

Codes distinguish between permanent and semipermanent pools. Permanent pools are located in the ground and cannot be disassembled. Semipermanent pools can be moved and reassembled at a new site. Both types of pool require easy access to water supply and drainage systems. In most cases, the pipes and fittings can be the same as those used in other water supply and drainage systems. Pools also require recirculation systems to sanitize the water. Use the installation techniques that you have already learned to install these components. Refer to your local code before installing swimming pools.

1.1.0 Sizing a Swimming Pool

Begin by calculating the pool's volume in gallons, also called its capacity. Refer to the design drawings for the pool's dimensions. For a rectangular pool, multiply the length, width, and height to obtain the volume. For a circular pool, multiply the height by the area of the circular base. Many pools are of a complex shape; break the volume down into smaller, simpler shapes. Calculate the volumes of the smaller shapes and add them together to get the total area in cubic feet (*Figure 1*). Once you have calculated the capacity, convert the cubic-foot measurement into gallons. Remember that there are 7.48 gallons in a cubic foot.

Once you have calculated a pool's capacity, you can determine its turnover rate and bathing load. The turnover rate is the time that it takes to cycle all the water in the pool through the filter. The bathing load is the maximum number of people allowed in the pool per hour. The turnover rate affects the bathing load (*Table 1*). Shorter turnover rates mean that the water will be cleaned more often, allowing a higher bathing load. If the bathing load is exceeded, people could become ill from contaminants in the water. Some local codes determine the bathing load based on the square footage of pool water surface area per person. For example, the shallow area may require 10 square feet per person and the deep area 25 square feet.

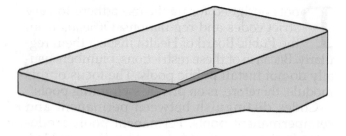

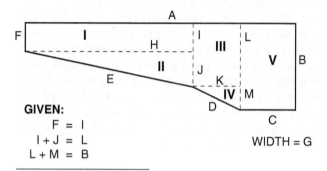

GIVEN:

$F = I$

$I + J = L$

$L + M = B$

$I = H \times F \times G$

$II = \frac{1}{2}(H \times J \times G)$

$III = L \times K \times G$

$IV = \frac{1}{2}(K \times M \times G)$

$V = B \times C \times G$

WIDTH = G

02410-14_F01.EPS

Figure 1 Calculating the area of a swimming pool with a complex shape.

For example, suppose you are installing a pool with a capacity of 55,000 gallons. If you install a pump that can handle 92 gallons per minute (gpm), a complete turnover will take 10 hours. At that turnover rate, the pool can safely handle no more than nine people per hour. If, on the other hand, you install a pump that is rated at 155 gpm, the pump can cycle the water in just six hours. The pool could then handle 40 people per hour.

Calculate the customer's needs when sizing the pool. High turnover rates are not always best. Larger pumps also mean larger pipes and filters to handle the increased load. The result is a greater expense for the customer. In the example above, the bathing load might never exceed 15 people. In that case, you can compromise on a 115-gpm pump that will allow up to 17 people in the pool per hour (*Table 1*). Proper sizing ensures that the water supply, filtering, and drainage systems will be adequate and safe.

The *International Swimming Pool and Spa Code*™ (ISPSC) specifies turnover rates by the category of the pool. For example, the turnover time for swimming pools at hotels and residential complexes is equal to 1½ times the average depth of the pool in feet, for a maximum of six hours. Wave-action pools installed in public recreation facilities, on the other hand, are required to have turnover times between one and two hours, depending on the capacity of the pool. Aboveground and in-ground residential pools are required to provide 100 percent turnover within a minimum of 12 hours.

Table 1 Calculating Turnover and Bathing Load Based on Capacity

Capacity to Waterline (gal)	Turnover (gpm)				Bathing Load per Hour at Turnover (Refiltration Rate)			
	6 hr	8 hr	10 hr	12 hr	6 hr (225 gal)	8 hr (400 gal)	10 hr (625 gal)	12 hr (900 gal)
31,600	88	66	53	44	23	10	5	3
38,400	105	80	64	54	23	12	6	4
55,000	155	115	92	77	40	17	9	5
80,900	255	170	135	115	60	25	13	7
119,500	335	250	200	165	89	37	19	11
158,700	440	330	265	220	117	50	25	15
207,600	575	435	345	290	154	65	33	19
254,300	710	530	425	355	188	80	41	24
305,700	850	640	510	425	226	96	49	28
422,400	1,170	880	705	585	314	133	68	39
557,600	1,550	1,160	930	775	413	179	89	52
632,100	1,750	1,320	1,050	880	468	197	101	58
703,900	1,950	1,460	1,170	975	521	220	112	65
883,000	2,450	1,840	1,470	1,220	654	276	142	82
1,074,900	2,990	2,240	1,790	1,490	796	336	172	99

1.2.0 Installing Water Supply and Recirculation Systems

Once the pool has been properly sized and built, install a suitable water supply system. Pools can draw from a public water supply or a private well. In either case, install a backflow preventer on the supply line to prevent an accidental cross-connection while the pool is being filled. If a garden hose is used, install a vacuum breaker on the hose bibb. If the fill line is connected directly to the water supply system, install a backflow preventer designed to prevent back siphonage. If a permanent fill spout is used, ensure that there is an air gap of at least 6 inches between the outlet and the water overflow line.

Well water must be free of turbidity, hardness, and harmful organisms. Refer to your local code for guidelines. Use only approved pipe and fittings on a pool's fresh-water supply line. In most cases, pipe that is suitable for use in other water supply systems is permitted.

A recirculation system cleans the pool water. It also makes up for water lost through splashing, seepage, and evaporation. Make-up water helps keep the system's pumps primed. Recirculation systems use inlet and outlet fittings, pumps, and filters. The requirements for each of these are discussed in more detail in the following sections. Requirements for recirculation systems vary. Refer to your local code before installing a recirculation system.

1.2.1 Inlet and Outlet Fittings

The following fittings are used to direct water into and out of a recirculation system:

- Vacuum fittings
- Sidewall drains
- Skimmers
- Water supply fittings

Size and locate these inlets and outlets to allow clean water to circulate uniformly through the pool. A good rule of thumb is one inlet for every 350 square feet of pool surface. The ISPSC specifies the installation of one outlet for each 300 square feet of pool surface.

Vacuum fittings (*Figure 2*) are used as connections for cleaning hoses. When water and debris are vacuumed from the pool bottom, they are pumped through the filter. These fittings should be installed 18 inches below the water surface and should be easily accessible.

Sidewall drains (*Figure 3*) are installed on the pool wall below the waterline. This location pre-

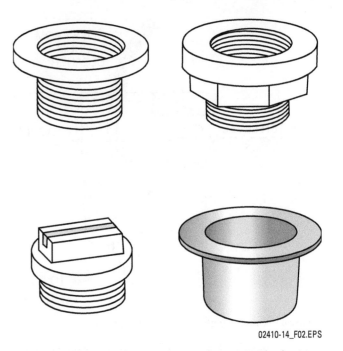

02410-14_F02.EPS

Figure 2 Plastic and chrome vacuum fittings (with plastic plug).

vents leaves or other objects from blocking the drain as the water enters.

Skimmers trap contaminants in a basket that can be removed for cleaning. The skimmer can be located in a gutter or behind an inlet on the pool wall (*Figure 4*). This allows easy access to the strainer or a vacuum hose connection. Skimmers can also be located on the pool deck to catch over-

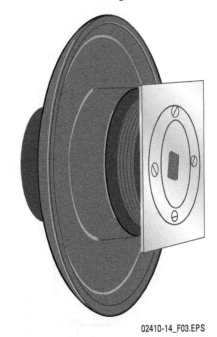

02410-14_F03.EPS

Figure 3 Sidewall drain.

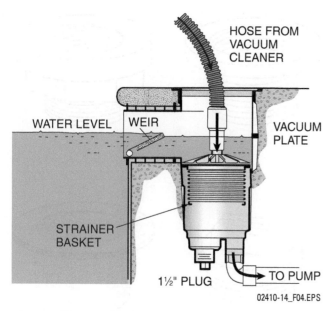

Figure 4 Skimmer installation in a pool wall.

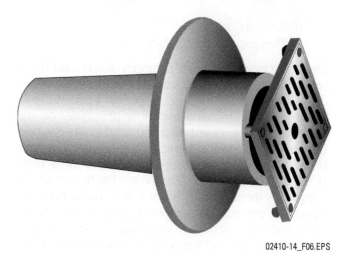

Figure 6 Water supply fitting.

flow (*Figure 5*). The ISPSC requires one skimmer per 400 square feet of pool surface area in a public pool, and one per 800 square feet for residential pools.

Once the water has been filtered, it returns to the pool through water supply fittings (*Figure 6*). Water supply fittings are adjustable and come in a variety of sizes. Install water supply fittings 15 feet apart in the pool walls. Install at least one fitting within 5 feet of each corner of the pool. For placement in pools with irregular shapes, refer to your local code.

1.2.2 Pumps

Codes require the installation of pumps to circulate the pool's water through filters and chemical feeders. Once the water has been filtered and treated, it is pumped back into the pool. Many recirculation systems use water-lubricated centrifugal pumps (*Figure 7*). These are made of stainless steel or plastic so that they will not rust. Pool pumps are designed to run continuously. They should be fitted with removable protective screens to trap debris and hair before it can enter and clog the pump.

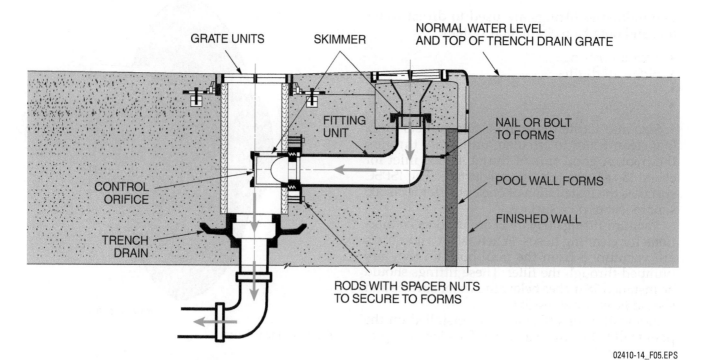

Figure 5 Skimmer installation on a pool deck.

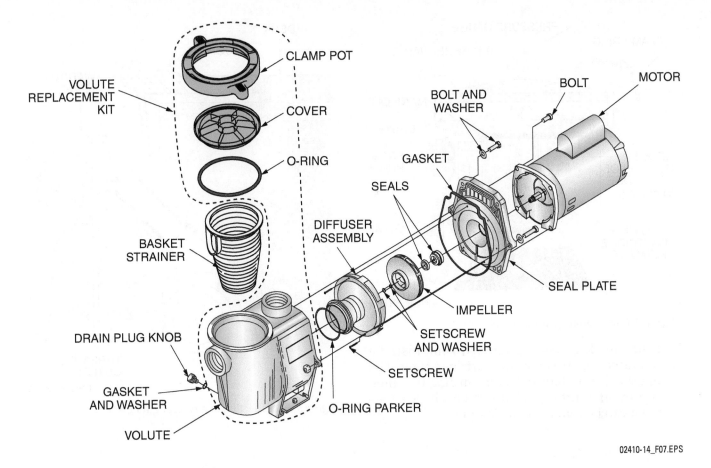

Figure 7 Water-lubricated centrifugal pump.

Install the pump, piping, and electrical systems where they can be reached easily for service. Secure the pump firmly to its base according to the installation instructions. Union connections allow the pump to be removed easily for repair or replacement. Install an interceptor ahead of the pump on the suction line to trap hair and sediment. The interceptor should have a screen that has at least five times the area of the suction pipe.

Most pool packages include a presized pump. For a site-built pool, you will need to size the pump to match the pool's capacity and turnover rate. Carefully set the pump's operating pressure. If the pressure is too high, water will move through the filter too fast. If the pressure is too low, the pump will not be able to move a sufficient volume of water. Filters, heaters, and other items will increase friction loss in the system. Size the pump accordingly. Install easily accessible shutoff valves on the pump's suction and discharge lines below the waterline. An emergency shutoff switch should be installed a minimum of 5 feet from the pool.

1.2.3 Filters

During normal use, pool water collects a wide variety of materials that degrade its quality and increase turbidity. The following types of materials are common:

- Hair
- Dirt
- Leaves
- Insects
- Suntan oils

Pool recirculation systems use filters to reduce turbidity. Most private pools use cartridge filters (see *Figure 8*). As the filter collects impurities, the pump has to work harder to circulate water through it. As a result, the system pressure increases. When the pressure reaches 10 to 12 pounds above the normal pressure, change the filters. Follow the manufacturers' instructions when replacing filters. Ensure that the replacement cartridge filters are the correct size.

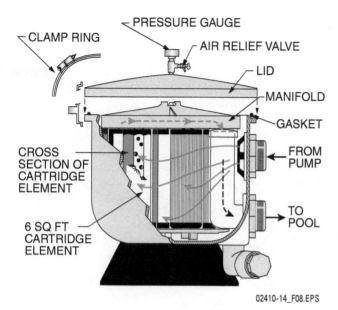

02410-14_F08.EPS

Figure 8 Cartridge-filter installation.

Hair can clog a recirculation system. Install a hair strainer upstream of the cartridge filter. The strainer will trap hair before it can clog the filter, pump, or other fittings. Hair strainers have a basket that can be removed for cleaning.

1.3.0 Installing Gutters and Drains

Gutters (*Figure 9*) are an effective way to drain water from the pool into the recirculation and drainage systems. Gutters are installed around the entire perimeter of the pool. Drains are located along the gutter at intervals. The drains connect to a main gutter line, which drains water to the filter. If the gutter line drains into the sewer system, install a backwater valve on the line (*Figure 10*). Backwater valves prevent sewage backflow into the pool's drainage system.

Deck drains catch the overflow and channel it to the filter. Ensure that the deck is sloped away from the pool. The slope prevents contaminated water from draining back into the pool. Ensure that the drainpipes connected to gutters and deck drains are pitched to provide easy discharge.

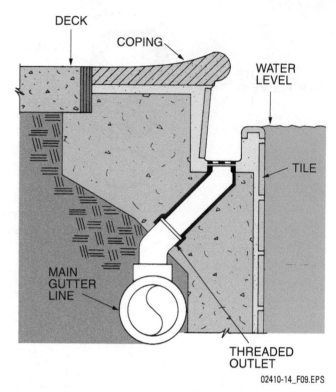

02410-14_F09.EPS

Figure 9 Cross section of a typical gutter drain.

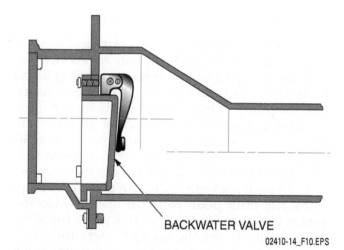

BACKWATER VALVE

02410-14_F10.EPS

Figure 10 Backwater valve.

Using Pumps and Skimmers to Ground a Pool

Local codes require protection against electrical shorts and accidental electrocution beyond the standard installation of a ground fault circuit interrupter (GFCI) to service pool pumps and skimmers. Pumps and skimmers can be used to ground plastic and fiberglass in-ground, aboveground, and prefabricated pools. Run solid copper grounding wire from at least four points on the metal frame of the pool to the designated grounding lug or connector on the pump and the skimmer, and ensure that all the wire is connected using slip bolts or other approved metallic connectors. If either the pump or the skimmer is double insulated, the grounding wire connected to the pump's or skimmer's electrical plug will then safely ground the pool. Otherwise, run another copper grounding wire from the pump or skimmer safely into the ground according to the local applicable code.

The main drain is designed to allow the entire pool to drain in no more than four hours (see *Figure 11*). Locate the main drain at the pool's lowest point. The main drain must have a grate with an opening that is at least four times the area of the drainage pipe. Attach the grate in such a way that tools are required to remove it. The grate will prevent objects and children from getting caught by the drain's suction. Many codes specify the installation of antivortex gratings on the main drain to prevent the formation of a vortex during the draining process by limiting the velocity of the draining water.

Once drained, the pool water can be disposed of in several ways. Pool water is waste water, so refer to your local code for preferred drainage methods. Pool water, filter backwash, and pool-deck drain water can be directed to the building drainage system. Connect the main drain to the drainage system, using an indirect waste pipe with an air gap. The main drain can also be connected directly to the sanitary sewer. Install a backwater valve to protect against contamination caused by a sewage backup.

Water can be channeled into a drain field or allowed to puddle on a properly sized and graded field. If the puddling method is used, the water must be able to percolate into the ground in less than 60 minutes. Fields must be at least 50 feet from wells used for potable water. Another option is to connect the drain to an irrigation sprinkler system. Codes prohibit backwash waste from entering an irrigation system.

Overflow Standpipes

Overflow standpipes prevent spillover in fountains and decorative pools. The pipe has a dome to keep out leaves and other floating debris. The dome can be removed to provide access to the standpipe interior. The number and diameter of the pipes depends on the design of the fountain or pool. Consult the design drawings before installing standpipes.

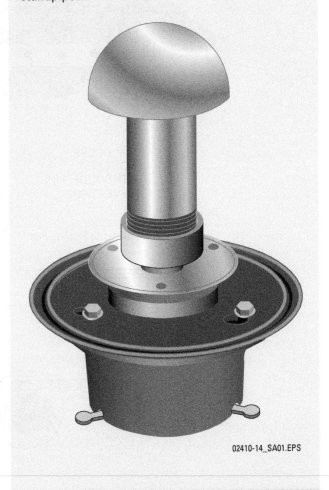

02410-14_SA01.EPS

SUMP INSTALLATION

SUPPLY TO POOL FROM FILTER

PLASTER POOL FLOOR

GRATE

RING

GUNITE

SUMP

2" PLUMBING SHOWN 1½" WITH REDUCER

1½"

COLLECTION TUBE

RETURN TO PUMP SUCTION

02410-14_F11.EPS

Figure 11 Main drain with a return line connection.

A main drain may also serve the recirculation system. If so, ensure that the drain's flow rate to the recirculation system is less than 2 feet per second. A higher flow rate would create dangerously high suction. Install a gate valve on the main drain in an accessible location. Use the valve to close off the drain from the recirculation system when draining the pool.

In some places, high groundwater levels can actually float an empty pool out of the ground. This situation is caused by hydrostatic pressure. Protect the pool against hydrostatic pressure by installing a relief valve in the main drain (*Figure 12*). The relief valve allows water to flow into the empty pool until the pressure is equalized. Another way to relieve hydrostatic pressure is to install a header pipe (refer to *Figure 12*). The header pipe runs from underneath the pool to a connection on the pool deck. A pump connected to the pipe can siphon out the groundwater until the pressure is relieved.

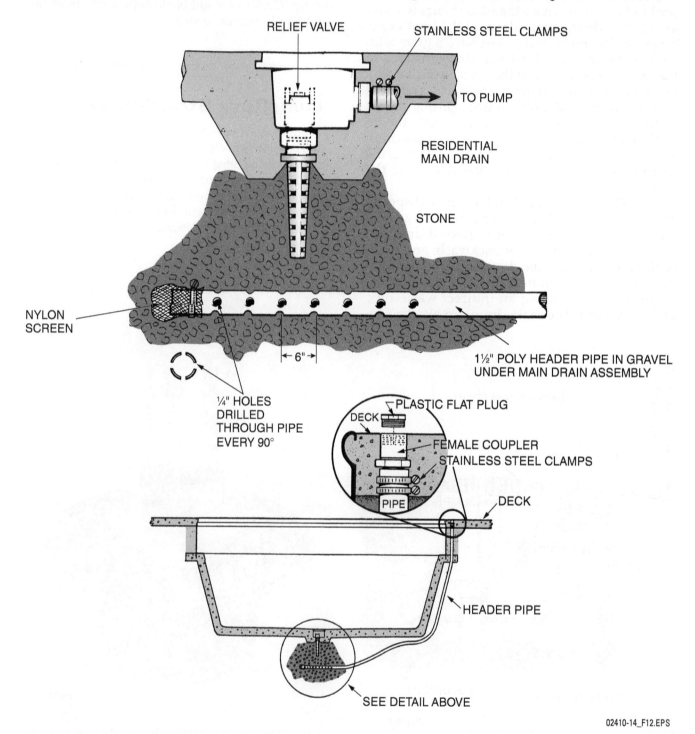

Figure 12 Relief-valve and header-pipe installations.

02410-14_F12.EPS

1.4.0 Maintaining Water Quality

Swimming pools can be contaminated by bacteria and other harmful organisms. Small amounts of chlorine in the water are enough to kill most organisms, so install an in-line chlorinator on the recirculation line. Chlorinators release a measured amount of chlorine into the water as it is pumped to the filter. The chlorine can be in tablet or stick form. Some large pools use chlorine gas under pressure. Another option is to place chlorine tablets in a floater, which is a device that floats in the pool and releases chlorine through vents on its underside (*Figure 13*). Test chlorine levels regularly. Replace tablets or sticks according to the manufacturers' instructions.

Acidity and alkalinity levels in the pool water must be correctly balanced. High acid levels can cause the following problems:

- Corrosion of iron and copper pipes
- Damage to pumps and other recirculation components
- Irritation of swimmers' skin and eyes
- Inability of the water to hold chlorine

On the other hand, high alkaline levels can cause the following problems:

- Formation of scale on plumbing components
- Clouding of pool water
- Prevention of chlorine from killing harmful organisms

The measure of the acidity or alkalinity levels is called the pH (potential hydrogen) value. High pH values indicate high alkaline content. Low pH values indicate high acid content. A value of 7.0 indicates a neutral solution. The proper pH range for a pool is from 7.4 to 7.8. Use muriatic acid to lower the pH value. Refer to your local code. Consult a local pool expert for advice on how to maintain a correct pH value.

Swimming pool water should be tested regularly. Test kits can be purchased that will give you a quick reference to check the chlorine level and pH value. More thorough tests are required to identify other health hazards. For example, most over-the-counter test kits do not test for fecal coliform, a dangerous type of bacteria. The public health department usually conducts such tests. However, customers often contact the pool installer first. Take the time to help the customer obtain more information about testing. Ensure that your level of service matches the quality of the components and workmanship so that the swimming pool will remain a source of recreation, not headaches.

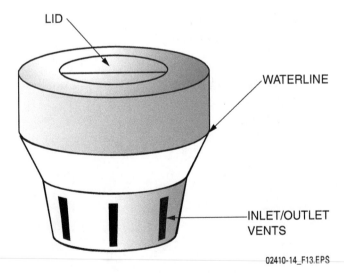

LID

WATERLINE

INLET/OUTLET VENTS

02410-14_F13.EPS

Figure 13 Floater chlorine dispenser.

Algae

Alga is a type of seaweed. Algae (the plural of alga) can form in a pool that has not been used for a long time. Algae cause the water to turn green and have an unpleasant odor. They can leave a floating scum on the water surface or the pool walls. Algae can grow if the chlorine level is below normal. Adding chlorine is often enough to kill algae. If it is not, you may need to call in a pool service professional.

Disinfecting Pools Using Methods other than Chlorine

Chlorine is a popular and widely used disinfectant in swimming pools, but it has drawbacks. Chlorine can cause allergic skin reactions, and its disinfecting properties can be lessened by imbalances in the pH level of the water. Pool manufacturers offer a variety of other methods for disinfecting pool water. Saltwater systems, for example, are less expensive than chlorine and eliminate the strong chemical odor associated with chlorine. However, saltwater systems require regular cleaning to prevent calcium buildup. Furthermore, saltwater systems are currently more expensive than chlorine systems, and must run all day, which results in greater energy consumption.

In copper-ion systems, a low-voltage electric current flows over copper or copper-alloy electrodes in the water. This releases electrically charged copper atoms, called ions, into the water. Copper ions are harmless to humans, but they prevent the buildup of algae and other organic materials in the water. Silver-ion systems use silver and silver-alloy electrodes to achieve the same results. Both copper-ion and silver-ion systems require little maintenance, although the electrodes have to be replaced every three to five years.

Ultraviolet light, which is used to disinfect potable water supplies, is also used to disinfect swimming pools. UV systems are generally less expensive than chlorine systems to install and operate.

Additional Resources

Basic Aboveground Pool Installation. 1995. Don Burger. Hesperia, CA: Superior Publishing.

The Pool Book: Building and Maintaining Swimming Pools & Spas. 1990. Gerard C. O'Connell. Tucson, AZ: Fisher Books.

International Plumbing Code®, Latest Edition. Falls Church, VA: International Code Council.

International Swimming Pool and Spa Code™, Latest Edition. Falls Church, VA: International Code Council.

1.0.0 Section Review

1. The calculation of bathing load is partly determined by the _____.

    a. filter size
    b. pool capacity
    c. pump type
    d. turnover rate

2. The minimum distance that an emergency-pump shutoff switch should be installed from the pool is _____.

    a. 3 feet
    b. 5 feet
    c. 7 feet
    d. 9 feet

3. If a pool's gutter line drains into the sewer system, the line should be fitted with a _____.

    a. skimmer
    b. backwater valve
    c. backflow preventer
    d. filter

4. The proper pH range for a swimming pool is between _____.

    a. 7.4 and 7.8
    b. 6.4 and 6.8
    c. 5.4 and 5.8
    d. 4.4 and 4.8

2.0.0 HOT TUB AND SPA SYSTEMS AND COMPONENTS

Objective

Identify hot tub and spa systems and their components.
   a. Identify hot tub and spa components.
   b. Explain how to install hot tubs and spas.

Performance Task

Identify the components of piping for a spa.

Trade Term

Hydrojet: A nozzle that sends a high-pressure stream of air and water into a spa.

The terms *hot tub* and *spa* are often used interchangeably. However, there are some important technical differences between them. You should be aware of these differences because they can affect the successful installation, use, and servicing of the hot tub or spa.

Hot tubs are filled with hot water and are typically made of molded fiberglass or acrylic, though wood is still sometimes used. The water in a hot tub circulates slowly through a heater. A spa uses jets, or air injectors, to provide additional massaging action. Spas are usually made of fiberglass, steel, or concrete.

This section reviews the components used in hot tubs and spas, including the components that are used in both and those that are used specifically in one or the other. It then summarizes the installation steps required for hot tubs and spas. As always, refer to the manufacturer's instructions and the local applicable code when installing hot tubs and spas. Proper installation is essential for ensuring that hot tubs and spas are safe and healthy for the people who use them.

2.1.0 Identifying Hot Tub and Spa Components

Except for the spa's air-injection ports, the components and installation requirements for hot tubs and spas are very similar. However, the differences could be important when ordering new

or replacement parts. Before installing a spa or hot tub, ensure that you and the customer are referring to the same type of installation. The following sections review key hot tub and spa components that plumbers are responsible for installing.

2.1.1 Hot Tub Components

In most hot tubs, water is circulated slowly through a natural gas or electric heater. Self-contained hot tubs that are purchased as complete units are typically equipped with electric heaters. In addition to the heater, the plumbing components of a typical hot tub include the pipe and fittings that connect to the water supply, and a pump to circulate the water to the heater. Approved piping materials for use in hot tubs as well as in spas include the following:

- Acrylonitrile butadiene styrene (ABS) pipe
- Chlorinated polyvinyl chloride (CPVC) pipe
- Copper or copper-alloy tubing
- Polyvinyl chloride (PVC) hose and pipe
- Stainless steel pipe

Direct-contact water heaters are often used in hot tubs because of their greater efficiency, smaller footprint, and reduced complexity when compared with other types of water heaters. Unlike conventional water heaters that are used to provide potable hot water, direct-contact water heaters do not use a heat exchanger to heat water. Instead, the water is pumped directly over the heating element. Heating elements in direct-contact water heaters must be resistant to corrosion from exposure to water and any chemicals in the water such as chlorine or ozone.

Some hot tubs use heat pumps to heat water. Heat pumps require the circulating pump to operate much more frequently than for a water heater, and they also take longer to heat the water. Furthermore, the temperature of heat-pump water heaters can be harder to regulate under certain conditions.

Some hot tubs are fitted with separate pumps for circulating and filtration, while others are fitted with a single pump that handles both functions. Unlike other types of pumps, which can dissipate the heat caused by their operation into the surrounding air or water, a hot-tub pump typically operates in temperatures that are at least as hot, if not hotter, than its typical operating temperature. In addition, the pump must be able to withstand the corrosive effects of both the heated water and any chemicals that are dissolved in the water. Therefore, it is important to ensure that a pump is specifically designed for use in high-temperature, corrosive environments such as a hot tub.

Pumps and heaters are typically operated by electronic controls that are sealed against damage from humidity. These controls may include sophisticated digital features such as password protection to guard against unauthorized use. Optional features include insulating covers and saltwater filtration.

2.1.2 Spa Components

Spa systems come in a variety of sizes and shapes. Most systems have the same basic components (*Figure 14*). A pump circulates the water through a filter that traps particles and organisms. The water is then pumped to a heater, which warms it to a preset temperature. A thermostat in the tub controls the heater. From the heater, the water circulates into the tub by way of a hydrojet. A hydrojet is a nozzle that squirts a mixture of air and water at high pressure into the tub (*Figure 15*). Some spas use jets of water with no air mixed in. As the water circulates in the tub, the pump draws it out through drains in the sides and bottom of the tub, and the cycle begins again.

An air blower provides the compressed air for the hydrojets. Install the air blower above the

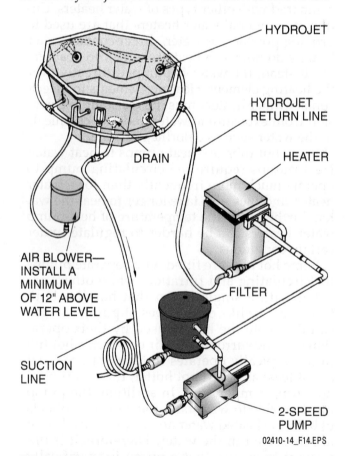

02410-14_F14.EPS

Figure 14 Typical spa installation.

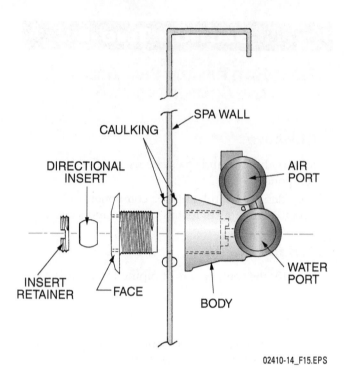

02410-14_F15.EPS

Figure 15 Exploded view of a hydrojet.

water level to prevent water from flooding the air-blower system. Refer to the manufacturer's instructions for the proper placement of the air blower. In most cases, 12 inches above the waterline is sufficient.

2.2.0 Installing Hot Tubs and Spas

The procedures for installing a hot tub or a spa are similar to those for a swimming pool. Most instructions come with detailed illustrations to help you install the components correctly. If you need additional information, contact the manufacturer or visit the manufacturer's website. Many companies have product information available to print or download.

2.2.1 Before Installing a Hot Tub or Spa

Many hot tubs and spas are installed indoors. They will have to be carefully located to fit in with existing supply and drainage systems. The wholesaler may deliver the complete hot tub or spa system directly to the job site, or you may have to pick it up yourself from a supplier or a contractor's warehouse. In either case, after you receive the system, install it according to the manufacturer's instructions, keeping the following guidelines in mind:

• Check the brand and model to ensure that the system meets the job specifications. Make sure that all components are included. Test-fit the components before final assembly.

- Check the condition of the tub and its components. Open all containers delivered to the job site while the delivery person is there. Doing this eliminates questions about who is at fault if damage is discovered later. If you pick up the system from a supplier or contractor's warehouse, inspect it before you leave. Check the contents of previously opened containers for missing items.
- Verify that the rough-in dimensions are correct. This is a critical step. Errors could have occurred in the initial rough-in. Changes in the building, such as wall placement and wall covering, could affect your work. You might have to install a hot tub or spa that is different than the one originally specified.
- Protect the tub and its components before installation. Ideally, you will be able to schedule delivery of the fixtures right before you are ready to install them. If that is not possible, store the fixtures in a clean, dry space that can be locked. After the components are installed, cut and fit cartons over them to protect surfaces from scratches and chips.
- Save all of the packing materials. You can use the package to protect the installed components from weather and other work. You can also use the package to return an unsuitable or damaged component.
- Save all of the manufacturer's installation and care instructions and warranty materials. Turn them over to the customer when the job is completed.
- Protect your customers. They expect to be the first to use their new hot tub or spa. Turn off the water supply and place a Do Not Use sign on the tub.

Follow this procedure on every installation. You will reduce (and even eliminate) mistakes, and save time and money.

> **NOTE**
> Turnover rates for hot tubs and spas vary with the manufacturer and the size of the hot tub or spa. Always refer to the manufacturer's instructions for guidance on calculating the appropriate turnover for the particular application.

2.2.2 Hot Tub and Spa Installation Procedures

Spa tubs may come with precut holes for the hydrojets. If not, follow the manufacturers' directions for cutting them. Carefully locate and size the holes. If the holes are too big, the hydrojet fixture will fit loosely on the tub wall. Oversized holes could cause excess vibration and wear. Seal the fitting on both the inside and the outside of the tub wall.

Always follow the manufacturer's instructions when positioning the required blocking, grout, or other supporting material for the tub body. Trial-fit the unit to ensure that the loadbearing points are properly supported. Fiberglass has a tendency to crack under stress, so you must correctly fit the framing around fiberglass units. Avoid bending the material during installation. Bending may cause cracks that will not become visible until later. After installation is complete, fill the tub to the overflow line and check for leaks. Turn off the water, drain the tub, and cover it with packing material to protect it.

Hydrostatic pressure can also float empty hot tubs and spas out of the ground. Install a header pipe under the tub (*Figure 16*). Use perforated pipe with ¼-inch holes every 90 degrees around the circumference of the pipe. Embed the pipe in gravel to allow water to flow into the pipe and to relieve the hydrostatic pressure. Plug the top of the header pipe when it is not in use, to keep the pipe clear of debris.

The Jacuzzi Brothers

The Jacuzzi™ name is synonymous with whirlpool baths. However, when the Jacuzzi brothers first came to California from Italy, they designed and built an airplane to carry mail for the US Post Office. The plane was very advanced for its day. It was a monoplane (meaning it had only one set of wings) and had an enclosed cabin for the pilot. At the time, most planes were biplanes, and the pilot sat in an open cockpit, exposed to the wind.

The Jacuzzi brothers also invented and marketed deep-well pumps. In the late 1950s, they turned their attention to hydrotherapy (curing physical ailments with water), developing a portable whirlpool pump for bathtubs. In 1968 Roy Jacuzzi invented the first self-contained whirlpool bath, and two years later the company developed its first spa model. Today, the company has manufacturing plants around the world, and the name Jacuzzi is synonymous with whirlpool baths.

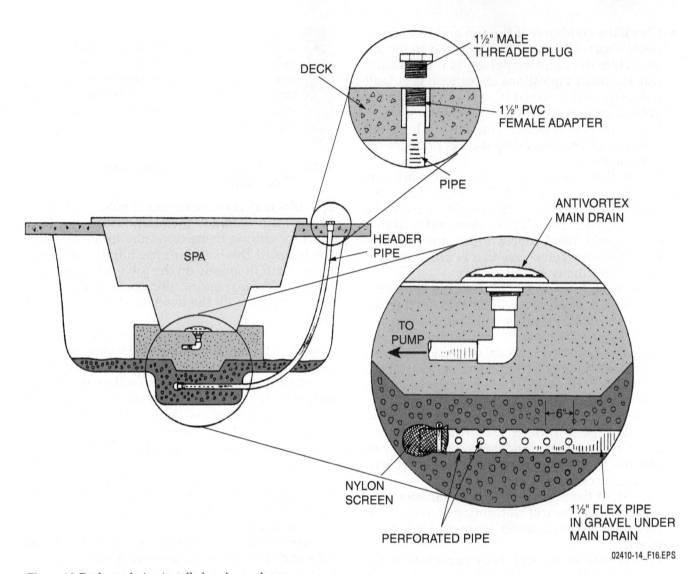

Figure 16 Perforated pipe installed underneath a spa.

If the main drain is directly connected to the sanitary drainage system, trap and vent the drain according to local code. If the drain empties into a floor drain, provide an air gap between the tub drain outlet and floor drain. Install the required number of skimmers per the local applicable code. The ISPSC, for example, requires the installation of one skimmer for every 150 square feet of water surface area in a hot tub or spa.

Refer to the installation instructions to ensure that the heaters and pumps are sized properly. If the heater will be turned off when the hot tub or spa is not in use, ensure that the heater is large enough to heat the water quickly. If the heater will be left on all the time, a smaller unit can be used. Spas with hydrojets require larger pumps than those with water jets that do not mix air into the stream.

Proper pH and chlorine levels are even more important in a spa or hot tub than in a swimming pool. The water is warmer, and most spas and hot tubs are not under direct sunlight. The combina-

tion of heat, humidity, and darkness is ideal for bacteria and fungus. Hot tubs and spas should be inspected regularly. Ensure that filters and chlorinators are well maintained. Provide the customer with a maintenance schedule. This information may also be included in the owner's manual.

> **CAUTION**
>
> Store chlorine in a well-ventilated area.

For public spas, local codes require the installation of shutoff switches and alarms on pumps to indicate when the pump has stopped. The alarm should generate an audible sound of no less than 80 decibels, and also have a visible light. Both the audible and visible alarms must operate continuously until turned off using the shutoff switch. Codes also require public spa facilities to display

Energy Conservation

Pools and hot tubs can be expensive to operate. When you install a unit, offer the customer some tips on how to save energy and keep costs down. The following are common suggestions:

- Run the filtration system only during off-peak hours.
- When leaving for more than a week, turn off the heater and extinguish its pilot light.
- If you use the pool or hot tub only on weekends, lower the thermostat during the week.
- For outdoor spas and hot tubs, erect windbreaks such as fences, hedges, or cabanas to minimize heat loss through the prevailing winds.
- At the beginning of the season, ensure that the pump is working before you add chemicals to the water.
- If the heater is more than five years old, consider purchasing a newer and more energy-efficient model.
- Backwash or clean the filter regularly to obtain maximum filtration.

a sign explaining the purpose of the alarm, and warning people to not use the spa when the alarm is activated. The sign must be visible from the spa. Ensure that air jets are fitted with backflow preventers according to the local applicable code.

Like swimming pools, hot tubs and spas lose water because of splashing and evaporation. They require make-up water to maintain the water level. If the water level falls too low, the pump may lose its prime and become damaged. Connecting hot tubs and spas to a water supply line will prevent this. Install a float valve to control the water level in the tub, and install a backflow preventer on the supply line to guard against back siphonage. If the tub is equipped with a fill pipe or faucet, install the fixture above the tub's flood-level rim.

Restrictions on Plastic Pipe

Refer to your local code before installing plastic water supply pipe. The code may restrict the use of plastics in commercial fire-rated buildings or in multistory plumbing installations.

Additional Resources

How to Build a Hot Tub. 1981. Carlton Hollander. New York: Sterling Publishing Company.

The Pool Book: Building and Maintaining Swimming Pools & Spas. 1990. Gerard C. O'Connell. Tucson, AZ: Fisher Books.

International Plumbing Code®, Latest Edition. Falls Church, VA: International Code Council.

International Swimming Pool and Spa Code™, Latest Edition. Falls Church, VA: International Code Council.

Pools & Spas, Second Edition. 2008. David Short and Fran J. Donegan. Mahwah, NJ: Creative Homeowner.

2.0.0 Section Review

1. Unless the manufacturer's instructions specify otherwise, air blowers are typically installed in a spa at _____.

    a. 12 inches above the waterline
    b. the waterline
    c. 12 inches below the waterline
    d. 12 inches above the base of the spa

2. When installing a header pipe under a spa, the circumference of the pipe should be perforated with _____.

    a. ⅛-inch holes every 90 degrees
    b. ⅛-inch holes every 45 degrees
    c. ¼-inch holes every 90 degrees
    d. ¼-inch holes every 45 degrees

SUMMARY

Swimming pools, hot tubs, and spas are among the most popular recreational fixtures that plumbers install. This module discussed the components and installation procedures for pools, hot tubs, and spas. Each type of fixture is equipped with water supply, drainage, and recirculation systems. These must be installed correctly for the system to work efficiently.

This module also explained how to calculate a pool's capacity, or volume, in gallons. This information is used to determine the pool's turnover rate and bathing load. The turnover rate will affect the bathing load. Frequent turnover cleans the water more often. As a result, more people per hour can use the pool. Turnover and bathing load affect the sizing of the pump and filter.

Pools use a variety of fittings to draw water from the pool and return it to the pool. Pumps draw water out of the pool to be filtered and chlorinated. Most private pools use cartridge filters. Install gutters around the perimeter to drain water into the recirculation and drainage systems. Deck drains can be installed to return pool overflow to the filter. Chlorinators can be installed in the recirculation system or allowed to float in the pool.

Hot tubs and spas differ in the construction materials used and the action of the water. Spas use jets with air injectors to provide high-speed streams of water. Spas are usually made from fiberglass, steel, or concrete, whereas hot tubs are often made of molded fiberglass or acrylic. Both spas and hot tubs use heaters to provide warm water.

Install hot tubs and spas according to the manufacturer's instructions. The process is similar to installing a swimming pool. Carefully locate and size holes for the hydrojet fixtures. To prevent a fiberglass tub from cracking, use proper blocking, grout, or other supporting material. Install backflow prevention devices if the hot tub or spa is connected to the water supply or drainage systems. These steps will help ensure customers will get years of satisfaction from their recreational installation.

1. Permanent pools are located in the ground and therefore cannot be _____.

   a. inspected
   b. disassembled
   c. cleaned
   d. regulated

2. A floater is a device that floats in the swimming pool and releases _____.

   a. chlorine
   b. bacteria
   c. alga
   d. alkaline

3. The purpose of the recirculation system in a swimming pool is to _____.

   a. measure the water's capacity
   b. release chlorine into the system
   c. increase the hydrostatic pressure
   d. sanitize the water

4. The capacity of a swimming pool is measured as _____.

   a. volume in gallons
   b. number of people it will hold
   c. square feet
   d. maximum depth

5. A residential swimming pool with a length of 20 feet, a width of 12 feet, and a height of 6 feet has a volume of _____.

   a. 1,440 gallons
   b. 10,771 gallons
   c. 748 square feet
   d. 25,832 liters

6. When using a permanent fill spout, the outlet and water overflow line must have an air gap of at least _____.

   a. 3 inches
   b. 4 inches
   c. 5 inches
   d. 6 inches

7. According to the ISPSC, a residential pool must have one skimmer per _____.

   a. 200 square feet
   b. 350 square feet
   c. 550 square feet
   d. 800 square feet

8. Pool pumps are designed to run _____.

   a. every 2 hours
   b. every 24 hours
   c. continuously
   d. occasionally

9. Pool recirculation systems use filters to reduce _____.

   a. pressure
   b. turbidity
   c. friction
   d. capacity

10. High acid levels in a pool can cause _____.

    a. scale to form on plumbing components
    b. ultraviolet radiation
    c. iron and copper pipes to corrode
    d. contamination in the leach field

11. Most hot tubs are made of acrylic or molded fiberglass, although some are made of _____.

    a. steel
    b. concrete
    c. wood
    d. plastic

12. The massaging action found in spas is produced by _____.

    a. air injectors
    b. hydrostatic pumps
    c. recirculation systems
    d. vacuum fittings

13. When installing hot tubs and spas, refer to your local code and the _____.

    a. customer
    b. plumber
    c. public health code
    d. manufacturer's instructions

14. To relieve hydrostatic pressure in a hot tub or spa, install a _____.

    a. loadbearing point
    b. perforated pipe
    c. hydrojet
    d. saltwater filtration system

15. A spa pump's audible alarm should be no less than _____.

    a. 50 decibels
    b. 60 decibels
    c. 70 decibels
    d. 80 decibels

Trade Terms Quiz

Fill in the blank with the correct term that you learned from your study of this module.

1. A drain installed on a swimming pool wall below the waterline is called a _____.

2. A _____ is a valve that prevents the backflow of sewage into a pool drain.

3. A pipe used to relieve hydrostatic pressure that runs underneath the pool to a pump connection on the pool deck is called a _____.

4. A _____ is a device that traps pool debris in a removable basket.

5. The maximum number of people allowed in a pool per hour, based on capacity and turnover rate, is called the _____.

6. A _____ is the force applied to the bottom of a pool, spa, or hot tub by high groundwater levels beneath it.

7. A pool drain that serves as a connection for cleaning hoses is called a _____.

8. A _____ is a device that releases chlorine through vents while floating in a pool.

9. The time required for all the water in a pool to cycle through the filter is called the _____.

10. A _____ is a fixture that allows water to flow from a recirculation system back into a swimming pool.

11. The volume of a pool in gallons is the pool's _____.

12. A _____ is a nozzle that sends a high-pressure stream of air and water into a spa.

Trade Terms

Backwater valve
Bathing load
Capacity
Floater

Header pipe
Hydrojet
Hydrostatic pressure
Sidewall drain

Skimmer
Turnover rate
Vacuum fitting
Water supply fitting

Trade Terms Introduced in This Module

Backwater valve: A valve that prevents the backflow of sewage into a pool drain.

Bathing load: The maximum number of people allowed in a pool per hour, based on capacity and turnover rate.

Capacity: The volume of a pool in gallons.

Floater: A device that releases chlorine through vents while floating in a pool.

Header pipe: A pipe used to relieve hydrostatic pressure. The pipe runs underneath the pool to a pump connection on the pool deck.

Hydrojet: A nozzle that sends a high-pressure stream of air and water into a spa.

Hydrostatic pressure: The force applied to the bottom of a pool, spa, or hot tub by high groundwater levels beneath it.

Sidewall drain: A drain installed on a swimming pool wall below the waterline.

Skimmer: A device that traps pool debris in a removable basket.

Turnover rate: The time required for all the water in a pool to cycle through the filter.

Vacuum fitting: A pool drain that serves as a connection for cleaning hoses.

Water supply fitting: A fixture that allows water to flow from a recirculation system back into a swimming pool.

EXERCISES FOR PERFORMANCE TASK 1

Use the drawing of a hypothetical swimming-pool installation in *Figure A-1* and the information in this module to answer the following questions. Round your answers up or down to the nearest whole number. Remember to show all your work.

1. Calculate the total pool volume in cubic feet:

   _____

2. Calculate the total gallon capacity of the pool:

   _____

3. Given that turnover is said to occur once every 10 hours, the pool's bathing load would be:

   _____

4. Given that turnover is said to occur once every six hours, the pool's bathing load would be:

   _____

5. Enter the turnover in gpm for Questions #3 and #4:

   _____ and _____

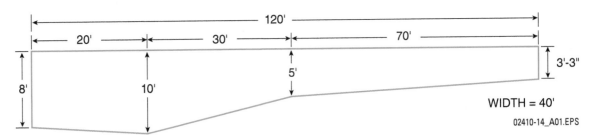

Figure A-1 Swimming pool installation.

Additional Resources

This module presents thorough resources for task training. The following resource material is suggested for further study.

Basic Aboveground Pool Installation. 1995. Don Burger. Hesperia, CA: Superior Publishing.

How to Build a Hot Tub. 1981. Carlton Hollander. New York: Sterling Publishing Company.

The Pool Book: Building and Maintaining Swimming Pools & Spas. 1990. Gerard C. O'Connell. Tucson, AZ: Fisher Books.

International Plumbing Code®, Latest Edition. Falls Church, VA: International Code Council.

International Swimming Pool and Spa Code™, Latest Edition. Falls Church, VA: International Code Council.

Pools & Spas, Second Edition. 2008. David Short and Fran J. Donegan. Mahwah, NJ: Creative Homeowner.

Figure Credits

Courtesy Jim Penney Photo, Module Opener

Pentair Pool Products, Figure 7

Courtesy of Jay R. Smith Mfg. Co. Montgomery, AL, Figure 10

Section Review Answer Key

Answer	Section Reference	Objective
Section One		
1. d	1.1.0	1a
2. b	1.2.2	1b
3. b	1.3.0	1c
4. a	1.4.0	1d
Section Two		
1. a	2.1.2	2a
2. c	2.2.2	2b

NCCER CURRICULA — USER UPDATE

NCCER makes every effort to keep its textbooks up-to-date and free of technical errors. We appreciate your help in this process. If you find an error, a typographical mistake, or an inaccuracy in NCCER's curricula, please fill out this form (or a photocopy), or complete the online form at **www.nccer.org/olf**. Be sure to include the exact module ID number, page number, a detailed description, and your recommended correction. Your input will be brought to the attention of the Authoring Team. Thank you for your assistance.

Instructors – If you have an idea for improving this textbook, or have found that additional materials were necessary to teach this module effectively, please let us know so that we may present your suggestions to the Authoring Team.

NCCER Product Development and Revision
13614 Progress Blvd., Alachua, FL 32615

Email: curriculum@nccer.org
Online: www.nccer.org/olf

❏ Trainee Guide   ❏ Lesson Plans   ❏ Exam   ❏ PowerPoints   Other _____

Craft / Level: _____   Copyright Date: _____

Module ID Number / Title: _____

Section Number(s): _____

Description: _____

_____

_____

_____

_____

Recommended Correction: _____

_____

_____

_____

_____

Your Name: _____

Address: _____

_____

Email: _____   Phone: _____

02411-14

Plumbing for Mobile Homes and Travel Trailer Parks

OVERVIEW

Plumbers install water supply and DWV systems in parks that serve mobile homes and travel trailers. Mobile homes are independent vehicles that can be connected to public utilities for extended periods of time using flexible pipe and quick-disconnect fittings. Travel trailers that are classified as independent vehicles are fitted with storage tanks for fresh water and wastewater. Plumbers also install sanitary dump stations in the parks to receive wastes from travel trailers' holding tanks. Travel trailer parks must also provide bathroom and laundry facilities.

Module Ten

Objectives

When you have completed this module, you will be able to do the following:

1. Describe water supply and DWV systems for mobile home parks.
   a. Describe water supply systems for mobile home parks.
   b. Describe DWV systems for mobile home parks.
2. Describe water supply and DWV systems for travel trailer parks.
   a. Describe water supply systems for travel trailer parks.
   b. Describe DWV systems for travel trailer parks.

Performance Task

Under the supervision of your instructor, you should be able to do the following:

1. Size the water supply and DWV systems for a mobile home park using information provided by your instructor.

Trade Terms

Dependent vehicle
Independent vehicle
Mobile home

Sanitary dump station
Tent trailer
Travel trailer

Industry-Recognized Credentials

If you're training through an NCCER-accredited sponsor, you may be eligible for credentials from NCCER's Registry. The ID number for this module is 02411-14. Note that this module may have been used in other NCCER curricula and may apply to other level completions. Contact NCCER's Registry at 888.622.3720 or go to **www.nccer.org** for more information.

Code Note

Codes vary among jurisdictions. Because of the variations in code, consult the applicable code whenever regulations are in question. Referring to an incorrect set of codes can cause as much trouble as failing to reference codes altogether. Obtain, review, and familiarize yourself with your local adopted code.

Contents

Topics to be presented in this module include:

Figures

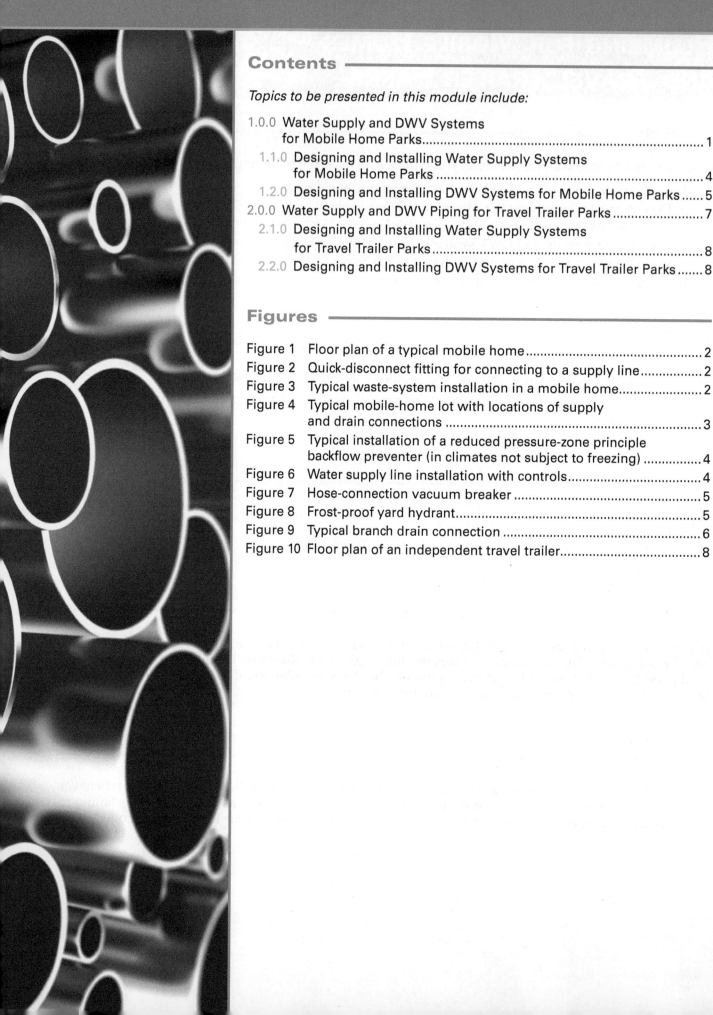

1.0.0 WATER SUPPLY AND DWV SYSTEMS FOR MOBILE HOME PARKS

Objective

Describe water supply and DWV systems for mobile home parks.

 a. Describe water supply systems for mobile home parks.
 b. Describe DWV systems for mobile home parks.

Performance Task

Size the water supply and DWV systems for a mobile home park using information provided by your instructor.

Trade Terms

Independent vehicle: A mobile home or recreational vehicle (RV) equipped with bathroom facilities. Independent vehicles can be connected to public utilities for long periods of time.

Mobile home: A portable dwelling built on a wheeled chassis that has no fixed foundation and that is typically no smaller than 8 feet wide by 32 feet long.

Travel trailer: A portable dwelling that is towed by a motorized vehicle.

A mobile home is a portable dwelling that is built on a chassis without a fixed foundation. Most codes further define a mobile home as a trailer that is at least 8 feet wide and 32 feet long (*Figure 1*). Owners are responsible for maintaining the mobile home's supply and drainage systems. Mobile homes are considered independent vehicles. This means they have bathroom facilities and can be connected to public utilities for extended periods of time, using flexible pipe and quick-disconnect fittings (*Figure 2*). Note that some codes classify mobile homes as permanent dwellings and travel trailers as temporary dwellings. Refer to the local applicable code and follow the classifications that apply to your jurisdiction.

Manufacturers install the plumbing systems in mobile homes. However, plumbers design and install water supply and drain, waste, and vent (DWV) systems in the parks that serve mobile homes. The two sets of systems must be compatible; otherwise, one or the other could be damaged when they are connected. The materials and equipment used are the same as those used in other installations, but the plumbing systems in mobile home parks may differ from familiar systems. Flow rates and pressures differ from other systems, and there are specialized methods for connecting the vehicles to the mains.

Construction standards for mobile homes vary widely. However, they must meet all of the local plumbing codes that are in force where they are installed, and they must use approved pipe and fittings. Their supply and discharge systems must be able to handle the pressures and flow rates of the main system. All connections to public utilities must be in good working order.

Piping for the water supply line must be at least ½ inch in diameter. Some codes require dielectric fittings on water supply lines. The drainage line must be at least 3 inches in diameter. Ensure that the drainage-line connections are both gastight and watertight. The lines must be as short as possible. The waste system can be vented using air admittance vents (*Figure 3*).

The HUD Code

In 1974 Congress passed a law providing for national construction and quality standards for all manufactured homes (a term that includes mobile homes, modular homes, and manufactured homes) in the United States. The National Home Construction and Safety Standards (commonly known as the HUD Code, after the US Department of Housing and Urban Development) is followed nationwide, and preempts all local building codes for manufactured housing. Each manufactured home must have a red HUD label, certifying that the home has been factory constructed, tested, and inspected to comply with strict, uniform federal standards.

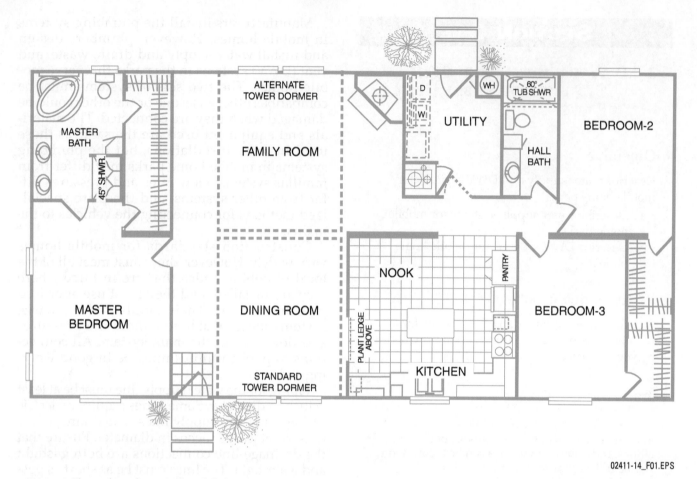

Figure 1 Floor plan of a typical mobile home.

Figure 2 Quick-disconnect fitting for connecting to a supply line.

Figure 3 Typical waste-system installation in a mobile home.

If the mobile home uses natural gas, refer to the local applicable code to select the gas piping. Size the gas system according to the anticipated use; the local code will provide tables that will allow the system size to be calculated. Do not install gas lines under a mobile home unless they are enclosed in a gastight conduit. These conduits are vented at both ends, which allows gas to escape without posing a risk. Each home lot in the park should have a gas shutoff valve. Ensure that the shutoff valve is located in the open air, not under the mobile home. Plug the lot's gas line when it is not in use.

Provide the owner with a maintenance schedule for the plumbing systems. Do not alter water service lines without referring to the local code. In some cases, prior approval from local code officials may be required. Before using a chemical drain cleaner, ensure that it is safe to use on the pipe material. Remember to flush sediment screens and water heaters regularly. These steps should help keep the mobile home's plumbing system trouble free.

When installing plumbing systems in a mobile home park, ensure safe and reliable connections between the main system and each house. The first step is to develop plans and specifications for the installations. Before laying out the components of the plumbing system, complete the plot plan. The plot plan must show the following:

- Individual lots in the park
- Fittings at each lot (see *Figure 4*)
- Nearby roads and utilities
- Park service buildings

Temporary Housing and Natural Disasters

Starting in September 2005, the federal government ordered nearly 25,000 mobile homes and over 114,000 travel trailers from both retailers and manufacturers to provide temporary housing for residents of the Gulf States who were left homeless by Hurricane Katrina. Federal, state, and local governments work hand in hand with builders of mobile and manufactured homes and travel trailers to respond to natural disasters—not only hurricanes, but tornadoes, floods, and wildfires— to provide immediate shelter solutions for those whose homes are damaged or destroyed.

- Specifications and layouts of supply and drainage lines
- Location of the fire protection system
- Site elevations and property lines

Once the plot plan has been drawn up, file copies with the local building officials. Contact the local building office for the proper procedure. Once the plans have been approved, the building officials will issue a permit to begin work. Be sure to follow all applicable local codes, and use approved piping materials and fittings. The following sections provide information on sizing and installing supply and drainage systems that are standard for most mobile home parks. Refer to the local applicable code for specific guidelines.

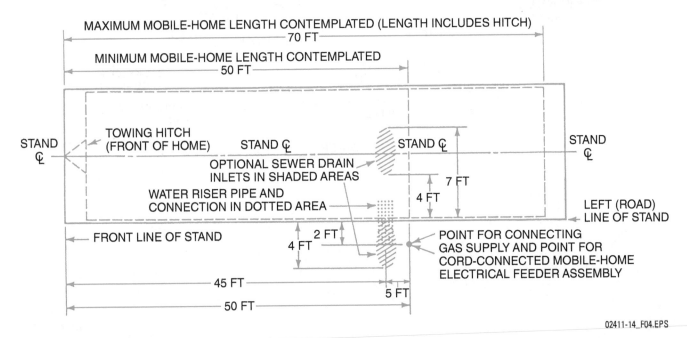

Figure 4 Typical mobile-home lot with locations of supply and drain connections.

1.1.0 Designing and Installing Water Supply Systems for Mobile Home Parks

Size each main supply line according to the number of mobile home lots to be serviced. Most codes require the system to provide at least 150 gallons of water per day to each home. The system must also provide water for the park's own needs, such as lawn care, fire protection, and laundry services. Connect the supply line to the public water supply or an approved private well. Most codes require the supply system to provide water at a minimum pressure of 20 pounds per square inch (psi). In climates that are not subject to freezing, install a reduced pressure-zone principle backflow preventer on the supply line (*Figure 5*); then install a shutoff valve on the supply side of the backflow preventer and a pressure-relief valve on the discharge side.

Install shutoff valves, backflow preventers, and pressure-relief valves on all branch lines. For protection against cold weather, use leak-proof valves and locate them in a freeze-proof enclosure. Install a pressure-reducing valve on each line to an individual lot, after the water meter but before the shutoff valve (*Figure 6*).

Codes require that each mobile home lot be connected to the supply line. The minimum size for this line is usually ¾ inch, but local requirements may vary. The local applicable code will

02411-14_F05.EPS

Figure 5 Typical installation of a reduced pressure-zone principle backflow preventer (in climates not subject to freezing).

specify the minimum rate of flow in gallons per minute (gpm) to be provided to each lot. If the water pressure in the system is low, resize the line to provide the required gpm.

Install a shutoff valve on each line serving an individual lot. Most codes require a backflow preventer or check valve and a relief valve on the discharge side of the shutoff valve (refer to *Figure 6*). Locate the valves near the service connection so they can be accessed easily. Terminate the water supply line on the left (curb) side of the mobile home lot.

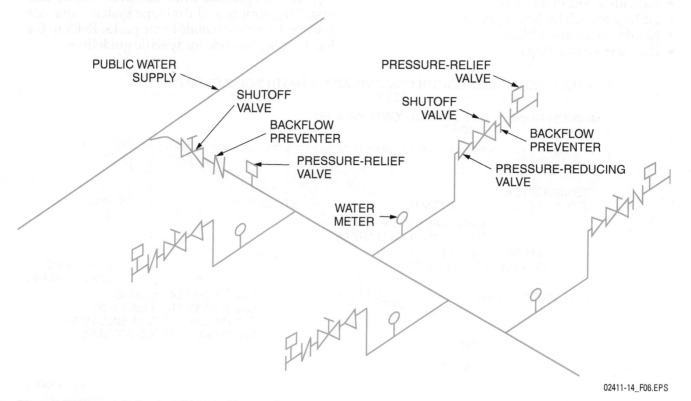

02411-14_F06.EPS

Figure 6 Water supply line installation with controls.

NCCER – *Plumbing Level Four* 02411-14

Install lawn and yard fixtures in accordance with local code. Provide hose-connection vacuum breakers (see *Figure 7*) for all hose bibbs. Yard hydrants should be frost proof in regions where freezing occurs (see *Figure 8*). The hydrant has a small drain at its base, below grade, which empties out of the drain into a gravel bed when the hydrant is shut off. The drain prevents water from freezing in the pipe and shattering it. Set the hydrant in concrete to prevent accidental damage. Refer to the local applicable code for other requirements related to lawn and yard fixtures.

Make every effort to keep the water supply piping as clean as possible before installing it. Sand, gravel, or mud trapped in the pipes will flow through the piping system, lodge in the valves and faucets, and damage the washers and O-rings; the faucets will then leak. Store pipe in a clean, dry area and cap all ends of pipe at the end of each workday. Be sure to flush the line as thoroughly as possible to remove any sand, mud, or gravel from the system.

1.2.0 Designing and Installing DWV Systems for Mobile Home Parks

The process for calculating fixture unit loads in a mobile home park is different from the process used to calculate loads in a building. Normally, the load can be found by totaling up each fixture in the system. In a mobile home park, each home lot has a drain line already installed, even though the number of fixtures in a home can vary widely. Therefore, assign a fixed number of drainage fixture units (DFUs) per lot. Many codes require an estimate of 15 DFUs per lot. Refer to the local applicable code for the correct DFU rating in the local area.

Use pipe of at least 4 inches in diameter for the branch that connects the home to the main drain line. The pipe slope should allow for a minimum flow rate of 2 feet per second. Ensure that all pipes

and fittings are gas- and watertight. The branch drain inlet should extend 3 to 6 inches above grade and should connect to the rear third of the mobile home (refer to *Figure 4*). To protect the inlet pipe from damage, set it in concrete. Install a gas-tight plug on the inlet when the lot is unoccupied. If the branch is longer than 30 feet, install a vent and a cleanout that extends above grade. Install a trap on the branch line (*Figure 9*).

> **WARNING!**
> Toxic and flammable vapors may be present when a DWV line connects to a sewer line. The public sewer is also a biohazard. Wear appropriate personal protective equipment when working with DWV and sewer lines.

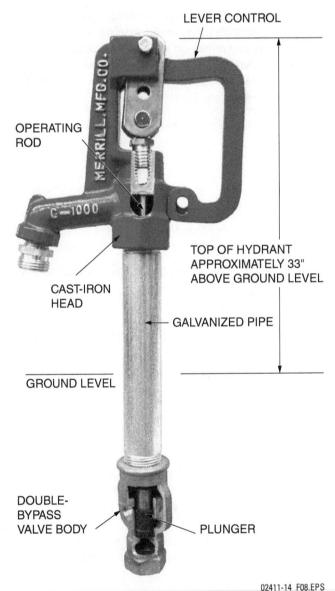

Figure 8 Frost-proof yard hydrant.

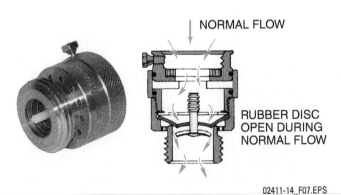

Figure 7 Hose-connection vacuum breaker.

02411-14_F07.EPS

02411-14_F08.EPS

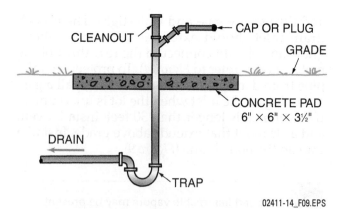

CLEANOUT

CAP OR PLUG

GRADE

CONCRETE PAD
6" × 6" × 3½"

DRAIN

TRAP

02411-14_F09.EPS

Figure 9 Typical branch drain connection.

Multisection Mobile Homes

Multisection mobile homes are made up of two or more separate components that are shipped separately and joined on site. They are becoming increasingly popular. In 2004, multisection mobile homes outsold single-section mobile homes almost three to one.

Size each main drainage line to accommodate all of the mobile homes and buildings that will empty into it. Consider the potential for future expansion of the park when sizing the line. Note that the park's service building must have its own connection. Install vents on each main drainage line according to the local applicable code. Each vent should have a minimum diameter of 4 inches and should be located no more than 5 feet downstream from the highest trapped branch on the line. Install relief vents every 200 feet along the main drainage line and cleanouts at regular intervals. Finally, connect the main drainage line to the public sewer according to local code.

> **WARNING!**
> Check the local code before modifying any building structural members. Improper modification can seriously weaken a building's support structure.

Additional Resources

Code Check Plumbing: A Field Guide to Plumbing, Fourth Edition. 2011. Redwood Kardon, Michael Casey, and Douglas Hansen. Newtown, CT: Taunton Press.

International Plumbing Code®, Latest Edition. Falls Church, VA: International Code Council.

The Manual for Manufactured/Mobile Home Repair and Upgrade. 2002. Mark N Bower. Aberdeen, SD: Aberdeen Home Repair.

Manufactured Housing Site Development Guide. 1993. Welford Sanders. Chicago, IL: American Planning Association.

NFPA 501, *Standard on Manufactured Housing*, Latest Edition. Quincy, MA: National Fire Protection Association.

1.0.0 Section Review

1. Gas lines installed under a mobile home must be _____.

   a. enclosed in a gastight conduit
   b. made from PVC or CPVC
   c. buried at least 6 feet under grade
   d. fitted with check valves at both ends

2. When connecting a mobile home to a main drain line, the diameter of the pipe used in the branch should be at least _____.

   a. 1 inch in diameter
   b. 2 inches in diameter
   c. 3 inches in diameter
   d. 4 inches in diameter

2.0.0 WATER SUPPLY AND DWV PIPING FOR TRAVEL TRAILER PARKS

Objective

Describe water supply and DWV systems for travel trailer parks.

   a. Describe water supply systems for travel trailer parks.
   b. Describe DWV systems for travel trailer parks.

Trade Terms

Dependent vehicle: An RV that is not equipped with bathroom facilities. Dependent vehicles are not intended to be connected to public utilities for an extended time.

Sanitary dump station: A container designed to receive wastes from a travel trailer holding tank.

Tent trailer: A collapsible travel trailer with canvas walls.

Travel trailers are among the most popular types of recreational vehicles (RVs). They are portable dwellings that are towed by a car or truck. Travel trailers come in a wide range of sizes and styles, and many can be collapsed for easy travel and storage. Collapsible trailers with canvas walls are also called tent trailers or pop-up trailers. Most travel trailers can be classified as dependent vehicles, meaning they do not have bathrooms and are not intended to be connected to public utilities for long periods of time. Some trailers are independent vehicles and are fitted with storage tanks for fresh water and wastewater (see *Figure 10*).

As with the systems that serve mobile homes, plumbers are responsible for designing and installing water supply and DWV systems in the parks that serve travel trailers. The plumbing systems used in travel trailers are basic and simple. Use flexible pipe and quick-disconnect fittings when filling or emptying the storage tanks. Codes only permit waste tanks to be emptied at approved dump stations in travel trailer parks. Like

Job-Site Office Trailers

Many construction projects require an on-site office. These offices are towed to the project site at the outset and towed away when the project has been completed. Some portable offices are small with very few amenities, and resemble travel trailers. Others are larger with more amenities, and are more similar to mobile homes. Regardless of the size of the office trailer, plumbers will be called on to connect their water supply and DWV systems once they have been situated on site. Be sure to correctly identify the trailer's inlet and outlet connections before making the connections.

02411-14_SA01.EPS

mobile homes, travel trailers are required to meet the local code of the areas where they are parked. Discourage owners from modifying the trailer's supply and waste lines. If modifications are required, the owner should hire a trained plumber.

The requirements for supply and drainage systems in travel trailer parks are very similar to those for mobile home parks. Travel trailer parks offer only basic facilities because trailers are there for only a short time. Most codes do not require these parks to have individual water and sewer connections for each lot. In most cases, trailers do not even have fixtures that need to be hooked up. Refer to the local applicable code for the standards for safe operation in the local area.

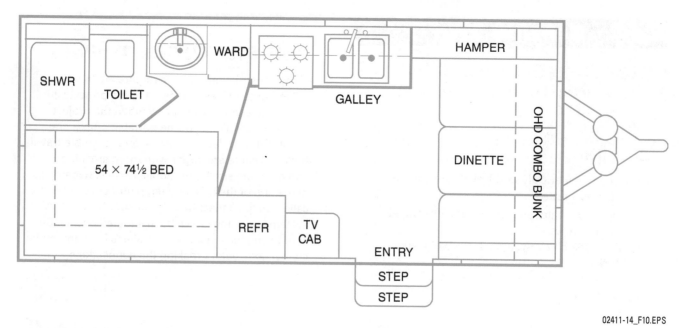

02411-14_F10.EPS

Figure 10 Floor plan of an independent travel trailer.

2.1.0 Designing and Installing Water Supply Systems for Travel Trailer Parks

In most local codes, the regulations governing water supply to mobile home parks also cover travel trailer parks. However, travel trailer parks do not have to supply water at the same rate as mobile home parks. Size the supply system so that it provides water to each branch at a rate of 8 gpm. Individual branches should be a minimum of ½ inch in diameter. Refer to the local applicable code's guidelines for mobile home parks when installing the main supply line. Use only piping materials and fittings that have been approved by the local applicable code.

2.2.0 Designing and Installing DWV Systems for Travel Trailer Parks

Plumbers may be called on to install sanitary dump stations in parks that serve independent vehicles. Dump stations receive wastes from a travel trailer's holding tank. Ensure that the dump tank meets the local code's environmental standards. The hose connection on the dump tank must provide a tight fit with the trailer's waste line. Empty the dump station regularly. The station can be connected to the public sewer, or it can be cleaned by a pumper truck. Some cities provide municipal dump stations that are connected to the public sewer system, making the city responsible for maintaining these systems.

RV Classification

For legal and insurance purposes, RVs are classified into two groups: motorized and towable. The motorized RV category is further broken down into three subcategories:

- Type A: Large motorhomes
- Type B: Van campers
- Type C: Small or mini motorhomes

Towable RVs are broken down into the following subcategories:

- Folding camping trailers
- Truck campers
- Fifth-wheel travel trailers
- Conventional travel trailers

Recreational Vehicles

According to a 2005 study by the University of Michigan, nearly 8 million households in the United States own an RV, an increase of 15 percent from 2000 and 58 percent from 1980. RV rentals are also a popular choice, growing by 36 percent in 2005 alone. People are clearly attracted to the affordability and flexibility of RV travel. In fact, the online auction house eBay reported that in 2004 "RV" was the most popular search term among its 125 million users. Other studies have shown that even with high gas prices, RVs are still the most economical mode of travel.

Travel trailer parks must provide bathroom and laundry facilities. The number and type of facilities depend on the type of trailers being served. If the park serves only independent trailers, minimal facilities are required. If the park serves either only dependent trailers or a mix of both types, then the park must offer services that meet its occupancy level.

Most codes require travel trailer parks to offer at least the following facilities for every 10 trailer lots:

- One laundry tray
- Separate bathroom facilities for men and women
- One shower per bathroom facility

Gray Water and Black Water

Most travel trailers are fitted with separate holding tanks for wastewater, called gray-water tanks and black-water tanks. Gray water is water from sinks, showers, and similar fixtures, while black water is the wastewater from toilets. Both tanks are discharged into the same receptacle at the travel trailer park. Typically, black water will be emptied first.

- One lavatory per bathroom facility
- Two water closets for each women's bathroom facility
- One water closet and one urinal for each men's bathroom facility

If the number of lots in the park is between multiples of 10, round up to the next multiple. For example, if the park has 32 lots, provide facilities for 40 lots. Refer to the local applicable code for specific requirements.

> **WARNING!**
>
> Incorrectly installed sanitary dump stations can cause and spread diseases such as typhoid, scarlet fever, infectious hepatitis, poliomyelitis, and dysentery. Remember, correct installations are essential to public health and safety.

Additional Resources

Code Check Plumbing: A Field Guide to Plumbing, Fourth Edition. 2011. Redwood Kardon, Michael Casey, and Douglas Hansen. Newtown, CT: Taunton Press.

International Plumbing Code®, Latest Edition. Falls Church, VA: International Code Council.

The Manual for Manufactured/Mobile Home Repair and Upgrade. 2002. Mark N Bower. Aberdeen, SD: Aberdeen Home Repair.

Recreation Vehicle Plumbing Systems. 2000. Reston, VA: Recreational Vehicle Industry Association.

2.0.0 Section Review

1. Size the water supply systems that serve travel trailers to provide water to each branch at a rate of _____.

   a. 6 gpm
   b. 8 gpm
   c. 10 gpm
   d. 12 gpm

2. In a travel trailer park, provide a complete set of facilities for every group of _____.

   a. 5 lots
   b. 10 lots
   c. 15 lots
   d. 20 lots

SUMMARY

Plumbers may be responsible for installing water supply and DWV systems for mobile home and travel trailer parks. Though such systems are built with commonly used materials and equipment, they differ from other types of plumbing systems in design and operation. This module discussed how to design and install plumbing systems that service mobile homes and travel trailers.

A mobile home is built on a chassis without a fixed foundation. It is designated as an independent vehicle, which means it can be semipermanently connected to public utilities. It requires water supply lines of at least ½ inch in diameter and drainage lines of at least 3 inches in diameter. Most travel trailers are designated as dependent vehicles, which means that they should not be connected to public utilities for an extended time.

This module stressed the importance of having safe and reliable connections in mobile home parks. Begin installation by developing the plumbing system's plans and specifications. File copies of the plans with the local building officials. Once the building officials issue a permit, size each main supply line to meet the number of lots to be serviced. The water supply should provide at least 150 gallons of water per home per day. Each mobile home lot must connect to the water supply. For DWV systems, refer to the local applicable code to determine the appropriate DFUs per home lot. Use pipe of at least 4 inches in diameter for branches. Slope the pipe to provide a minimum flow rate of 2 feet per second, and install vents as required by code.

Travel trailer park systems are similar to mobile home park systems. However, individual water and sewer connections may not be required for each lot. The supply system should be sized so that it provides water at a rate of at least 8 gpm. Use pipe with a minimum diameter of ½ inch for individual branches, and install sanitary dump stations to store wastes from trailer holding tanks. Install bathroom and laundry facilities to meet the needs of vehicles. Design and install park systems with the same attention to detail that you would give to any plumbing system.

1. A mobile home is defined by most codes as a trailer that has a width of at least 8 feet and a length of at least _____.

   a. 18 feet
   b. 24 feet
   c. 32 feet
   d. 48 feet

2. A mobile home is typically connected to public utilities using flexible pipe and _____.

   a. quick-disconnect fittings
   b. flared fittings
   c. compression fittings
   d. threaded brass fittings

3. Some codes require the installation of dielectric fittings in mobile home park _____.

   a. gas supply lines
   b. drainage lines
   c. water supply lines
   d. water supply and drainage lines

4. Most codes specify that water supply mains in a mobile home park should be sized to provide each home with a daily minimum of _____.

   a. 100 gallons
   b. 150 gallons
   c. 200 gallons
   d. 250 gallons

5. The water supply line serving a mobile home should be provided with a reduced pressure-zone principle _____.

   a. relief valve
   b. surge protector
   c. shutoff valve
   d. backflow preventer

6. Yard hydrants with a drain at the base that empties the pipe when the hydrant is shut off are described as _____.

   a. autodraining
   b. freeze free
   c. frost proof
   d. discharge type

7. Many codes require drainage systems for mobile home parks to be sized using a per-unit allowance of _____.

   a. 11 DFUs
   b. 15 DFUs
   c. 18 DFUs
   d. 24 DFUs

8. A vent and cleanout that extends above grade should be installed on any drainage line longer than _____.

   a. 30 feet
   b. 50 feet
   c. 80 feet
   d. 100 feet

9. A vent on a main drainage line should be located downstream from the highest trapped branch by a distance of no more than _____.

   a. 2 feet
   b. 5 feet
   c. 7 feet
   d. 10 feet

10. Water supply and waste lines of a trailer should only be modified by _____.

    a. the original manufacturer
    b. the trailer owner
    c. a factory-certified technician
    d. a trained plumber

Trade Terms Quiz

Fill in the blank with the correct term that you learned from your study of this module.

1. A(n) _____ is a portable dwelling that is towed by a motorized vehicle.

2. A portable dwelling built on a wheeled chassis that has no fixed foundation and that is typically no smaller than 8 feet wide by 32 feet long is called a(n) _____.

3. A(n) _____ is an RV that is not equipped with bathroom facilities and is not intended to be connected to public utilities for an extended time.

4. A collapsible travel trailer with canvas walls is called a(n) _____.

5. A(n) _____ is a container designed to receive wastes from a travel trailer holding tank.

6. A mobile home or RV equipped with bathroom facilities and which can be connected to public utilities for long periods of time is called a(n) _____.

Trade Terms

Dependent vehicle
Independent vehicle
Mobile home

Sanitary dump station
Tent trailer
Travel trailer

Trade Terms Introduced in This Module

Dependent vehicle: An RV that is not equipped with bathroom facilities. Dependent vehicles are not intended to be connected to public utilities for an extended time.

Independent vehicle: A mobile home or recreational vehicle (RV) equipped with bathroom facilities. Independent vehicles can be connected to public utilities for long periods of time.

Mobile home: A portable dwelling built on a wheeled chassis that has no fixed foundation and that is typically no smaller than 8 feet wide by 32 feet long.

Sanitary dump station: A container designed to receive wastes from a travel trailer holding tank.

Tent trailer: A collapsible travel trailer with canvas walls.

Travel trailer: A portable dwelling that is towed by a motorized vehicle.

Additional Resources

This module presents thorough resources for task training. The following resource material is suggested for further study.

Code Check Plumbing: A Field Guide to Plumbing, Fourth Edition. 2011. Redwood Kardon, Michael Casey, and Douglas Hansen. Newtown, CT: Taunton Press.

International Plumbing Code®, Latest Edition. Falls Church, VA: International Code Council.

The Manual for Manufactured/Mobile Home Repair and Upgrade. 2002. Mark N Bower. Aberdeen, SD: Aberdeen Home Repair.

Manufactured Housing Site Development Guide. 1993. Welford Sanders. Chicago, IL: American Planning Association.

Recreation Vehicle Plumbing Systems. 2000. Reston, VA: Recreational Vehicle Industry Association.

NFPA 501, Standard on Manufactured Housing, Latest Edition. Quincy, MA: National Fire Protection Association.

Figure Credits

Courtesy of S4Carlisle Publishing Services, Module Opener, Figure 7

Oak Creek Homes , Figure 1

Courtesy of Stacy Sabers, Figure 5

Courtesy of Merrill Mfg. Co., Figure 8

Northwood Manufacturing, Figure 10

Courtesy of **GeneralContractor.com**, SA01

Answer	Section Reference	Objective
Section One		
1. a	1.0.0	1
2. d	1.2.0	1b
Section Two		
1. b	2.1.0	2a
2. b	2.2.0	2b

NCCER CURRICULA — USER UPDATE

NCCER makes every effort to keep its textbooks up-to-date and free of technical errors. We appreciate your help in this process. If you find an error, a typographical mistake, or an inaccuracy in NCCER's curricula, please fill out this form (or a photocopy), or complete the online form at **www.nccer.org/olf**. Be sure to include the exact module ID number, page number, a detailed description, and your recommended correction. Your input will be brought to the attention of the Authoring Team. Thank you for your assistance.

Instructors – If you have an idea for improving this textbook, or have found that additional materials were necessary to teach this module effectively, please let us know so that we may present your suggestions to the Authoring Team.

NCCER Product Development and Revision

13614 Progress Blvd., Alachua, FL 32615

Email: curriculum@nccer.org
Online: www.nccer.org/olf

❏ Trainee Guide   ❏ Lesson Plans   ❏ Exam   ❏ PowerPoints   Other _____

Craft / Level: _____   Copyright Date: _____

Module ID Number / Title: _____

Section Number(s): _____

Description: _____

_____

_____

_____

_____

Recommended Correction: _____

_____

_____

_____

_____

Your Name: _____

Address: _____

Email: _____   Phone: _____

02412-14

Introduction to Medical Gas and Vacuum Systems

OVERVIEW

Medical gas and vacuum systems are installed, tested, and serviced by plumbers who are specially trained and certified. This module covers the system requirements and professional qualifications required by code. It describes common types of medical gas and vacuum systems, and introduces the safety requirements for installing, testing, and servicing these systems. It covers special requirements for storing and preparing system components, labeling and identification requirements, techniques for joining copper tubing and fittings, and safety tests required by code.

Module Eleven

Trainees with successful module completions may be eligible for credentialing through the NCCER Registry. To learn more, go to **www.nccer.org** or contact us at **1.888.622.3720**. Our website has information on the latest product releases and training, as well as online versions of our *Cornerstone* magazine and Pearson's product catalog.

Your feedback is welcome. You may email your comments to **curriculum@nccer.org**, send general comments and inquiries to **info@nccer.org**, or fill in the User Update form at the back of this module.

This information is general in nature and intended for training purposes only. Actual performance of activities described in this manual requires compliance with all applicable operating, service, maintenance, and safety procedures under the direction of qualified personnel. References in this manual to patented or proprietary devices do not constitute a recommendation of their use.

Objectives

When you have completed this module, you will be able to do the following:

1. Identify the building system requirements for medical gas and vacuum systems as outlined in the latest edition of NFPA 99, *Health Care Facilities Code,* and the professional qualifications as outlined in the latest edition of ASSE/IAPMO/ANSI Series 6000, *Professional Qualifications Standard for Medical Gas Systems Personnel.*
   a. Explain NFPA 99 building system categories.
   b. Explain ASSE/IAPMO/ANSI Series 6000 professional qualifications.
2. Describe the various medical gas and vacuum systems.
   a. Identify medical gas and vacuum systems.
   b. Identify control panels and alarm systems.
   c. Identify zone valves and station outlets/inlets.
3. Identify the safety issues related to medical gas and vacuum system installation.
   a. Identify personal protective equipment.
   b. Identify fire extinguishers and establish a fire watch.
   c. Explain the use of shutdown and hot-work permits.
   d. Recognize life safety for the end user.
4. Identify the materials and tools for medical gas and vacuum systems and their storage and handling requirements.
   a. Explain how materials and tools are stored and used.
   b. Identify labels and lettering used on medical gas and vacuum systems.
5. Identify the equipment required for brazing copper tube with and without purging.
   a. Explain how to braze a joint with a purge.
   b. Explain how to braze a joint without a purge.
6. Describe the process for testing and verifying a medical gas and vacuum system.
   a. Explain how to conduct installation tests.
   b. Explain how to conduct system verification tests.

Performance Tasks

Under the supervision of your instructor, you should be able to do the following:

1. Braze copper tube with purging.
2. Braze copper tube without purging.

Trade Terms

Anesthetic
Area alarm
Bulk oxygen system
Category 1 system
Category 2 system
Category 3 system
Category 4 system
Charged
Dehydrated
Flowmeter
Hot work

Initial blowdown test
Initial cross-connection test
Initial piping-purge test
Initial pressure test
Instrument air
Life safety
Local alarm
Master alarm
Medical air USP
Medical gas system
Medical-surgical vacuum system

Medical vacuum system
Noncancelable
Oxygen-enriched atmosphere
Pharmaceutical grade
Respiration
Standing pressure test
Standing vacuum test
Transducer
Verifier

Industry-Recognized Credentials

If you're training through an NCCER-accredited sponsor, you may be eligible for credentials from NCCER's Registry. The ID number for this module is 02412-14. Note that this module may have been used in other NCCER curricula and may apply to other level completions. Contact NCCER's Registry at 888.622.3720 or go to **www.nccer.org** for more information.

Code Note

Codes vary among jurisdictions. Because of the variations in code, consult the applicable code whenever regulations are in question. Referring to an incorrect set of codes can cause as much trouble as failing to reference codes altogether. Obtain, review, and familiarize yourself with your local adopted code.

Contents

Topics to be presented in this module include:

Contents (continued)

Figures and Tables

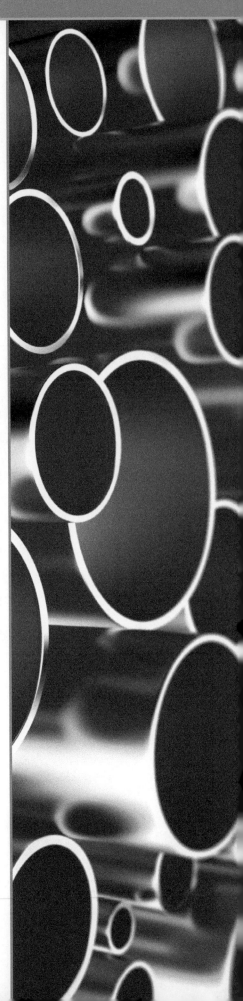

1.0.0 BUILDING SYSTEM REQUIREMENTS AND PROFESSIONAL QUALIFICATIONS

Objective

Identify the building system requirements for medical gas and vacuum systems as outlined in the latest edition of NFPA 99, *Health Care Facilities Code*, and the professional qualifications as outlined in the latest edition of ASSE/IAPMO/ANSI Series 6000, *Professional Qualifications Standard for Medical Gas Systems Personnel*.

a. Explain NFPA 99 building system categories.
b. Explain ASSE/IAPMO/ANSI Series 6000 professional qualifications.

Trade Terms

Anesthetic: A gas that is inhaled to inhibit sensitivity to pain.

Category 1 system: A medical gas system in which equipment or system failure is likely to cause serious injury or even death.

Category 2 system: A medical gas system in which equipment failure is likely to cause minor injury.

Category 3 system: A medical gas system in which equipment failure is unlikely to cause injury but may cause some discomfort to patients.

Category 4 system: A medical gas system in which equipment failure would not affect patient care at all.

Flowmeter: A meter that measures and indicates the volumetric flow rate of a gas or a liquid.

Medical gas system: Equipment and piping that supplies nonflammable medical gases under pressure, along with the associated pumps and compressors, filters and purifiers, valves, alarms, gauges, and controls.

Medical vacuum system: Equipment and piping that uses negative pressure to remove waste gases, liquids, and solids, along with the associated pumps and compressors, filters and purifiers, valves, alarms, gauges, and controls.

Respiration: The breathing cycle.

Verifier: A person who inspects medical gas and vacuum systems and verifies that they meet all applicable standards and codes as well as the manufacturers' specifications.

Hospitals, doctors' and dentists' offices, clinics, and other medical facilities are equipped with special piping systems that supply oxygen, nitrous oxide, helium, carbon dioxide, and on-site compressed air under pressure. These nonflammable gases are used for many things, including the following:

- Drying medical instruments
- Operating pneumatic tools such as booms and drills
- Assisting patients with respiration (breathing)
- Delivering an anesthetic (a gas that helps patients temporarily lose their ability to feel pain)
- Calibrating medical devices that are used for patient respiration and anesthetic delivery

The equipment and piping that supply these gases under pressure, along with their associated pumps and compressors, filters and purifiers, valves, alarms, gauges, and controls, are called medical gas systems. Medical gas systems are often paired with systems that use negative pressure to remove waste gases, liquids, and solids. These systems are called medical vacuum systems. In some ways medical gas and vacuum systems are similar to the way water supply systems and drain, waste, and vent (DWV) systems work, but with gas instead of water.

Medical gas and vacuum systems are installed, tested, and serviced by plumbers who are specially trained and certified to work with these systems. In this section, you will learn about the three main categories of medical gas systems. You will also learn about the professional qualifications that are required to install, inspect, and verify (test) medical gas and vacuum systems.

Always consult your local applicable code before installing, testing, or servicing medical gas and vacuum systems. Never work on a medical gas or medical vacuum system without proper training, certification, and supervision according to your local applicable code. Medical gas can injure and even kill people if the delivery and waste removal systems malfunction or are installed or used improperly. Safety, for yourself and others, must be your primary concern when working around medical gases and medical gas and vacuum systems.

> **WARNING!**
> Never use a medical gas system to supply flammable gases such as natural gas, hydrogen, or acetylene. Medical gas systems are designed for use only with nonflammable gases.

1.1.0 Understanding NFPA 99 Building System Categories

The requirements for medical gas and vacuum systems are outlined in the National Fire Protection Association's model code NFPA 99, *Health Care Facilities Code*. The 2012 edition of NFPA 99 classifies medical gas and vacuum systems into four broad categories. The category to which a medical gas system belongs depends on the health and safety risks that the system presents to patients and medical personnel. The facility owner is responsible for determining the specific category to which medical gas and vacuum systems belong. This is determined by conducting a risk assessment for each system.

The four categories for medical gas and vacuum systems are as follows:

- Category 1 systems, in which equipment or system failure is likely to cause serious injury or even death.
- Category 2 systems, in which equipment failure is likely to cause minor injury.
- Category 3 systems, in which equipment failure is unlikely to cause injury but may cause some discomfort to patients.
- Category 4 systems, in which equipment failure would not affect patient care at all.

NFPA 99 provides detailed design requirements for Category 1, 2, and 3 systems. The installation and testing requirements for each of the three systems is similar, but each system has its own specific requirements as well. Plumbers who install and service medical gas and vacuum systems need to be familiar with all of these requirements to ensure that systems are installed and tested correctly.

This module covers the procedures for installing and testing Category 1, 2, and 3 systems. The following sections provide a brief overview of the three categories covered in the *Health Care Facilities Code* and describe how each is used.

1.1.1 Category 1 Systems

A medical gas and vacuum system is designated a Category 1 system if the failure of the entire system or any part of it could lead to serious injury or death. Category 1 systems are considered essential for providing life support to patients. Therefore, codes require them to be capable of operating or being available at all times. Examples of medical facilities that require Category 1 systems include ambulatory surgical centers with operating rooms, critical care centers, and surgeon's offices equipped with general anesthesia.

1.1.2 Category 2 Systems

A medical gas and vacuum system is designated a Category 2 system if the failure of the entire system or any part of it could lead to minor injury. Category 2 systems do not provide critical life support. Therefore, codes require these systems to be highly reliable, but they also permit them to have brief periods of downtime. Examples of medical facilities that require Category 2 systems include outpatient clinics that provide sedation, and general care rooms.

1.1.3 Category 3 Systems

A medical gas and vacuum system is designated a Category 3 system if the failure of the entire system or any part of it could lead to discomfort, but not to injury. Category 3 systems support patients' needs, but failure of the system would not affect their care. Therefore, codes permit Category 3 systems to have the same reliability as other building systems. Examples of medical facilities that require Category 3 systems include dental offices without general anesthesia, and basic care rooms.

1.2.0 Understanding ASSE/IAPMO/ANSI Series 6000 Professional Qualifications

Medical gas and vacuum systems are installed, inspected, and verified by professionals who have been specially trained and certified to perform these tasks. The minimum training standards that are used today were originally developed by the American Society of Sanitary Engineering (ASSE) in 1987 and adopted by the International Association of Plumbing and Mechanical Officials (IAPMO) and the American National Standards Institute (ANSI). The standard is called ASSE/IAPMO/ANSI Series 6000, *Professional Qualifications Standard for Medical Gas Systems Personnel*. The standard is called a series because it is actually a group of several related standards, one for each of the medical gas specialties. They include:

- Standard #6010: *Medical Gas Systems Installers*
- Standard #6020: *Medical Gas Systems Inspector*
- Standard #6030: *Medical Gas Systems Verifiers*
- Standard #6040: *Medical Gas Systems Maintenance Personnel*

This section provides an overview of the education, training, and general knowledge requirements for each of these categories. Note that the ASSE/IAPMO/ANSI Series 6000 standards provide the minimum requirements for people who work with medical gas and vacuum systems. Local codes may have additional requirements for training and certification above and beyond these standards. Be sure to review and understand the requirements that apply to where you work. Never attempt to install, inspect, test, or perform maintenance on a medical gas or medical vacuum system if you have not received the proper training and certification.

1.2.1 Medical Gas Systems Installer

Plumbers who install equipment and piping for medical gas and vacuum systems, including brazers, must follow the requirements established in ASSE/IAPMO/ANSI Standard #6010 when working in the field. Installers are required to know the applicable codes, laws, and regulations that apply to installing medical gas and vacuum systems where they work. Medical gas systems installers are required to recertify when there are changes to the local applicable code that apply to medical gas and vacuum systems. Brazers must recertify every six months by submitting an affidavit stating that they have brazed according to the requirements laid out in ASTM (American Society for Testing and Materials) B88, *Standard Specification for Seamless Copper Water Tube*, during the previous six months. The company's designated supervisor must sign the affidavit. Otherwise, the brazer must begin the certification process from the start.

Medical gas systems installers may be required to report to the project architect or engineer of record, to job inspectors, the owner of the facility where the systems are being installed, and to the authority having jurisdiction (AHJ). They are also required to demonstrate that they understand the principles, components, and operation of the various types of medical gas and vacuum systems. This includes not only knowledge of the local code requirements, but also detailed knowledge of manufacturers' specifications and requirements for locating and ventilating medical gas and vacuum systems, and ensuring that system components are accessible for service and repair.

Medical gas systems installers should be able to describe basic test procedures and explain how test components are used, but they are only allowed to perform tests associated with the installation of medical gas and vacuum systems.

Verification tests are performed by qualified verifiers. When installation tests are complete, the installer provides detailed written reports on each system test to the facility owner or the AHJ documenting the results of the test.

> **NOTE**
>
> Never allow noncertified personnel to work on the installation of a medical gas and vacuum system under your supervision. You will lose your certification.

Brazers can be qualified under ASSE/IAPMO/ANSI #6010 if they demonstrate the ability to braze copper tube according to the requirement in NFPA 99. They must know how to clean the copper tube and fittings, ensure proper joint clearance and overlap, perform an internal purge, monitor the flow rate of purge gas, and use filler metal properly. The techniques for brazing with and without a purge are covered in detail later in this module.

To be certified as a medical gas systems installer, a plumber must first be able to demonstrate at least four years' experience installing plumbing or other piping systems. Then, the plumber must attend an approved training course of at least 32 hours that covers the ASSE/IAPMO/ANSI #6010 standard and the applicable sections of NFPA 99, and must take and pass a written exam and a practical exam. To become recertified, a medical gas systems installer must attend a refresher course of at least four hours and take an exam on the contents of the latest edition of NFPA 99.

1.2.2 Medical Gas Systems Inspector

Plumbers who inspect medical gas and vacuum system installations are required to follow the requirements established in ASSE/IAPMO/ANSI Standard #6020. As with installers, inspectors are required to know the applicable codes, laws, and regulations that apply to installing medical gas and vacuum systems where they work. They are also required to demonstrate that they understand the principles, components, and operation of the various types of medical gas and vacuum systems. This includes not only knowledge of the local code requirements, but also detailed knowledge of manufacturer's specifications and requirements for locating and ventilating medical gas and vacuum systems, and ensuring that system components are accessible for service and repair.

To qualify as a medical gas systems inspector, a plumber must be able to describe the test procedures required for system components, including (but not limited to) the following:

- Brazed joints
- Welded joints
- Alarm systems
- Medical air system components
- Medical gas manifolds and outlets
- Medical vacuum system components
- Valves used in medical gas and vacuum systems

To be certified as a medical gas systems inspector, a plumber should have at least two years' documented experience as a plumbing and/or mechanical inspector, or as a qualified medical gas systems installer. Candidates are required to take a training course of at least 24 hours on the ASSE/IAPMO/ANSI #6020 standard and the applicable sections of NFPA 99, as well as on applicable sections of NFPA 55, *Compressed Gases and Cryogenic Fluids Code*. In addition, candidates must take and pass a written exam and a practical exam. To become recertified, a medical gas systems inspector must attend a refresher course of at least four hours and take an exam on the contents of the latest edition of NFPA 99.

1.2.3 Medical Gas Systems Verifier

Plumbers who inspect medical gas and vacuum systems and verify that they meet all applicable standards and codes, as well as the manufacturers' specifications, are called medical gas systems verifiers. Verifiers are required to follow the requirements established in ASSE/IAPMO/ANSI Standard #6030. Verifiers must be able to demonstrate knowledge of applicable codes, laws, and regulations, as well as the principles, components, operation, and manufacturers' specifications of the various types of medical gas and vacuum systems.

Codes require verifiers to have no affiliation with either the installers or the health care facilities in which medical gas and vacuum systems are installed. This ensures that the verification tests are done without bias. A medical gas and vacuum system cannot be used until the verifier submits a detailed report approving the system for use. The report is submitted after a series of tests are performed satisfactorily. These tests are discussed elsewhere in this module. The report includes information about any repairs or servicing that was performed, the function of alarm and equipment warning systems, and the circumstances surrounding any retests that were required. Copies of the report are provided to the facility's owner and to the inspector or AHJ.

Medical gas systems verifiers are required to have extensive knowledge about testing the components of medical gas and vacuum systems. This includes the items that installers and inspectors are familiar with, as well as the following:

- Carbon dioxide, oxygen, nitrous oxide, and dew-point analyzers
- Flowmeters
- Gas sampling equipment
- Particulate filters
- Pressure and vacuum gauges

To be certified as a medical gas systems verifier, a plumber should have at least two years' documented experience as a verifier as well as appropriate certificates of general and professional liability insurance. Candidates are required to take a training course of at least 32 hours on the ASSE/IAPMO/ANSI #6030 standard and the applicable sections of NFPA 99, NFPA 55, and the Compressed Gas Association's (CGA) standard M-1, *Guide for Medical Supply Systems at Consumer Sites*. Like installers and inspectors, candidates for the medical gas systems verifier certification must take and pass a written exam and a practical exam. Recertification requires attending a refresher course of at least four hours and taking an exam on the contents of the latest edition of NFPA 99.

1.2.4 Medical Gas Systems Maintenance Personnel

Plumbers who maintain medical gas and vacuum systems that have already been installed, tested, and verified are called medical gas systems maintenance personnel. Maintenance personnel are required to follow the requirements established in ASSE/IAPMO/ANSI Standard #6040. They must also be able to demonstrate knowledge of applicable codes, laws, and regulations, as well as the principles, components, operation, and manufacturers' specifications of the various types of medical gas and vacuum systems.

Medical gas systems maintenance personnel are required to be able to identify system components and describe how they operate, as well as their performance characteristics and safety requirements. They are also required to demonstrate their familiarity with a variety of system components, and describe the various categories of medical gas and vacuum systems identified in NFPA 99. Their knowledge of system maintenance must include the following:

- Requirements of the local applicable code
- System operating requirements
- Locations of system components, ventilation, and access points

Maintenance personnel are required to identify and describe the safety hazards to the system, to patients, and to themselves resulting from improper system maintenance. They are required to describe maintenance and test procedures for medical gas and vacuum systems, as well as the safety precautions that must be taken when working in confined spaces and shutting down electrical power. ASSE/IAPMO/ANSI #6040 also requires maintenance personnel to be able to document and record the status of system components such as valves, connections, alarms, and gas source equipment.

To be certified as a medical gas systems maintenance personnel, a plumber should have at least one year's experience working for a health care facility on the maintenance of its medical gas and vacuum systems at the time of application for certification. Candidates are required to take a training course of at least 32 hours on the ASSE/IAPMO/ANSI #6040 standard and the applicable sections of NFPA 99, and take and pass both a written and a practical exam. Recertification requires attending a refresher course of at least four hours and taking an exam on the contents of the latest edition of NFPA 99.

Additional Resources

ASSE/IAPMO/ANSI Standard #6010, *Medical Gas Systems Installers*, Latest Edition. Washington, DC: American National Standards Institute.

ASSE/IAPMO/ANSI Standard #6020, *Medical Gas Systems Inspector*, Latest Edition. Washington, DC: American National Standards Institute.

ASSE/IAPMO/ANSI Standard #6030, *Medical Gas Systems Verifiers*, Latest Edition. Washington, DC: American National Standards Institute.

ASSE/IAPMO/ANSI Standard #6040, *Medical Gas Systems Maintenance Personnel*, Latest Edition. Washington, DC: American National Standards Institute.

ASTM B88, *Standard Specification for Seamless Copper Water Tube*, Latest Edition. West Conshohocken, PA: ASTM International.

CGA M-1, *Guide for Medical Supply Systems at Consumer Sites*, Latest Edition. Chantilly, VA: Compressed Gas Association.

NFPA 55, *Compressed Gases and Cryogenic Fluids Code*, Latest Edition. Quincy, MA: National Fire Protection Association.

NFPA 99, *Health Care Facilities Code*, Latest Edition. Quincy, MA: National Fire Protection Association.

1.0.0 Section Review

1. Medical gas and vacuum systems that are *not* permitted to have brief periods of downtime are considered _____.

    a. Category 1 systems
    b. Category 2 systems
    c. Category 3 systems
    d. Category 4 systems

2. Knowledge of the requirements in CGA M-1, *Guide for Medical Supply Systems at Consumer Sites*, is required of _____.

    a. medical gas systems inspectors
    b. medical gas systems maintenance personnel
    c. medical gas systems installers
    d. medical gas systems verifiers

SECTION TWO

2.0.0 TYPES OF MEDICAL GAS SYSTEMS

Objective

Describe the various medical gas and vacuum systems.

  a. Identify medical gas and vacuum systems.
  b. Identify control panels and alarm systems.
  c. Identify zone valves and station outlets/inlets.

Trade Terms

Area alarm: An alarm that continually monitors medical gas and vacuum systems that are located in a specific area, such as at a treatment room.

Bulk oxygen system: The equipment used to store oxygen and distribute it to the supply line at system pressure.

Instrument air: Compressed air that is used to power pneumatic tools and to clean instruments; not to be confused with medical air USP.

Local alarm: An alarm that continually monitors one system and is located near the system being monitored.

Master alarm: An alarm that continually monitors the gas supply source and mainline pressure for all systems and is directly connected to each of the devices it monitors.

Medical air USP: Medical air that has been certified as pharmaceutical grade according to the standards of the United States Pharmacopeial Convention.

Medical-surgical vacuum system: A special type of medical vacuum system designed to suction away bodily fluids during surgery.

Oxygen-enriched atmosphere: An atmosphere that contains more than 23.5 percent oxygen by volume.

Pharmaceutical grade: Meeting the specifications for medical purity established by the United States Pharmacopeial Convention.

Transducer: An electrical device that converts an analog signal, such as line pressure, into an electrical signal that can be read by an electrical device such as an alarm circuit board.

Medical gas systems are designed to distribute nonflammable gases from the source, or point of origin, to outlets at the point of use. Medical vacuum systems, on the other hand, are designed to remove waste gases, liquids, and solids from inlets at the point of use and carry them away for safe disposal. The equipment, valves, controls, and other components of medical gas and vacuum systems vary depending on the type of gas the system is designed for. In this section, you will learn about the most common types of medical gas and vacuum systems used in health care facilities. This section also covers the control panels, alarm systems, zone valves, and system inlets and outlets that are commonly used to operate the various systems.

The shape and size of medical gas system outlets and vacuum system inlets vary depending on the manufacturer (*Figure 1*). Outlet shapes can also vary depending on the type of gas. This prevents the accidental connection of tools and equipment to the wrong type of gas. When using a flowmeter to measure the flow rate of gases in a system, therefore, a plumber needs to carry a variety of adapters designed for use with the particular system being tested (*Figure 2*). Never use an adapter on a medical gas system outlet or vacuum system inlet for which the adapter is not designed. You can damage the system, which could result in a gas leak that can cause injury or even death.

2.1.0 Identifying Medical Gas and Vacuum Systems

Medical gas and vacuum systems can be designed to deliver and carry away a wide variety of nonflammable gases that are used for patient respiration, clinical applications, power for pneumatic devices, and calibration. This section describes the features of five medical gas and vacuum systems:

- Oxygen
- Nitrous oxide
- Medical air
- Vacuum
- Waste anesthetic gas disposal (WAGD)

When installing, testing, or servicing the components of a medical gas and vacuum system designed to deliver or dispose of a certain type of gas, never use fittings and components designed

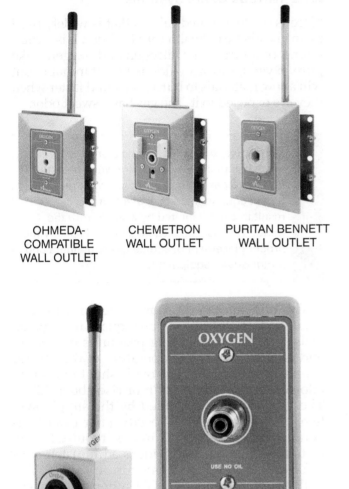

OHMEDA-
COMPATIBLE
WALL OUTLET

CHEMETRON
WALL OUTLET

PURITAN BENNETT
WALL OUTLET

BRITISH STANDARD
OUTLET

DISS-COMPATIBLE
CEILING OUTLET

02412-14_F01.EPS

Figure 1 Common medical gas system outlets.

for a different type of gas. Some fittings are universal, meaning they are designed for use with all types of gas. Unless universal fittings are specifically called for in the system specifications, use fittings and components that are designed for use with that type of gas. Codes specifically prohibit the use of adapters or conversion fittings (sometimes called "cheater" fittings) to adapt one gas-specific fitting to another.

2.1.1 Oxygen Systems

Oxygen is a colorless, odorless, nonflammable gas. Humans and most other living creatures require oxygen to sustain life. Around 21 percent of the

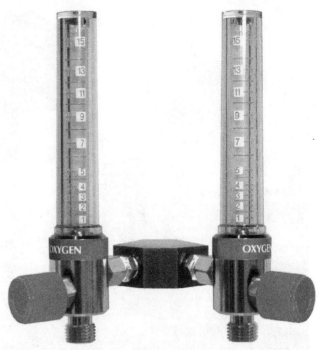

02412-14_F02.EPS

Figure 2 Oxygen flowmeters attached to a dual-port outlet.

air we breathe is made up of oxygen. Health care facilities use pure oxygen to help patient respiration before, during, and after medical procedures. The gas is delivered under pressure to hoses that are placed into a patient's nostrils, masks that are worn over the nose and mouth, and hoods that are placed over the head.

Although oxygen itself is not flammable, oxygen aids the combustion of other flammable gases, liquids, and solids. A material that catches fire in an oxygen-enriched atmosphere, which NFPA 99 defines as an atmosphere that contains more than 23.5 percent oxygen by volume, will burn faster and hotter than it would in regular air. Safety precautions against the accidental release of oxygen must be taken when installing, testing, using, and servicing oxygen systems, in order to prevent the risk of fire.

Oxygen for medical gas systems can be provided as a gas or as a compressed liquid. Both types are stored in tanks (*Figure 3*). Pressure regulators attached to the oxygen tanks reduce the gas to the pressure required by the system. The equipment used to store oxygen and distribute it to the supply line at system pressure is called the bulk oxygen system. From there, the gas flows through manifolds (*Figure 4*) and piping to reach the various outlets where it is to be used. The standards for bulk oxygen systems are covered in NFPA 55, *Compressed Gases and Cryogenic Fluids Code*.

Reproduced with permission from NFPA's *Medical Gas and Vacuum Systems Installation Handbook*, 2012 edition, Copyright © 2011, National Fire Protection Association. This reprinted material is not the complete and official position of the NFPA on the referenced subject, which is represented only by the standard in its entirety.

02412-14_F03.EPS

Figure 3 Liquid oxygen tanks in a bulk oxygen system.

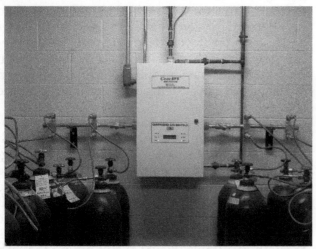

02412-14_F04.EPS

Figure 4 Manifold in a medical gas system.

2.1.2 Nitrous Oxide Systems

Nitrous oxide is a medical gas that is widely used as an anesthetic in health care facilities. It is a compound of oil-free, dry nitrogen and oxygen. Like pure oxygen, nitrous oxide is not flammable but will allow materials to burn hotter and faster when they are exposed to it. The gas has a sweet odor.

> **WARNING!**
>
> Nitrous oxide supplies can be stored near oxygen supplies, but take care when placing these two gas sources in a room with motor-driven equipment. A leak of either gas can result in a fire if ignited by a spark from the motor. The local applicable code will specify the proper placement of source gas containers and motor-driven equipment.

The typical operating pressure for a nitrous oxide system is 5 pounds per square inch gauge (psig) below the operating pressures of oxygen and medical air. The temperature of the central supply system for nitrous oxide should never be allowed to fall below –20°F or rise above 125°F. The specifications provided by the supply system's manufacturer may specify a narrower temperature range. A WAGD inlet is required to be installed where nitrous oxide is administered, to safely carry away the waste nitrous oxide. You will learn more about WAGD systems elsewhere in this module.

> **WARNING!**
>
> When nitrous oxide is used in an anesthesia machine that blends the gas with other gases, never allow the pressure of the nitrous oxide to exceed the pressure of the other gases. In the event of backflow caused by reduced pressure in the other line, nitrous oxide could be allowed to flow through oxygen lines, potentially exposing patients to lethal doses of the gas.

2.1.3 Medical Air Systems

Medical air is purified atmospheric air (78 percent oil-free, dry nitrogen, 21 percent oxygen, and trace amounts of many other gases). It is used to provide support for patient respiration and can be supplied from cylinders, storage tanks, or compressors. Medical air can also be supplied by blending special types of compressed oxygen and oil-free, dry nitrogen. Medical air is sometimes confused with instrument air, which is compressed air that is used to power pneumatic tools and to clean instruments. NFPA 99 classifies instrument air as a medical support gas distinct from medical air.

Medical air must be treated before it can be used. The treatment process removes water and oil from the air. Water in medical gas piping and fittings can freeze when exposed to cold gas, slowing and even blocking the gas flow. Oil can pose a fire hazard if it is exposed to an oxygen-enriched atmosphere. Unlike other medical gases, medical air systems are not required to be vented to the outside.

NFPA 99 requires medical air to be produced at pharmaceutical grade, which means that the gas meets the specifications for medical purity established by the United States Pharmacopeial Convention (USP). The USP publishes the standards for all medicines, health care products, food ingredients, and dietary supplements sold in the United States. The set of standards is called the *United States Pharmacopeia* (also abbreviated USP). Pharmaceutical-grade medical air is therefore called medical air USP. The USP standards for medical air require the following:

- It must be supplied from cylinders, bulk containers, or medical air compressors, or be blended from USP-grade oxygen and oil-free, dry nitrogen
- It must be free of liquid hydrocarbons and have no more than 25 parts per million (ppm) gaseous hydrocarbons
- It must have no more than 6.85×10^{-7} pounds per cubic yard particulate matter larger than one micron across at standard atmospheric pressure

In most cases, ambient (ordinary) air meets or exceeds the USP standards for medical air, so very little needs to be done to prepare air to serve as medical air beyond drying and filtering it at the supply source. Your local applicable code will specify the requirements for medical air where you work. Medical air compressors (*Figure 5*) cannot be used to provide air for applications other than patient respiration and calibrating respiratory equipment. Misuse of medical air compressors can damage the compressors as well as the devices to which they are connected.

Codes require a minimum of two compressors on a Category 1 medical air system. Locate medical air system compressors indoors in a dedicated area equipped with proper ventilation and all required utility connections. The temperature of the equipment space should be maintained according to the manufacturers' specifications. Backflow preventers should be fitted to prevent backflow due to pressure differentials between compressors that are operating and those that are not. Each compressor should be fitted with a manual shutoff valve to isolate it from the medical air system without loss of system pressure. Pressure-relief valves should be set at 50 percent above the line pressure. Always follow the manufacturer's specifications to ensure the materials used on components and pipes are compatible with the compressor.

Medical air compressors that use a reciprocating (piston) engine are required to have an inspection hole that is at least 1.5 times the size of the piston shaft so that a failure in the crankcase seal can be easily spotted. Otherwise, crankcase oil can contaminate the medical air and cause a health or fire risk.

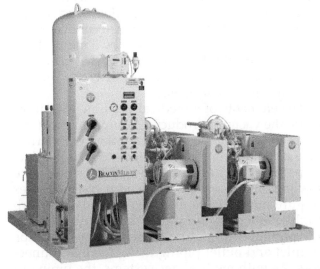

02412-14_F05.EPS

Figure 5 Medical air compressor.

Duplex Systems

Some medical gas systems are designed so that the failure of a single device on the system does not require the entire system to be shut down. Instead, if a device fails, it can be isolated with valves and the gas routed around the device to continue serving other devices further down the line. This is called duplexing. The system designer determines whether to install a duplex arrangement on a particular system.

Codes require medical air system piping to be installed so that continuous service can be maintained if a component fails. The piping should be sized to serve the peak calculated demand with one compressor offline. Codes also require a minimum of two air dryers, filters, and regulators on each medical air system. Dryers, filters, and regulators must be equipped with manual shutoff valves upstream and check valves downstream to allow for servicing. The valves can be installed to allow the component to be either isolated or bypassed, depending on the system design.

2.1.4 Vacuum Systems

Medical vacuum systems (*Figure 6*) use negative pressure to provide suction to remove waste gases away from patient treatment areas to a place where they can be vented safely. A special type of vacuum system is the medical-surgical vacuum system, which is designed to suction away bodily fluids during surgery. Category 3 medical vacuum systems are further classified as either "wet" or "dry" systems. Wet systems remove liquids, air-gas mixtures, and solids from the work area. Dry systems trap liquids and solids before the service inlet and allow only air-gas mixtures to pass through the inlet. The inlets used in Category 3 vacuum systems are called service inlets, while the outlets used in Category 3 medical gas systems are called service outlets. The different types of inlets and outlets are discussed in the section *Station Outlets/Inlets*.

The elements of a medical vacuum system include the vacuum-producing equipment, controls for pressure and operation, shutoff valves, alarms, gauges, and piping that extends to station inlets in patient care areas. Pumps are used to create low pressure in a medical vacuum system. Codes require pumps to be mounted on pads to absorb high-frequency vibrations that could damage the piping, fittings, and valves. Connect pumps to inlet and outlet piping using flexible connectors. As with medical gas systems, the piping in

a Category 1 system should be installed to ensure continuous service in the event of a component failure.

Pump motors must be fitted with dedicated disconnect switches, motor starters, overload protection, backup transformers or voltage controls, and automatic restart. Control circuits for vacuum pumps should be wired so that the failure of a pump will not interrupt the operation of the other pumps. Medical vacuum systems are required to have their exhaust lines run outdoors at least 10 feet from any doors, windows, or other openings or air intakes. They must be located at a different level than air intakes to avoid accidental re-ingestion of waste gases into the building. The end of the exhaust piping should be turned to face down toward the ground, and screened. This will prevent entry by pests, dirt, and rain or snow. Refer to your local applicable code for procedures applicable to your area.

Install exhaust piping to avoid dips or loops that would allow condensate oil to collect. Install a drip leg and valved drain at the lowest point of the exhaust line. Codes allow the exhaust piping from multiple pumps to be joined through a manifold into a single exhaust line, provided the line is sized properly according to the local applicable code. This can save time and money during the installation process.

Reproduced with permission from NFPA's *Medical Gas and Vacuum Systems Installation Handbook*, 2012 edition, Copyright © 2011, National Fire Protection Association. This reprinted material is not the complete and official position of the NFPA on the referenced subject, which is represented only by the standard in its entirety.

02412-14_F06.EPS

Figure 6 Medical vacuum system.

2.1.5 Waste Anesthetic Gas Disposal (WAGD) Systems

Waste anesthetic gas disposal (WAGD) is the capture and removal of vented gases exhaled by patients when under anesthesia. WAGD is sometimes called scavenging or evacuation, though NFPA 99 discourages the use of those terms because medical professionals use them to refer to other things. WAGD systems are considered by code to be part of the medical-surgical vacuum system. Depending on the system design, WAGD may share the same piping as the medical-surgical vacuum system, or it may be a separate system. However, NFPA 99 encourages the use of separate WAGD systems for various reasons, including the following:

- The maximum safe negative pressure for patient respiratory systems is much lower than the minimum negative pressure of a standard medical-surgical vacuum system.
- The alarm systems required for medical-surgical vacuum systems cannot be used to monitor WAGD systems.
- High airflows between the two systems may cause problems for vacuum pumps.

> **WARNING!**
>
> Never label WAGD terminals "suction" or use suction terminals for WAGD.

WAGD systems may be located adjacent to the sources for medical air compressors, medical-surgical vacuums, and instrument air. Ensure that pumps and motors that provide the source negative pressure for WAGD systems are adequately ventilated. NFPA 99 distinguishes between WAGD systems based on the horsepower of the engine or motor producing the low pressure in the system. Systems that use an engine or motor with a total power output of less than 1 horsepower may be located near the inlet(s) they serve. Systems with engines or motors that produce 1 horsepower or more are considered central systems and must be installed according to the same local applicable code requirements used for locating central supply systems.

Vacuum pumps used to provide WAGD service are required to meet the same criteria as pumps used for medical-surgical vacuum systems. They must be made from materials that are inert (meaning they won't react chemically) in the presence of medical gases. If a WAGD system is to be connected to another vacuum system, the connection must be at least 5 feet from any vacuum inlet. Install WAGD exhaust systems using the same requirements that apply to medical-surgical vacuum systems.

The location of WAGD inlets will vary depending on the needs of the facility. The growing use of injected, rather than gas, anesthetics means that patient treatment rooms may not require WAGD piping. However, NFPA 99 recommends that system designers designate at least one location in the facility where WAGD will be provided, in case it is needed.

2.2.0 Identifying Control Panels and Alarm Systems

Medical gas and vacuum systems are fitted with controls and alarms that allow operators to monitor and control their operation. They are also equipped with alarm systems that indicate when line pressure rises above or falls below the specified range. Certified medical gas and vacuum systems installers must install control panels and alarm systems. The local applicable code will specify the minimum standards for control panel and alarm systems in medical gas and vacuum systems. Additional requirements may be specified in the system design specifications.

This section provides a general overview of control panels and alarm systems that are commonly used on medical gas and vacuum systems. The controls and layout of the control and alarm panels will vary depending on the manufacturer. Always follow the manufacturer's instructions when installing, testing, and servicing a control panel or alarm panel. Follow all required safety precautions such as turning off the power and disconnecting alarms. Ensure that the medical gas or vacuum system can continue to operate safely and reliably during servicing.

2.2.1 Control Panels

Manufacturers provide control panels for medical gas and vacuum systems as pre-assembled units that can be roughed in and joined to the system piping according to instructions provided by the manufacturer (*Figure 7*). Control panels may include the following components:

- Gauges that indicate supply and outlet pressures
- Outlet pressure regulator
- Supply-valve open/close switch
- Gas outlet for testing and piping to additional outlets

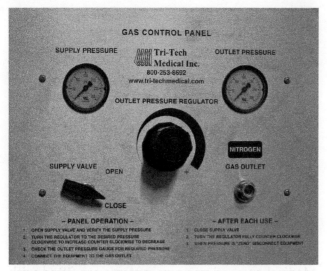

Figure 7 Typical medical gas system control panel.

The location of a medical gas and vacuum system control panel will be specified in the design drawings. When installing the control panel box, ensure that it is supported on rigid mounting on the sides or the top and bottom. The outlet and inlet extensions on the box should be brazed to the system lines using the techniques discussed in the sections on brazing elsewhere in this module.

Medical gas and vacuum system sources such as compressors, pumps, and air-drying equipment are also equipped with control panels that are specifically used to control those devices. These control panels are attached to the source equipment and are not installed by the plumber. All controls must be checked and tested to ensure they are working properly before the control panel can be put into service.

2.2.2 Alarm Systems

Alarm systems are used to warn operators when part of a medical gas and vacuum system is functioning incorrectly, or is no longer functioning. Alarms provide both a visual warning (for example, a flashing light) and an audible warning (for example, a horn or buzzer) when there is a problem. Alarm systems also indicate when a system has switched to the reserve gas supply, which requires the primary supply to be recharged. NFPA 99 specifies three types of alarms for Category 1 and Category 2 medical gas and vacuum systems:

- **Master alarms**, which continually monitor the gas supply source and mainline pressure for all systems and are directly connected to each of the devices they monitor (*Figure 8*).
- **Area alarms**, which continually monitor medical gas and vacuum systems that are located in a specific area, such as at a treatment room (*Figure 9*).

Figure 8 Typical master alarm panel.

Where Do You Need an Alarm System?

NFPA requires the installation of alarm systems in every critical care or vital life-support area, but the definition of those areas can vary from facility to facility. Thanks to advances in medical technology, many activities that required doctors to administer medical gas can now be done without the use of any medical gas at all. That can make it hard for designers to know where to place alarm systems.

The NFPA's *Medical Gas and Vacuum System Installation Handbook* offers designers a new criterion for determining whether to install a medical gas or medical vacuum system. According to NFPA, the designer should ask: "If gases were to fail in the area or room during a procedure, would the procedure have to terminate prematurely? Is it possible that patient outcomes would be negatively impacted?" If the answer is yes, then the room can be used to provide critical care that includes the use of medical gas and vacuum systems, and thus the room warrants the installation of an alarm system.

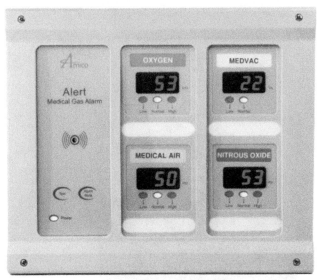

ANALOG AREA ALARM PANEL

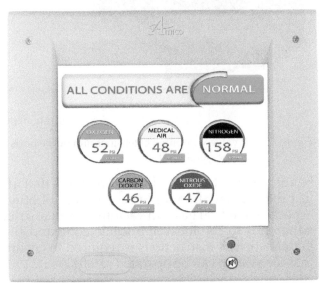

DIGITAL AREA ALARM PANEL

02412-14_F09.EPS

Figure 9 Typical area alarm panels.

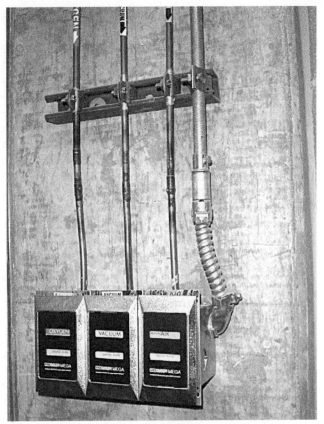

02412-14_F10.EPS

Figure 10 Typical local alarm panel.

- Local alarms, which continually monitor one system and are located near the system being monitored (*Figure 10*).

Alarm systems in Category 3 medical gas and vacuum systems are different from those used in Category 1 and Category 2 systems. These alarms are designed to indicate high and low pressures and changeovers to secondary systems on oxygen and nitrous oxide systems. Pressure switches and sensors for Category 3 alarm systems are fitted at the source equipment and downstream from shutoff valves.

NFPA 99 requires all master, area, and local alarms to have visual indicators for each component or system being monitored, and visual alarm indicators that must be manually switched off when the problem is resolved. Alarm systems must also have audible alarms that can be turned off at the discretion of the operator, and which produce a sound of at least 80 decibels at 3 feet. However, if a new alarm condition occurs after the audible alarm has been manually turned off, the audible alarm must be able to override the shutoff and provide an audible warning of the new problem.

There must be a way to identify whether a warning light has failed and when the alarm has been disconnected to a device being monitored. All indicators must be clearly labeled, as well as the area of coverage of the alarm panel itself. When room numbers or names change, the labels on the alarms must also be changed to ensure that the alarm coverage is properly identified.

Electrical power for alarms and sensors should be provided by the same electrical branch that powers the air compressor system. Always follow the local applicable electrical code when installing alarm wiring, including protection against fire. Alarm systems should also be equipped with an automatic restart capability after a power loss of 10 seconds or longer.

Codes require the installation of two separate master alarm systems, one in the nurse's station and the other in another location that is constantly staffed, such as a security office, maintenance office, or telephone switchboard. This is to ensure that there is always someone monitoring the alarm system who knows what to do in the event of an alarm situation. Many new codes now allow the installation of a central computer system for one of the master alarm panels. The signals from one master alarm panel cannot be relayed to another.

Area alarms monitor medical gas, medical-surgical vacuum, and WAGD systems where anesthesia and critical care services are provided. Area alarms provide visual and audible warning signals for high and low line pressure, low medical-surgical vacuum, and low WAGD vacuum. Area alarms are placed in areas that are constantly staffed, such as nurses' stations.

Local alarms are installed near air compressors, medical-surgical vacuum pumps, WAGD systems, instrument air, and proportioning equipment to monitor their functions. They are designed to signal the failure of the main compressor, vacuum pump, or proportioning system; high carbon monoxide levels; high dew point in the mechanical air system; and abnormally high temperature in the discharge air. Master alarm panels are required to have at least one warning signal to indicate an alarm condition with source equipment.

Alarm panels may include a power supply, line pressure gauges, an alarm circuit board, and transducers that convert line pressure into electrical signals that can be read by the alarm circuit board. Displays indicate the pressures on the various lines and the status of the line. An audible alarm speaker will sound if the status becomes abnormal. The alarm control panel will include buttons to test and silence the alarm, and keys to program the various functions. Sensors for area alarms should be located in critical care areas alongside zone-valve boxes.

2.3.0 Identifying Zone Valves and Station Outlets/Inlets

Zone valves are quarter-turn ball valves designed to isolate station outlets and inlets in a medical gas and vacuum system (*Figure 11*). Station outlets and inlets are connected to terminal points along the medical gas and medical vacuum lines to which instruments and devices can be connected. In this section, you will learn how zone valves and station outlets and inlets work, and you will

02412-14_F11.EPS

Figure 11 Ball valve with fitted extensions and purge ports used for gas shutoff.

be introduced to the code requirements that affect their installation and use. As with all other aspects of medical gas and vacuum systems, strict adherence to the local applicable code and the manufacturer's instructions is required to ensure that the system can be operated safely.

2.3.1 Zone Valves

NFPA 99 does not define the term *zone*, but in practice a zone is considered to be one branch of outlets or inlets serving a designated area. A zone valve, therefore, is designed to quickly isolate a set of outlets or inlets serving a zone from the rest of the system in the event of a malfunction or an emergency. Codes require the installation of zone valves on all station outlets and inlets. Zone valves are accessible through zone valve boxes (*Figure 12*), which may house a single valve or multiple valves depending on the design of the system.

NFPA 99 requires zone valves to be placed in easily accessible areas where they are visible at all times. They must also be installed so that there is a wall between the valve and the outlets or inlets that it controls. This is to allow a person to operate the valve from behind a safe barrier that protects the person against fire, smoke, and escaped gas. If a zone has more than one operating room, a separate shutoff valve is required for each room. Zone valves must be installed on the same floor as the zone it controls. Closing a zone valve should not affect the supply of medical gas or vacuum to other zones in the system. Never place zone valves in series.

Install a pressure or vacuum indicator gauge on the outlet or inlet side of the zone valve. The zone valve should be located upstream of all gas delivery columns, hoses, booms, medical instrument pendants, and other equipment served by the outlets and inlets in the particular zone. Additional in-line shutoff valves may be installed to disconnect gas and vacuum service to individual rooms or areas within a zone.

In certain types of facilities where security may be an issue, such as in pediatric hospitals or psychiatric wards, zone-valve boxes may be modified to protect against vandalism or damage. Acceptable forms of modification include lockable valve-box doors or location of the valve box inside a locked room.

2.3.2 Station Outlets/Inlets

Medical gas outlets and vacuum system inlets may be located on walls or ceilings depending on the design of the system and the equipment that the outlet is designed to serve. To help prevent accidental use of the wrong gas, the shape of the outlets and inlets for each gas type are mechanically different. This way the wrong hose cannot be connected (refer to *Figure 1*). Codes require outlets and inlets to be placed high enough so that they will not damage equipment connected to them, or be damaged by furniture or equipment in the room such as the headboard of a patient bed. Each outlet or inlet must be clearly labeled according to the labeling standards discussed in the section *Using Labels and Identification*. The local applicable code will specify the dimensions of the copper tubes used on system inlets. Never allow a station outlet to be used for anything other than the delivery of medical gas.

When locating station outlets and inlets, be sure to account for the equipment that will be con-

nected to them, and allow sufficient space between outlet or inlet and the space above and below the outlet or inlet. WAGD inlets should be provided wherever anesthetic gas is administered. They should be clearly noninterchangeable with other medical gas and vacuum system outlets or inlets and clearly marked for WAGD service.

SINGLE-ZONE
VALVE BOX

MULTIPLE-ZONE
VALVE BOX

02412-14_F12.EPS

Figure 12 Zone valve boxes.

Additional Resources

NFPA 55, *Compressed Gases and Cryogenic Fluids Code*, Latest Edition. Quincy, MA: National Fire Protection Association.

NFPA 99, *Health Care Facilities Code*, Latest Edition. Quincy, MA: National Fire Protection Association.

United States Pharmacopeia, Latest Edition. Rockville, MD: United States Pharmacopeial Convention, Inc.

2.0.0 Section Review

1. Category 3 medical vacuum systems are classified as either _____.

   a. medical air or instrument air systems
   b. outlet or inlet systems
   c. high-pressure or low-pressure systems
   d. wet or dry systems

2. Alarms that continually monitor medical gas and vacuum systems that are located in a specific area are called _____.

   a. local alarms
   b. area alarms
   c. master alarms
   d. Category 3 alarms

3. NFPA 99 requires that zone valves be installed so that the valve and the outlets or inlets that it controls are separated by _____.

   a. at least one floor
   b. a wall
   c. fireproofing
   d. at least 3 feet

SECTION THREE

3.0.0 MEDICAL GAS SYSTEM INSTALLATION SAFETY

Objective

Identify the safety issues related to medical gas and vacuum system installation.

a. Identify personal protective equipment.
b. Identify fire extinguishers and establish a fire watch.
c. Explain the use of shutdown and hot-work permits.
d. Recognize life safety for the end user.

Trade Terms

Hot work: Work performed on heated metal.

Life safety: Essential for ensuring the safety of patients.

The safety of patients and medical personnel is a primary concern when working with medical gas and vacuum systems. Exposure to gas in sufficient quantities can cause injury or even death. The pneumatic tools that use instrument air can injure patients and medical personnel if the pressure is too high or too low. The failure of a pump, compressor, or other mechanical device at the system source can cause a fire. Improper exhaust can cause suffocation. As a plumber, you know how to work safely, but medical gas and vacuum systems require additional safety precautions to ensure the system is not damaged and that patients and medical personnel remain safe. Whether installing, testing, or servicing components in a gas or vacuum system, it is important to follow the safety requirements provided in the local applicable code.

In this section, you will learn about the personal protective equipment that should be used when installing, testing, and servicing medical gas and vacuum systems. You will also learn about proper fire prevention and fire safety techniques that should be followed when working with these systems. This section also covers permits that are required for shutting down an existing medical gas or vacuum system, and for performing hot work, which, as its name suggests, is work performed on metal while it is being heated. Finally, this section addresses the important issue of life safety for users of medical gas and vacuum systems. Always refer to your local applicable code for the applicable standards that apply to medical gas and vacuum system installations in your area. Remember that the standards in NFPA 99 are minimum standards established for safe operation, and that local codes may be more stringent. When working with medical gas, strive to be as safe as possible.

3.1.0 Using Personal Protective Equipment

As you are aware, personal protective equipment (PPE) is designed to protect you from injury. Many plumbers are injured on the job because they do not use PPE. Working around medical gas and vacuum systems requires the use of PPE for brazing and working with metals. PPE will greatly reduce your chances of getting hurt, but the best PPE is of no use unless you do the following:

- Inspect PPE regularly, and replace any item that is damaged or worn.
- Care for it properly.
- Use it properly when it is needed.
- Avoid altering or modifying PPE in any way.

Wear eye protection (*Figure 13*) when installing, testing, and servicing medical gas and vacuum systems. There are different types of eye protection. Safety goggles give your eyes the best protection from all directions. Regular safety glasses will protect you from objects flying at you from the front, such as large chips, particles, sand, or dust. Face shields offer better protection than safety glasses, and protect your entire face. You can add side shields for further protection. You may substitute prescription safety glasses if they provide the same protection as regular safety glasses. If you are welding, you must use safety goggles with tinted lenses or a welding hood.

02412-14_F13.EPS

Figure 13 Tinted safety glasses.

Tinted lenses protect your eyes from the bright flame when brazing copper pipe. Tinted eye protection should have a tint value of 3 or 4.

Wear heavy-duty gloves to protect your hands from cuts and burns when brazing pipe and fittings (see *Figure 14*). Work gloves are usually made of cloth, canvas, or leather, but they may also be made of metal mesh or Kevlar®, which protects against metal cuts. Never wear cloth gloves around rotating or moving equipment. Inspect your gloves every day to ensure that they are in good condition. Immediately replace gloves that are worn, torn, or no longer fit. Wear long sleeves to protect your arms.

In addition to a visual test, an effective way to check rubber gloves is to conduct an air test. The following steps are outlined in 29 *CFR* (*Code of Federal Regulations*) 1910.137:

Step 1 Hold the glove by its cuff, flip it several times to make a seal, and roll the glove toward its fingers. An air pocket will form inside the glove.

Step 2 Hold the rolled portion of the glove tightly. Inspect the inflated exterior of the glove for cracks or degradation of the insulating material surface. Forcing air into the glove will expose damage to the insulation that cannot be seen during visual inspections.

Step 3 Inspect the glove for holes in the insulating material. Hold the glove close to your ear. If you hear air escaping from the glove or if the glove will not hold pressure, the glove is damaged. Damaged gloves must be removed from service.

> **CAUTION**
>
> Protective equipment that is good one day may be damaged the next. Always do a daily visual inspection of PPE to prevent injuries from defective or worn equipment.

> **WARNING!**
>
> Always thoroughly wash and disinfect your hands if you have been exposed to sewage, even if you were wearing gloves. Doing so can prevent disease or illness. In general, it is best to use an alcohol gel instead of an antibacterial cleaner. Recent research indicates that overuse of antibacterial cleaners may make bacteria more resistant to antibacterial agents.

3.2.0 Using Fire Extinguishers and Establishing a Fire Watch

Always have an approved fire extinguisher available before starting your work. You are not expected to be an expert firefighter, but you may experience a fire and find it necessary to protect your safety and the safety of others. You need to know the locations of fire fighting equipment when performing hot work, as well as which equipment to use on different types of fires. Remember that only qualified personnel are authorized to fight fires.

Most companies ensure that new employees know where fire extinguishers are kept. If you have not been told, be sure to ask. Also find out your company's policy for reporting fires and their fire safety procedures. The telephone number of the nearest fire department should be clearly posted in your work area. If your company has a company fire brigade, learn how to contact them. Know what type of extinguisher to use for different kinds of fires and how to use them. Make sure all extinguishers are fully charged. Never remove the tag from an extinguisher, as it shows the date the extinguisher was last serviced and inspected (see *Figure 15*).

02412-14_F14.EPS

Figure 14 Work gloves.

Four classes of fuels can be involved in fires: liquids (code A), gases (code B), ordinary combustibles (code C), and combustible metals (code D). Each class of fuel requires a different method of fire fighting and a different type of extinguisher. The label on a fire extinguisher clearly shows the class of fire for which it can be used (*Figures 16* and *17*).

When you check the extinguishers in your work area, you will see that some are rated for more than one class of fire. You can use an extinguisher coded ABC to fight a class A, B, or C fire. If the extinguisher has only one code letter, it cannot be used on any other class of fire, even in an emergency. You could make the fire worse and put yourself in great danger.

Post a fire watch when you are brazing or cutting. One person other than the welding or cutting operator must constantly scan the work area for fires. Fire-watch personnel should have ready access to fire extinguishers and alarms and know how to use them. Welding and cutting operations should never be performed without a fire watch. The area where welding is done must be monitored afterwards until there is no longer a risk of fire.

3.3.0 Using Shutdown and Hot-Work Permits

When working on medical gas and vacuum systems, it is important to complete and submit an appropriate permit before shutting the system down or performing hot work. The permit form should be completed and submitted to the authorized facility representative, typically a director of facility services, for approval. The permit request must be approved before the work can take place. Refer to the local applicable code for the standards that apply in your area. In some cases, you may also be required to post notices to inform facility staff that the permitted work will be performed. These notices will allow people to stay safely away from the work area and also to avoid using the system being worked on.

Shutdown permits (*Figure 18*) include a checklist of the systems that will be shut down during the work. The health care facility will specify the date and time of the shutdown on the permit. The contractor will be asked to identify the location and expected duration of the shutdown. If a detailed description of the work to be performed is requested, write out the steps.

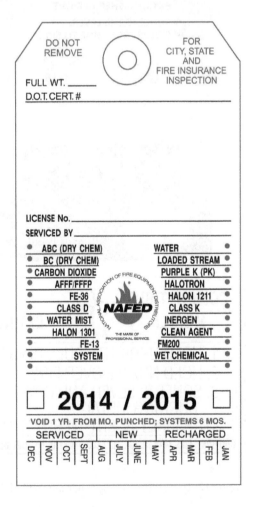
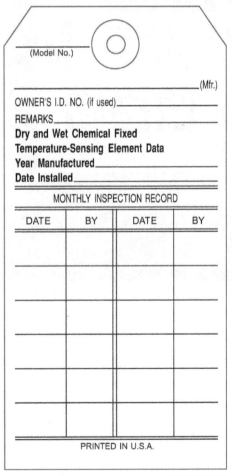

02412-14_F15.EPS

Figure 15 Fire-extinguisher tag.

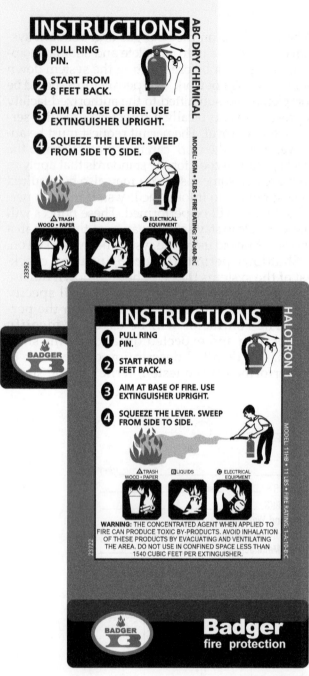

Figure 16 Typical fire extinguisher labels (1 of 2).

02412-14_F16.EPS

COMBUSTIBLE METALS | **D** | **FIRE EXTINGUISHER**

1. HOLD UPRIGHT - PULL RING PIN.
2. HOLD NOZZLE OVER FIRE.
3. SQUEEZE LEVER AND COVER ALL BURNING METAL
4. REAPPLY AGENT TO HOT SPOTS.

CAUTION: FIRE MAY RE-IGNITE, ALLOW METAL TO COOL BEFORE CLEANUP.

02412-14_F17.EPS

Figure 17 Typical fire extinguisher labels (2 of 2).

Spring Hills Health Care Facility
Utility System Shutdown Permit

Personnel performing installation, maintenance, or repair work on any utility system supplying a SHHC building will be required to apply for a written shutdown permit and receive approval for said shutdown. This permit must be completed and submitted to the Director of Facilities Services or his/her representative in a timely manner as indicated by the attached notification matrix. Utility systems will not be shut down under any circumstance without a properly completed and approved permit. All departments effected must receive prior notification. A separate shutdown permit must be submitted for each unrelated utility system manipulation.

Note: All valves for any Infrastructure system are to be opened and/or closed by Facility Services personnel only unless specific written consent is given.

Project Name:

Check the box of the utility system the shutdown is being requested for:

	Electric - Normal Power		Fire Alarm System
	Electric - Emergency Power		Mechanical Air
	Chilled Water		Sanitary Sewer
	Medical Gas: Specify_____		Exhaust System
	Medical Vacuum		Air Handlers
	Steam and Condensate		Communications - Nurse Call
	Heating Hot Water		Communications - Telephone
	Domestic Hot Water		Communications - Public Address
	Domestic Cold Water		Communications - Data
	Fire Sprinkler System		Air Distribution
	Fire Suppression		Other:
	Tube System		Other:

1. Date of Shutdown: _____   2. Time of Shutdown: _____

3. Duration of Shutdown: _____   4. Specific Location of Shutdown: _____

5. Provide a detailed description of the work: (note - attach map & POA if appropriate)

_____

_____

_____

Page 1 of 3

02412-14_F18A.EPS

Figure 18A Example of a shutdown permit. (1 of 3)

Spring Hills Health Care Facility
Utility System Shutdown Permit

6. List all building(s), floor(s), room(s), and department(s) that will be impacted by the shutdown.

7. Will mitigation measures be required to accommodate utility system loss during the shutdown? If yes, attach mitigation plan.

8. Are any special inspections or testing required? If yes, list inspector, company, and contact phone number.

9. Are there any post disruption/connection procedures required? (such as chemicals, flushing, air elimination, etc.) Please describe.

10. Are any other permits or notifications required for this work?

    a. Hot Permit - _____   d. ILSM - _____

    b. Over Ceiling Permit- _____   e. Fire Department - _____

    c. ICRA Permit - _____   f. Other - _____

11 . Has a contingency plan for a prolonged outage or other factors been developed? If yes, please attach plan or describe. Please include backup supplies.

Page 2 of 3

02412-14_F18B.EPS

Figure 18B Example of a shutdown permit. (2 of 3)

Spring Hills Health Care Facility
Utility System Shutdown Permit

I have received, read, and fully filled out the **Spring Hills Health Care Facility Utility System Shutdown Permit.** I will ensure all personnel are properly trained, have appropriate equipment, and shall be responsible for the successful shutdown and reactivation of the said utility systems.

_____
General Contractor / Company Name

_____
Date Request Submitted

_____
General Contractor / Personnel Name

_____
Contact Telephone Number

_____
General Contractor / Company Name

_____
Date Request Submitted

_____
General Contractor / Personnel Name

_____
Contact Telephone Number

_____
Subcontractor / Company Name

_____
Date Request Submitted

_____
Subcontractor / Personnel Name

_____
Contact Telephone Number

_____
Subcontractor / Company Name

_____
Date Request Submitted

_____
Subcontractor / Personnel Name

_____
Contact Telephone Number

REQUIRED APPROVALS:

_____
Facility Services Approval

_____
Date Approved

_____
Safety Approval

_____
Date Approved

_____
Infection Control Approval

_____
Date Approved

_____
Project Management Approval

_____
Date Approved

Reason for Disapproval:

Page 3 of 3

02412-14_F18C.EPS

Figure 18C Example of a shutdown permit. (3 of 3)

Describe any special inspections or testing that will be required, contingency plans, and mitigation methods if required. The form must be signed and dated by the general contractor and the personnel performing the work. The facility will designate the appropriate people to countersign and approve the shutdown permit. All affected organizational units, such as departments, that are affected by the shutdown must be notified prior to the shutdown. This can be done by posting notices of the service interruption (*Figure 19*).

Shutdowns in medical gas and vacuum systems are required when tying in new zones to an existing system or modifying existing piping. Before the shutdown, the health care facility is responsible for identifying the systems to be shut down and making all preparations to ensure continued facility operation during the shutdown. This may include arranging for alternative gas-supply cylinders, suction equipment, and other necessary tools and equipment. The health care facility is responsible for posting notices of the shutdown throughout the facility.

Prior to the shutdown, the contractor should work with facility staff to identify areas and systems that will be affected by the shutdown, including alarm systems. The installer is responsible for scheduling the necessary personnel, tools, and equipment for the job. Immediately prior to the shutdown, the installer notifies the facility representative. The installer completes the work in a timely fashion and notifies the facility representative when the work has been completed. The local applicable code will specify the necessary test and repair equipment that will be used when shutting systems down.

Hot-work permits (*Figure 20*) are required when brazing of copper tube is to be performed. The requirements for securing a hot-work permit are similar to those used for shutdown permits. Follow all safety requirements for hot work as specified in the local applicable code. Occupational Safety and Health Administration (OSHA) regulations restrict hot work in confined or enclosed spaces; adjacent to spaces that have combustible materials; near fuel tanks; or on pipelines that may contain fuel residue. For example, al-though you are permitted to take a torch into a confined or enclosed space, the fuel bottle cannot be brought inside. A torch cannot be left unattended in a confined or enclosed space; when you leave, you must take the torch with you.

Always follow the appropriate safety guidelines when performing hot work. Be sure to wear appropriate personal protective equipment. Always have a standby person present during hot work. The standby person should know whom to contact in case of emergency and how to disconnect the power.

3.4.0 Understanding Life Safety for the End User

Life safety systems are medical gas and vacuum systems that are essential for ensuring the safety of patients. These systems typically include the medical gases themselves, patient alarms, ventilation systems, and vacuum pumps. The National Fire Protection Association describes life safety requirements in NFPA 101, *Life Safety Code*. As with NFPA 99, NFPA 101 classifies life safety according to the level of hazard of the health and safety risks posed by the system to patients and medical personnel. All systems required for patient life safety must meet the requirements of NFPA 101 and the local applicable code for medical gas and vacuum systems.

Life safety branches of a medical gas and vacuum system are specially designated to supply power for lighting and other equipment that is essential for ensuring the safety of patients. Life safety systems are designed to automatically switch to backup power sources if electrical power to the facility is interrupted. Master, zone, and local alarms are all required to be connected to life safety branches. NFPA 101 also provides the requirements for penetrations through-interior fire-rated or smoke barrier walls.

> **WARNING!**
>
> NFPA 101, *Life Safety Code*, prohibits the installation of medical gas and vacuum piping in exit enclosures.

SPRING HILLS HEALTH CARE FACILITY

IMPORTANT NOTICE
SERVICE INTERRUPTION

SERVICE AFFECTED: _____

DATE _____

TIME: _____

AREA(S) AFFECTED: _____

REMARKS: _____

_____

_____

FOR INFORMATION, CALL : _____

_____

DATE ISSUED: _____

BY: _____

TITIE: _____

DISTRIBUTION: _____

_____

02412-14_F19.EPS

Figure 19 Example of a service interruption notice.

Spring Hills Health Care Facility
Brazing/Welding/Hot-Work Permit

Valid from _____ to _____. _____   Master Card No. _____
    (am/pm) (am/pm)   DATE

1. Work Description
   Equipment Location or Area _____
   Work to be done:

2. Gas Test

☐ None Required				
☐ Instrument Check	Test Results	Other Tests		Test Results
☐ Oxygen 20.8% Min				
☐ Combustible % LFL				

Gas Tester Signature   Date   Time

3. Special Instructions   ☐ None   ☐ Check with issuer before beginning work

4. Hazardous Materials   ☐ None   What did the line/equipment last contain?

5. Personal Protection   ☐ Standard Equipment: welder's hood with long sleeves; cutting goggles

    ☐ Goggles or Face Shield   ☐ Respirator   ☐ Forced Air Ventilation

    ☐ Standby Man   ☐ Other, specify: _____

6. Fire Protection   ☐ None Required   ☐ Portable Fire Extinguisher

    ☐ Fire Watch   ☐ Fire Blanket   ☐ Other, specify: _____

7. Condition of Area and Equipment

Required
Yes No   THESE KEY POINTS MUST BE CHECKED

Yes	No	THESE KEY POINTS MUST BE CHECKED
		a. Lines disconnected & blanked or if disconnecting is not possible, blinds installed?
		b. Lines steamed, purged, or otherwise properly cleaned of combustibles
		c. Area and equipment satisfactorily clean of oil or combustibles?
		d. Trenches, catch basins, & sewer connections properly covered or sealed?
		e. Immediate area and /or area under the work barricaded or roped off?
		f. Adjoining equip. & operations checked to have any effect on the job?
		g. Area fire suppression (fire water and sprinkler system) in service?

Comments

02412-14_F20A.EPS

Figure 20A Example of a hot-work permit. (1 of 2)

Spring Hills Health Care Facility
Brazing/Welding/Hot-Work Permit

8. Approval	Permit Authorization			Permit Acceptance		
	Area Supv.	Date	Time	Maint. Supv./Engineer Contractor Supv.	Date	Time
Issued by						
Endorsed by						
Endorsed by						

9. Individual Review

I have been instructed in the proper Hot Work Procedures

	Signed	Signed
Persons Authorized to Perform Hot Work	_____	_____
	_____	_____
	_____	_____
	_____	_____
	_____	_____
Fire Watch	_____	_____

10. Job Completion

☐ Yes ☐ No Is the work on the equipment completed?

☐ Yes ☐ No Has the worksite been cleaned and made safe?

Worker answering above questions _____

Issuer's Acceptance _____

Forward to Production Superintendent within 7 days of job completion

02412-14_F20B.EPS

Figure 20B Example of a hot-work permit. (2 of 2)

Additional Resources

NFPA 99, *Health Care Facilities Code*, Latest Edition. Quincy, MA: National Fire Protection Association.

NFPA 101, *Life Safety Code*, Latest Edition. Quincy, MA: National Fire Protection Association.

3.0.0 Section Review

1. The tint value of eye protection used for brazing should be at least _____.
   a. 3
   b. 4
   c. 5
   d. 6

2. The four classes of fuels that can be involved in fires are liquids, gases, combustible metals, and _____.
   a. untreated organic materials
   b. fuels
   c. plastics
   d. ordinary combustibles

3. On behalf of the contractor, shutdown permits must be signed and dated by _____.
   a. the on-site supervisor
   b. the personnel performing the work
   c. the general contractor and the personnel performing the work
   d. the head of the contractor's legal department

4. The National Fire Protection Association specifies the requirements for medical gases, patient alarms, ventilation systems, and vacuum pumps that affect the safety of patients in _____.
   a. NFPA 150, *Standard on Fire and Life Safety in Animal Housing Facilities*
   b. NFPA 72, *National Fire Alarm and Signaling Code*
   c. NFPA 99, *Health Care Facilities Code*
   d. NFPA 101, *Life Safety Code*

4.0.0 MEDICAL GAS SYSTEM MATERIALS, TOOLS, STORAGE, AND HANDLING

Objective

Identify the materials and tools for medical gas and vacuum systems and their storage and handling requirements.

a. Explain how materials and tools are stored and used.
b. Identify labels and lettering used on medical gas and vacuum systems.

Trade Terms

Charged: Filled with an inert gas such as oil-free, dry nitrogen.

Dehydrated: A condition in which all water has been removed.

Medical gas and vacuum systems use copper tube and fittings. Plumbers use brazing to connect tube and fittings in these systems. Although you are familiar with how to cut and join copper tube and fittings, this section builds on that knowledge with information that is specific to the proper assembly of medical gas and vacuum systems. Although the techniques may be familiar, medical gas and vacuum systems have specific safety requirements and special fittings that differ from other types of distribution systems that use copper.

For medical gas systems, NFPA 99 specifies the use of Type L hard-drawn seamless copper medical gas tube that has been manufactured according to ASTM B819, *Standard Specification for Seamless Copper Tube for Medical Gas Systems*. If the system operating pressure will be above 185 pounds per square inch (psi) and the pipe size is above 3 inches OD (outside diameter), Type K tube should be used. For medical-surgical vacuum systems, use hard-drawn seamless copper that has been manufactured to any of the following standards:

- ASTM B 819 medical gas tube.
- Type K, L, or M water tube manufactured according to ASTM B88, *Standard Specification for Seamless Copper Water Tube*.

- ACR (air conditioning and refrigeration) tube manufactured according to ASTM B280, *Standard Specification for Seamless Copper Tube for Air Conditioning and Refrigeration Field Service.*

NFPA 99 also allows the use of stainless steel tube or pipe in medical-surgical vacuum systems. Codes specify hard-drawn copper tube because soft-drawn tube can be easily kinked, which can restrict gas flow in the system. Codes require medical gas and vacuum system tubing to be at least ½ inch in diameter and that it be labeled properly by the manufacturer. The installer is required to provide documentation that shows the piping materials comply with the local applicable code.

This section covers the proper storage and use of materials and tools for installing medical gas and vacuum systems, and the proper use of labels and identification. Be sure to follow your local applicable code when using materials and tools to install medical gas and vacuum systems. Remember, NFPA 99 establishes the minimum safe standard for this work; the requirements in your local applicable code may be more strict. Always follow the manufacturers' instructions when using tools and equipment, and wear appropriate personal protective equipment when installing medical gas and vacuum systems.

4.1.0 Storing and Using Materials and Tools

The tools required for installing medical gas and vacuum systems are the same basic tools that plumbers carry in their tool kits. However, the specific safety requirements of medical gas and vacuum systems require extra care in preparing and using tools. All materials and tools should be clean and free of oil or grease. Oil and grease can pose a fire hazard when exposed to oxygen in a medical gas system.

In addition to materials and tools for brazing copper tube, you will also need the following tools when working on medical gas and vacuum systems:

- Tubing cutter
- Purge regulator
- Abrasive pads
- Lint-free white cloth wipes
- Oxygen analyzer
- Adapters for fitting pressure gauges on each type of station outlet and inlet used throughout the system
- Leak-detection solution
- Nitrogen purge alarm

4.1.1 Storing Materials and Tools

Because of the strict safety requirements for medical gas and vacuum systems, plumbers need to take extra care to prevent contamination of tools, equipment, and materials that are used in medical gas and vacuum system installation, testing, and servicing. Codes require tools and equipment to be stored separately from those used on other types of plumbing installations. For example, it is a code violation to store water pipe and medical gas tubing in the same storage trailer.

4.1.2 Preparing Pipe and Fittings

Prior to the publication of ASTM standard B819, plumbers had to clean medical gas and vacuum system tubing and fittings on the job site. Now, manufacturers prepare copper tube to meet the standard's requirements for cleanliness. Once the tube has been formed, the inside of the tube is dehydrated. This means that any trace of water is removed from the inner surface. The tube is then charged (filled) with oil-free, dry nitrogen gas to a low pressure in order to prevent copper atoms on the inner surface from reacting with oxygen in the atmosphere to form a layer of copper oxide.

Once the tube has been charged, the ends are capped or plugged to keep the oil-free, dry nitrogen inside. Caps are fitted on the outside of tube and are easily visible. Plugs fit flush with the end of the pipe and are harder to see. While codes may not specify the use of caps instead of plugs, it is good practice to use capped pipe to aid in remembering to remove the caps during the installation process. After the tube has been cut, the pieces are recapped.

> **CAUTION**
>
> Remember to remove internal plugs from copper tube when installing them. Leaving plugs in installed pipe can cause damage to the system when the system is tested.

Components such as valves and fittings are sealed in bags at the manufacturer's plant and must remain sealed until they are prepared for installation. Be sure to keep the documentation that comes with these sealed bags, as they are the only proof that they were delivered sealed. On-site cleaning of medical gas tube, fittings, valves, and other system components is permitted only under certain special conditions which are specified in the local applicable code. Codes prohibit any other types of cleaning of medical gas and vacuum system tube, fittings, valves, or components on the job site. This is because their cleanliness cannot be guaranteed outside of the controlled conditions of the manufacturer's facility, and also because the chemicals used to clean them represent a potential environmental hazard.

4.1.3 Cutting and Reaming

Cut copper tubing with a handheld tube cutter, a hacksaw, or a midget cutter. The handheld tube cutter (*Figure 21*) is preferred because it makes a cleaner joint and leaves no metal particles. Use a tube cutter that is the right size for the copper being cut, and make sure that the proper cutting wheel is in place.

After cutting, ream all cut tube ends to the full inside diameter of the tube. Reaming removes the small burr created when the tube is cut. Burrs left on tubing can cause the tube to corrode. A tube that is reamed correctly provides a smooth inner surface for better flow. Remove burrs on the outside of the tube to ensure a good fit. Tools used to ream tube ends include the reaming blade on the tube cutter, files (round or half round), a pocketknife, and a deburring tool. If the tube becomes deformed, use a sizing tool to bring the tube back to roundness. *Figure 21* shows an example of a deburring blade on a tube cutter. A variety of models are available, and the cutting sizes range from $\frac{1}{8}$ inch to $4\frac{1}{8}$ inches OD.

To use the handheld tube cutter, follow these steps (see *Figure 22*):

Step 1 Place the tube cutter on the tube at the point where you want to cut. Tighten the knob, forcing the cutting wheel against the tube.

Step 2 Make the cut by rotating the cutter around the tube under constant pressure.

Step 3 Use the built-in deburring blade to remove any burrs from inside the tube.

02412-14_F21.EPS

Figure 21 Handheld tube cutter.

CUTTING

REAMING

02412-14_F22.EPS

Figure 22 Using a handheld tube cutter.

4.1.4 Cleaning the Joints

The exterior surfaces of tube ends must be cleaned with nonabrasive pads. Never use steel wool, sanding paper or cloth, or wire brushes to clean copper tube, fittings, valves, or components used in medical gas and vacuum systems.

When the surfaces have been cleaned, wipe them with a clean, lint-free white cloth. Then inspect the interiors of tube, fittings, valves, and components to ensure they are free of contamination, obstructions, and debris. Never use contaminated items in a system designed for oxygen service. Codes require joints to be brazed within eight hours of cleaning.

4.2.0 Using Labels and Identification

Codes require all piping, valves, station outlets and inlets, and alarm panels to be clearly labeled (*Figure 23*). This is done to ensure that medical gas and vacuum systems are not accidentally cross-con-

nected during installation, testing, and servicing. Labeling is also used to help identify problems in the event of a system failure or an emergency. This section discusses the requirements for labeling medical gas and vacuum system piping, valves, and station outlets and inlets. The labeling of master-, zone-, and local-alarm panels is discussed in the section *Identifying Control Panels and Alarm Systems*. As with all medical gas and vacuum system requirements, refer to your local applicable code for the labeling and identification standards that are enforced in your area.

Labels may be stencils or adhesive markings. The colors for the lettering and background indicate the type of gas or vacuum line. NFPA 99 defines the standard colors for use on labels to help identify the different systems. *Table 1* lists the standard designation colors for medical gas and vacuum systems.

4.2.1 Labeling of Copper Tube

Manufacturers use special markings and labels to indicate that copper tube is suitable for medical gas and vacuum applications. Tube must be labeled with one of the following manufacturer's marks:

- OXY
- MED
- OXY/MED
- OXY/ACR
- ACR/MED

02412-14_F23.EPS

Figure 23 A properly labeled medical air compressor line.

Tube labels must follow the patterns described in *Table 1*. Labels must also include the following information:

- Name of the gas or vacuum system, or the appropriate chemical symbol
- Correct color code
- Operating pressure, if different from the standard gauge pressure for that type of gas

Labels on tube cannot be more than 20 feet apart. Each system must have at least one label per room, on each side of every wall or partition that the pipe penetrates. Vertical risers must be labeled at least once on every floor. Tube that is

Lettering on Medical Gas and Vacuum System Tubing

The manufacturer applies lettering to medical gas and vacuum system tubing. The color of the markings indicates the type of copper tube used. Blue lettering indicates Type L tube. Green lettering indicates Type K tube.

Table 1 Standard Designation Colors for Medical Gas and Vacuum Systems

Gas Service	Abbreviated Name	Colors (Background/Text)	Standard Gauge Pressure	
			kPa	psi
Medical air	Med air	Yellow/black	345–380	50–55
Carbon dioxide	CO_2	Gray/black or gray/white	345–380	50–55
Helium	He	Brown/white	345–380	50–55
Nitrogen	N_2	Black/white	1100–1275	160–185
Nitrous oxide	N_2O	Blue/white	345–380	50–55
Oxygen	O_2	Green/white or white/green	345–380	50–55
Oxygen/carbon dioxide mixture	$O_2/CO_2 n\%$ (n = % of CO_2)	Green/white	345–380	50–55
Medical–surgical vacuum	Med vac	White/black	380 mm to 760 mm (15 in. to 30 in.) HgV	
Waste anesthetic gas disposal	WAGD	Violet/white	Varies with system type	
Other mixtures	Gas A%/Gas B%	Colors as above Major gas for background/minor gas for text	None	
Nonmedical air (Category 3 gas-powered device))		Yellow and white diagonal stripe/black	None	
Nonmedical and Category 3 vacuum		White and black diagonal stripe/black boxed	None	
Laboratory air		Yellow and white checkboard/black	None	
Laboratory vacuum		White and black checkerboard/black boxed	None	
Instrument air		Red/white	1100–1275	160–185

labeled OXY is suitable for use with other non-flammable medical gas. This is because OXY tubing has been manufactured according to the standards of ASTM B88 for medical use.

4.2.2 Labeling of Shutoff Valves

Codes require shutoff valves to be labeled with the name or chemical symbol for the gas or vacuum system on which it is used, and the rooms or areas served. Valves must also be labeled with a warning notice that says the valve shall not be operated except in an emergency. Source, main line, riser, and service valves must all be labeled with the name of the line they are serving, and the area or areas served by that line. If the operating pressure of the line is different from the standard gauge pressure, the pressure on the line must also be labeled on the valve.

4.2.3 Labeling of Station Outlets and Inlets

As with valves, station outlets and inlets should be labeled with the name or chemical symbol for the gas or vacuum symbol that the line serves. If the operating pressure of the line is different from the standard gauge pressure, the line pressure must also be labeled on the station outlet or inlet.

Additional Resources

ASTM B88, *Standard Specification for Seamless Copper Water Tube*, Latest Edition. West Conshohocken, PA: ASTM International.

ASTM B280, *Standard Specification for Seamless Copper Tube for Air Conditioning and Refrigeration Field Service*, Latest Edition. West Conshohocken, PA: ASTM International.

ASTM B819, *Standard Specification for Seamless Copper Tube for Medical Gas Systems*, Latest Edition. West Conshohocken, PA: ASTM International.

NFPA 99, *Health Care Facilities Code*, Latest Edition. Quincy, MA: National Fire Protection Association.

4.0.0 Section Review

1. Codes require that the diameter of medical gas and vacuum system tubing should be at least _____.

    a. 1 inch
    b. ¾ inch
    c. ½ inch
    d. ¼ inch

2. The minimum number of labels that each system must have per room is _____.

    a. one
    b. two
    c. three
    d. four

5.0.0 BRAZING AND LINE PURGING OF MEDICAL GAS SYSTEMS

Objective

Identify the equipment required for brazing copper tube with and without purging.

a. Explain how to braze a joint with a purge.
b. Explain how to braze a joint without a purge.

Performance Tasks

Braze copper tube with purging.
Braze copper tube without purging.

Brazing is a heating process similar to soldering that uses the combustion of oxygen and a fuel gas, often acetylene, to heat filler metals that melt between 1,100°F and 1,500°F. Plumbers use brazing to create joints that are very strong. Plumbers use brazing torch kits (*Figure 24*) to braze joints. A brazing torch kit consists of a burner wand connected by a hose to pressurized oxygen and fuel gas tanks via a regulator (to the oxygen tank) and valve (to the fuel gas tank). Codes require brazers to be qualified according to the standards outlined in the local applicable code.

5.1.0 Brazing a Joint With a Purge

To ensure that a brazed joint is strong enough to withstand the positive and negative pressures in a medical gas and vacuum system, socket-type brazed joints are required. A socket joint involves coupling one end of a length of copper tube to another (*Figure 25*).

To ensure that the piping system will maintain its integrity in the event of a fire, use a brazing alloy that has a melting temperature greater than 1,000°F. Be sure to use filler metals that are compatible with the type of tube being brazed. When brazing copper-to-copper joints, use a copper-phosphorus or copper-phosphorus-silver filler metal without flux. Use flux sparingly to avoid contaminating the inside of the tube. Never field-braze cast metals. Joints that are brazed in place must be accessible for inspection.

Joints in medical gas systems must be continuously purged with oil-free, dry nitrogen during the brazing process. This is to prevent the formation of copper oxide on the insides of the joints. To further protect against contamination, always cap or plug all pipe openings to maintain the oil-free, dry nitrogen atmosphere inside the pipe prior to purging.

When purging, monitor the source gas supply. An audible alert is required to indicate low supply. Keep the flow rate of the oil-free, dry nitrogen purge gas low enough that it won't produce positive pressures in the system. Use a pressure regulator and a flowmeter to control the flow of the gas. Never use just a pressure regulator by itself.

To braze copper tube in medical gas and vacuum systems using a nitrogen purge, follow these steps:

Step 1 Insert the tube ends fully into the depth of the fitting socket or to a depth specified in the local applicable code.

Step 2 Make sure a discharge opening is present on the side of the joint that is opposite the point where purge gas will be introduced. Introduce the purge gas to the joint.

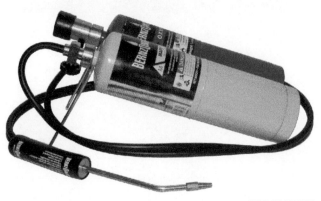

02412-14_F24.EPS

Figure 24 Typical brazing torch kit.

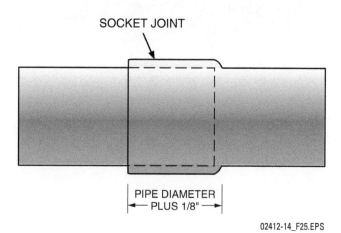

SOCKET JOINT

PIPE DIAMETER
PLUS 1/8"

02412-14_F25.EPS

Figure 25 A coupled joint.

Step 3 Insert an oxygen analyzer into the end of the tube to confirm that the pipe is free of oxygen.

Step 4 If flux is required, heat the joints slowly until the flux becomes liquid.

Step 5 After the flux has liquefied, or if flux is not required, quickly heat the joint to the temperature required for brazing, being careful to keep the joint from overheating.

Step 6 Maintain the flow of the oil-free, dry nitrogen purge gas until the brazing is complete and the joint is cool to the touch. Then cap or plug the purge discharge opening to prevent contamination and to maintain the internal nitrogen atmosphere.

5.2.0 Brazing a Joint Without a Purge

When completing the installation of medical gas and vacuum systems, the last connecting joints on both systems will need to be brazed without a purge. This is commonly called a dirty joint. The steps for brazing copper tube in a medical gas and vacuum system without using a nitrogen purge are similar to the steps for soldering copper tube that you are already familiar with. The differences are that brazing takes place at a much higher temperature (between 1,100°F and 1,500°F), and the flux material used for brazing, if it is required, is different from solder. To braze copper tube for a medical gas and vacuum system once the pipe has been cut, reamed, and cleaned, follow these steps:

Step 1 Insert the tube ends fully into the depth of the fitting socket or to a depth specified in the local applicable code.

Step 2 If flux is required, heat the joints slowly until the flux becomes liquid.

Step 3 After the flux has liquefied, or if flux is not required, quickly heat the joint to the temperature required for brazing, being careful to keep the joint from overheating.

Step 4 Allow the joint to cool until the filler metal solidifies.

Step 5 Once the filler metal has solidified, wipe it clean with a soft, wet cloth to remove any flux.

Additional Resources

Brazing, Second Edition. Mel M. Schwartz. 2003. Novelty, OH: ASM International.

The Copper Tube Handbook. 2010. New York: Copper Development Association.

NFPA 99, *Health Care Facilities Code*, Latest Edition. Quincy, MA: National Fire Protection Association.

5.0.0 Section Review

1. Copper tube used in medical gas and vacuum systems is filled with oil-free, dry nitrogen to prevent the insides of joints from being contaminated with _____.

   a. carbon dioxide
   b. copper oxide
   c. sulfur dioxide
   d. ferrous oxide

2. The last connecting joint on a medical gas or vacuum system is commonly called a(n) _____.

   a. clean joint
   b. open joint
   c. dirty joint
   d. close joint

6.0.0 TESTING AND VERIFYING MEDICAL GAS SYSTEMS

Objective

Describe the process for testing and verifying a medical gas and vacuum system.

a. Explain how to conduct installation tests.
b. Explain how to conduct system verification tests.

Trade Terms

Initial blowdown test: An installation test in which oil-free, dry nitrogen gas is blown through the distribution piping to remove dirt and other particulate matter.

Initial cross-connection test: An installation test performed to ensure that medical gas and vacuum systems are not cross-connected.

Initial piping-purge test: An installation test performed to purge the system of any remaining dirt and other particulate matter.

Initial pressure test: An installation test performed to ensure that there is no pressure loss in the distribution piping.

Noncancelable: A signal that cannot be turned off manually.

Standing pressure test: An installation test performed to ensure that the distribution piping in a medical gas system will not experience excessive pressure loss when under service pressure.

Standing vacuum test: An installation test performed to ensure that vacuum system piping in a medical vacuum system will not experience excessive vacuum loss when in use.

To minimize the risk of injury or death resulting from the failure of a component, all medical gas and vacuum systems and their components are tested thoroughly. Two separate sets of tests are conducted on each system. The installer conducts the first set of tests while the system is being installed, and the second set is conducted by a verifier following the completion of the installer's tests. This ensures that the tests are conducted impartially.

The various installer and verifier tests are detailed in the following sections. Never attempt to perform a test for which you are not qualified. Always follow the manufacturers' instructions and your local applicable code when testing medical gas and vacuum systems and their components. Wear appropriate personal protective equipment at all times when testing, and ensure that all safety guidelines are observed. These procedures will not only prevent damage to the system, but will also protect your health and safety as well as the health and safety of others.

6.1.0 Conducting Installation Tests

During the installation process for Category 1, 2, and 3 medical gas and vacuum systems, the installer is required to perform and document a series of tests of the systems and their components. Tests are conducted using oil-free, dry nitrogen gas. Tests of manufactured assemblies should be performed after the installation of the distribution piping, but before the system verification tests. The tests must also be performed before installing any manufactured assemblies that are supplied through flexible hoses or tubing, and at each station outlet/inlet on the installed manufactured assemblies that are supplied through copper tubing. The following sections explain the purpose of each test and provide the general steps for performing the test. Always follow the manufacturers' instructions and your local applicable code when conducting tests.

6.1.1 Initial Blowdown Test

The first installation test is called the initial blowdown test. In this test, oil-free, dry nitrogen gas is simply blown through the distribution piping to remove dirt and other particulate matter. Conduct the initial blowdown test after the distribution piping has been installed, but before installing other system components such as station outlet/inlet rough-in assemblies, pressure and vacuum alarms and indicators, pressure-relief valves, and manifolds.

6.1.2 Initial Pressure Test

To ensure that there is no pressure loss in the distribution piping, the installer performs an initial pressure test of each section of piping. Initial pressure tests are conducted following the initial blowdown test and after station outlet/inlet rough-in assemblies have been installed, but before installing system components that could be damaged by the test pressure. These components include pressure and vacuum alarms and indicators and pressure-relief valves. Follow these steps when conducting the initial pressure test:

Step 1 Close the source shutoff valve and ensure that it remains closed throughout the test.

Step 2 Pressurize the system to 1.5 times the system working pressure but not less than 150 psig.

Step 3 Maintain test pressure while examining each joint for evidence of leakage. Use an ammonia-free leak detectant that is safe for use with oxygen systems.

Step 4 Repair or replace any leaks according to your local applicable code, and then retest the system.

> **WARNING!**
>
> Leaks represent fire hazards as well as potential health hazards. Studies have found that excessive exposure to nitrous oxide and other medical gases can cause illness.

6.1.3 Initial Cross-Connection Test

Medical gas and vacuum systems must never be cross-connected. To ensure this, installers perform an initial cross-connection test on each system after the initial blowdown and pressure tests. The steps for this test are as follows:

Step 1 Reduce the pressure in all piping systems to the standard atmospheric pressure of 14.7 psi. Ensure that the test gas source is disconnected from all piping systems except for the system being tested.

Step 2 Pressurize the system with oil-free, dry nitrogen to 50 psig.

Step 3 Inspect each station outlet/inlet in all installed systems to ensure that only those on the system being tested are discharging gas.

Step 4 Repeat the steps for each medical gas and vacuum system in turn. While conducting the tests, confirm that the station outlets/inlets are labeled and identified properly.

Step 5 If you detect any cross-connection, or if any of the outlets/inlets are incorrectly labeled or identified, correct the problem before proceeding to test the next system.

6.1.4 Initial Piping-Purge Test

Following the completion of cross-connection tests, the system is purged to remove any remaining dirt and other particulate matter from the lines. This test is called the initial piping-purge test. It provides a more thorough cleaning than the initial blowdown test. Follow these steps to conduct the initial piping-purge test:

Step 1 Using manufacturer-approved adapters that are designed for use with the particular system being tested, purge each outlet with intermittent high-volume bursts of oil-free, dry nitrogen. Begin the purge at the system outlet/inlet closest to the zone valve and continue outward until reaching the furthest system outlet/inlet in the zone.

Step 2 Continue the blasts until the lines are free of debris. You can determine this by holding a clean white cloth up to the outlet.

Step 3 When the nitrogen blasts cause no discoloration of the cloth, the line is clean.

6.1.5 Standing Pressure Test for Positive-Pressure Medical Gas Piping

To ensure that the distribution piping in a medical gas system will not experience excessive pressure loss when under service pressure, installers conduct a 24-hour standing pressure test following the initial pressure test. Standing pressure tests are conducted after the station outlet valve bodies, faceplates, and other distribution system components have been installed. The source valve must remain closed during the standing pressure test. The steps for this test are as follows:

Step 1 Pressurize the positive-pressure portion of the system with oil-free, dry nitrogen to 20 percent above the normal operating line pressure for the system and allow the system to stand for 24 hours.

Step 2 At the end of that time, inspect the system pressure gauge to ensure that the test pressure has not changed as a result of leaks. Changes in pressure caused by fluctuations in the ambient temperature are permitted, however.

Step 3 Locate any leaks and conduct repairs or replacements as required by the local applicable code.

Codes require the AHJ to witness the standing pressure test. Following the successful completion of the test, the AHJ will complete a form confirming the test has been performed and completed according to the local applicable code, and will provide the form to the verifier prior to the start of the system verification tests.

6.1.6 Standing Vacuum Test for Vacuum Piping

As with positive-pressure medical gas systems, negative-pressure vacuum distribution piping must also undergo a 24-hour test to ensure it will not experience excessive vacuum loss when in use. The standing vacuum test is conducted following the successful completion of the standing pressure test and the installation of all vacuum system components. Follow these steps to conduct a standing vacuum test:

Step 1 Ensure that the source of the test vacuum is disconnected from the piping system.

Step 2 Reduce the pressure in the vacuum piping until it reads between 12 inches of mercury and a full vacuum. Allow the system to stand for 24 hours.

Step 3 At the end of that time, inspect the system pressure gauge to ensure that the test vacuum has not changed as a result of leaks. Changes in vacuum caused by fluctuations in the ambient temperature are permitted, however.

Step 4 Locate any leaks and conduct repairs or replacements as required by the local applicable code.

As with the standing pressure test for positive-pressure medical gas piping, codes require the AHJ to witness the standing vacuum test. Following the successful completion of the test, the AHJ will complete a form confirming the test has been performed and completed according to the local applicable code, and will provide the form to the verifier prior to the start of the system verification tests.

6.2.0 Conducting System Verification Tests

When the installer tests have been completed to the satisfaction of the AHJ, a second set of tests is performed on the medical gas and vacuum systems. These tests are performed by a verifier, a technically competent and experienced professional who has been specially certified by ASSE to perform these tests. The verifier must not be affiliated with the contractor that installed the systems. The verifier can be a member of the staff of the facility in which the systems have been installed, if properly trained and certified. Verification tests can be conducted using either nitrogen gas or the gas for which the system is designed to use in operation. Verifiers test each terminal connection point individually.

The following sections explain the purpose of each verification test and provide the general steps for performing the test. Always follow the manufacturers' instructions and your local applicable code when conducting tests.

6.2.1 Standing Pressure Test

To ensure that the distribution piping is free from leaks, the verifier first conducts a standing pressure test. This test is similar to the pressure test conducted by the installer, but instead of lasting 24 hours, this test lasts 10 minutes. Follow these steps:

Step 1 Pressurize the system with oil-free, dry nitrogen or source gas until operating pressure is reached.

Step 2 Close the source valve and all zone valves. Let the system stand for 10 minutes.

Step 3 At the end of that time, inspect the system pressure gauge to ensure that the test pressure has not changed as a result of leaks. Changes in pressure caused by fluctuations in the ambient temperature are permitted, however.

Step 4 Locate any leaks and conduct repairs or replacements as required by the local applicable code.

6.2.2 Cross-Connection Test

Next, the verifier conducts a cross-connection test to verify that the distribution piping remains free of any cross-connections. This test is performed following the closure of the walls and the installation of all system components. There are two types of cross-connection tests: the individual pressurization method and the pressure-differential method. The choice of test depends on the type of system as well as on manufacturer and code requirements.

To conduct a cross-connection test using the individual pressurization method, follow these steps:

Step 1 Reduce the pressure in all piping systems to the standard atmospheric pressure of 14.7 psi. Ensure that the test gas source is disconnected from all piping systems except for the system being tested.

Step 2 Pressurize the system with oil-free, dry nitrogen to 50 psig.

Step 3 Inspect each station outlet/inlet in all installed systems to ensure that only those on the system being tested are discharging gas.

Step 4 Disconnect the test gas source and reduce the system pressure to atmospheric pressure.

Step 5 Repeat the steps for each medical gas and vacuum system in turn until all systems have been tested for cross-connection.

To conduct a cross-connection test using the pressure-differential method, follow these steps:

Step 1 Reduce the pressure in all medical gas piping systems to the standard atmospheric pressure of 14.7 psi.

Step 2 Increase the test gas pressures in all systems according to *Table 2* and maintain these pressures for the duration of the test. If a system operates at a nonstandard operating pressure, test the system at a pressure that is at least 10 psig higher or lower than other systems being tested.

Step 3 Ensure that vacuum systems are in operation during this test so that they can also be tested at the same time.

Step 4 Use a test gauge to ensure that the test pressure at each station outlet is the correct pressure specified in *Table 2*. Ensure that the test gauges have been properly calibrated using the pressure indicator for the source pressure regulator.

Step 5 Identify each station outlet by label and, if required, by color marking. Ensure that the pressure indicated at each outlet matches the pressure specified in *Table 2*.

6.2.3 Valve Test

Verifiers test the system shutoff valves to ensure that they are functioning and that they are labeled properly. Refer to the manufacturers' instructions for testing the valves, as the test procedure can vary. Record the rooms or areas that are con-

Table 2 Test Pressures for Pressure Differential Cross-Connection Tests

Medical Gas	Pressure (Gauge)
Gas mixtures	140 kPa (20 psi)
Nitrogen/instrument air	210 kPa (30 psi)
Nitrous oxide	275 kPa (40 psi)
Oxygen	345 kPa (50 psi)
Medical air	415 kPa (60 psi)
Sy stems at nonstandard pressures	70 kPa (10 psi) greater or less than any other system HgV vacuum
Vacuum	510 mm (20 in.) HgV
WAGD	38 0 mm (15 in.) HgV (if so designed)

Reproduced with permission from NFPA 99-2012: *Health Care Facilities*, Copyright © 2011, National Fire Protection Association. This reprinted material is not the complete and official position of the NFPA on the referenced subject, which is represented only by the standard in its entirety.

trolled by each valve. This information is used to help verify the labeling of the valves.

6.2.4 Alarm Tests

To ensure that the master and area alarms installed in medical gas and vacuum systems are functioning properly and labeled correctly, the verifier tests them after completing the cross-connection and valve tests. An alarm test is also required when connecting a new piping system to an existing one, or when extending an existing system. Codes require the facility to maintain permanent records of all alarm tests on file. Use oil-free, dry nitrogen gas, system gas, or operating vacuum for these tests as required. If the alarm system is computer controlled, include the computer system in the alarm tests.

Test the master alarm to ensure that the audible and noncancelable (meaning that it cannot be turned off manually) visual signals are activated when the pressure in the main line fluctuates at least 20 percent above or below the normal operating pressure. Verify the operation of all master-alarm signals during this test.

Next, test the area alarms for individual positive- and negative-pressure systems to ensure that they are operating as designed. The area alarms in a positive-pressure piping system should activate if the pressure fluctuates at least 20 percent above or below the normal operating pressure. In a vacuum system, the alarm should activate if the vacuum falls below a gauge pressure of 12 inches of mercury.

6.2.5 Piping Purge Tests

Verifiers perform a piping purge test on medical gas and vacuum systems to remove any remaining dirt and particulate matter from systems' lines. This may be the first test conducted by a verifier. Follow these steps:

Step 1 Using an appropriate adapter supplied by the facility or the system's manufacturer, subject each outlet to a nitrogen gas purge of at least 8 scfm.

Step 2 Interrupt the purge several times during the test to create intermittent spurts of gas. This helps dislodge particulates still in the lines.

Step 3 Continue the test until the gas output leaves no discoloration on a white cloth that is placed loosely over the adapter. Always use an appropriate adapter for a purge test.

6.2.6 Piping Particulate Test

To verify that distribution piping is completely dirt-free, the purge test is followed by a particulate test. Follow these steps:

Step 1 Filter no less than 35 cubic feet of oil-free, dry nitrogen through a clean, white 0.45-micron filter at a flow rate of no less than 3.5 scfm.

Step 2 Test 25 percent of the zones at the outlets that are the farthest away from the source.

Step 3 If the filter collects more than 0.001 gram of matter from any outlet, test the most remote outlet in every zone.

6.2.7 Piping Purity Test

In addition to being free of particulate matter, medical gas distribution piping must also be free of excessive water vapor and hydrocarbons such as petroleum and natural gas. Follow these steps to perform a piping purity test:

Step 1 Test the outlet that is farthest from the gas source for total nonmethane hydrocarbons. Compare these results to the quality of the source gas. If using system gas as the source gas, conduct the test at the source equipment.

Step 2 The difference in purity between the source gas and outlet gas should not exceed 5 ppm (parts per million) of nonmethane and halogenated hydrocarbons (hydrocarbons that contain halogen).

Step 3 Measure the moisture concentration at the outlet to ensure that it does not exceed 500 ppm or an equivalent pressure dew point of 10°F at a pressure of 50 psig.

6.2.8 Final Tie-In Test

The final tie-in test is designed to ensure that the distribution system is free of any leaks at the connection point, and that no new contamination has been added. This test must be performed before connecting any new, extended, or additional piping. Follow these steps:

Step 1 Before joining new work to an existing system, conduct a leak test on each joint in the final connection. Use system gas at the designated operating pressure, and test for leaks using an approved leak detectant that is safe for use with oxygen and that contains no ammonia.

Step 2 Test vacuum joints using an ultrasonic leak detector or other approved method as specified by the manufacturer or the local applicable code.

Step 3 Test pressure joints following the brazing and leak testing of the final connection.

6.2.9 Operational Pressure Test

Positive- and negative-pressure distribution piping must be free from excessive loss of pressure or vacuum when operating. The operational pressure test is performed to ensure this. Perform pressure tests at each station outlet/inlet or terminal where connections are made, using either system gas or operating vacuum, depending on the system. NFPA 99 specifies the following delivery and draw requirements:

- Gas outlets with a pressure of 50 psig (including oxygen, nitrous oxide, medical air, and carbon dioxide) are required to deliver 3.5 scfm with a maximum pressure drop of 5 psi and a static pressure of between 50 to 55 psi.
- Support gas outlets are required to deliver 5 scfm with a maximum pressure drop of 5 psig and a static pressure of 160 to 185 psi.

- Medical-surgical vacuum inlets are required to draw 3 scfm without reducing the vacuum pressure below 12 inches of mercury at any adjacent station inlet.
- Oxygen and medical air outlets that serve critical care areas are required to permit transient flow rates of 6 scfm for 3 seconds.

6.2.10 Medical Gas Concentration Test

Verifiers conduct a medical gas concentration test to ensure that the correct concentration of system gas is available at each outlet. Once the system has been purged using system gas, analyze each gas source and outlet using appropriate measuring instruments operated according to the manufacturers' instructions. Refer to *Table 3* for the allowable gas concentrations for each type of gas.

6.2.11 Medical Air Purity Test

Test the quality and purity of medical air by taking a sample at the system's sample port. This test must be conducted before the source valve is opened. Refer to *Table 4* for the maximum allowable impurities in medical air.

6.2.12 Labeling

Verifiers ensure that proper labeling has been affixed to all distribution piping, system outlets/inlets, shutoff valves, master and area alarms, and source equipment. The style and placement of labels is discussed elsewhere in this module.

6.2.13 Source Equipment Verification

Finally, the verifier tests the complete gas supply, medical air compressor, and proportioning systems to ensure they are functioning correctly.

Table 3 Allowable Gas Concentrations

Medical Gas	Concentration
Oxygen	≥ 99% oxygen
Nitrous oxide	≥ 99% nitrous oxide
Nitrogen	≤ 1% oxygen or ≥ 99% nitrogen
Medical air	19.5%–23.5% oxygen
Other gases	As specified by ±1%, unless otherwise specified

Reproduced with permission from NFPA 99-2012: *Health Care Facilities*, Copyright © 2011, National Fire Protection Association. This reprinted material is not the complete and official position of the NFPA on the referenced subject, which is represented only by the standard in its entirety.

Table 4 Medical Air Contaminant Limits

Parameter	Limit Value
Pressure dew point	2° C (35° F)
Carbon monoxide	10 ppm
Carbon dioxide	500 ppm
Gaseous hydrocarbons	25 ppm (as methane)
Halogenated hydrocarbons	2 ppm

Reproduced with permission from NFPA 99-2012: *Health Care Facilities*, Copyright © 2011, National Fire Protection Association. This reprinted material is not the complete and official position of the NFPA on the referenced subject, which is represented only by the standard in its entirety.

These tests are performed once the interconnecting lines, accessories, and source equipment have all been installed.

To test a gas supply source, follow these steps:

Step 1 Ensure that the changeover from primary to secondary supply and the reserve are operating according to the manufacturer's instructions.

Step 2 Test the reserve unit's actuation and pressure-monitoring switches and signals, if installed.

Step 3 Test the bulk-supply and master signal-panel installations to ensure they are functioning correctly. Test the signal panels when changing or replacing storage units as well.

The medical air compressor system is tested to ensure air quality and to ensure that the alarm sensors are calibrated and installed according to the manufacturer's instructions. Test the medical air compressor system at the sample port. To test a medical air compressor system, follow these steps:

Step 1 Check the operation of system control sensors and controls including dew point, air temperature, and air quality, before allowing the system to go into service.

Step 2 Verify the quality of medical air being delivered by the compressor air supply when installing new components, and before patient use. Conduct air quality tests following normal operation of the medical air source system and with the source valve closed under a simulated load for at least 12 hours elapsed (not aggregate) time.

Step 3 Simulate loading by venting air continuously at a rate that is approximately 25 percent of the system's capacity.

Step 4 Create a demand of approximately 25 percent of the rated compressor capacity to cause the compressors to cycle and the dryers to operate.

Like the medical air compressor system, the proportioning system and its reserve are tested to ensure that they are functioning properly and that the alarms are calibrated and installed correctly. Conduct the test at the proportioning system's sample port. Finally, test the medical-surgical vacuum systems according to the manufacturer's instructions prior to putting it into service.

Additional Resources

ASSE/IAPMO/ANSI Standard #6010, *Medical Gas Systems Installers*, Latest Edition. Washington, DC: American National Standards Institute.

ASSE/IAPMO/ANSI Standard #6030, *Medical Gas Systems Verifiers*, Latest Edition. Washington, DC: American National Standards Institute.

NFPA 99, *Health Care Facilities Code*, Latest Edition. Quincy, MA: National Fire Protection Association.

6.0.0 Section Review

1. The tests that the AHJ is required by code to witness are the installer's _____.

   a. initial cross-connection and initial piping-purge tests
   b. initial blowdown and initial pressure tests
   c. standing pressure and standing vacuum tests
   d. medical gas concentration and medical air purity tests

2. In an operational pressure test, the maximum allowable pressure drop of a gas outlet with a pressure of 50 psig is _____.

   a. 20 psi
   b. 15 psi
   c. 10 psi
   d. 5 psi

SUMMARY

Health care facilities are equipped with special piping systems called medical gas and vacuum systems that supply oxygen, nitrous oxide, helium, carbon dioxide, and on-site compressed air under pressure. Medical gas systems consist of the equipment and piping that supply these gases under pressure as well as their associated pumps and compressors, filters and purifiers, valves, alarms, gauges, and controls. Medical gas systems are often paired with systems that use negative pressure to remove waste gases, liquids, and solids. These systems are called medical vacuum systems.

The requirements for medical gas and vacuum systems are outlined in NFPA 99, *Health Care Facilities Code*. NFPA 99 classifies medical gas and vacuum systems into four broad categories depending on the health and safety risks that the system poses to patients and medical personnel. Medical gas and vacuum systems are installed, inspected, verified, and maintained by people who have been specially trained and certified according to ASSE/IAPMO/ANSI Series 6000, *Professional Qualifications Standard for Medical Gas Systems Personnel*.

The most commonly used gases in medical gas and vacuum systems include oxygen, nitrous oxide, and medical air. Vacuum systems may be used for waste anesthetic gas disposal (WAGD). The equipment, valves, controls, and other components of medical gas and vacuum systems vary depending on the type of gas the system is designed to deliver or dispose of. Medical gas and vacuum systems are fitted with controls and alarms that allow operators to monitor and control their operation. They are also equipped with alarm systems that indicate when line pressure rises above or falls below the specified range.

Medical gas and vacuum systems require specific safety precautions that are spelled out in the local applicable code. Working around medical gas and vacuum systems requires the use of PPE for brazing and working with metals. Have an approved fire extinguisher available and post a fire watch when brazing or cutting. Permits are required before a system can be shut down or hot work can be performed.

The copper tube and fittings in medical gas and vacuum systems are joined by brazing. There are specific safety requirements that must be observed when brazing medical gas and vacuum system components. All materials and tools should be clean and free of oil or grease. Tools and equipment used for medical gas and vacuum systems must be stored separately from those used on other plumbing installations. Copper tube is dehydrated and charged with oil-free, dry nitrogen gas to prevent the formation of copper oxide. Care must be taken when cutting, reaming, and joining copper tube and fittings to ensure that they remain clean. Manufacturers use special markings and labels to indicate that copper tube is suitable for medical gas and vacuum applications; plumbers must apply these labels properly and according to the local applicable code.

To ensure that brazed joints are strong enough to withstand the positive and negative pressures in a medical gas and vacuum system, socket-type brazed joints are required. Brazing can be done with and without a nitrogen purge. When a medical gas and vacuum system is complete, it is thoroughly tested by the installer and then by a verifier to ensure the system is ready for use. Only when a medical gas and vacuum system passes all these tests is it considered safe for use to provide medical care to patients.

1. The special piping system used in hospitals, clinics, and doctors' offices to supply non-flammable gases is called a(n) _____.
   a. anesthetic supply system
   b. medical gas system
   c. clinical gas system
   d. respiratory-assist supply network

2. The requirements for installing, testing, and maintaining medical gas and vacuum systems are outlined in the model code designated as _____.
   a. NFPA 101
   b. NFPA 563, Part B
   c. NFPA 157.3
   d. NFPA 99

3. A surgeon's office equipped to administer general anesthesia would require a medical gas and vacuum system classified as _____.
   a. Category 1
   b. Category 2
   c. Category 3
   d. Category 4

4. A Category 2 medical gas and vacuum system is classified as one that if it failed, _____.
   a. could result in serious injury or death
   b. could result in discomfort, but not injury
   c. would not affect patient care
   d. could result in minor injury

5. To be certified for work with medical gas and vacuum systems, maintenance personnel must pass written and practical exams after completing a training course lasting _____.
   a. 8 hours
   b. 16 hours
   c. 32 hours
   d. 48 hours

6. Fittings designed for use with all types of gas in a medical gas system are called _____.
   a. standard fittings
   b. universal fittings
   c. cheater fittings
   d. adapter fittings

7. To meet the requirements of NFPA 99, medical air must be produced at a grade called _____.
   a. ultrarefined
   b. medical-surgical
   c. pharmaceutical
   d. premium

8. The abbreviation WAGD stands for _____.
   a. waste anesthetic gas disposal
   b. weighted average gas dispenser
   c. wide-area gas delivery
   d. water-aided gas dispersion

9. The location of a medical gas and vacuum system control panel will be specified in the _____.
   a. manufacturer's product manual
   b. local applicable code
   c. design drawings
   d. local building code

10. NFPA 99 specifies that audible alarms must produce a sound of 80 decibels at _____.
    a. 3 feet
    b. 12 feet
    c. 25 feet
    d. 50 feet

11. Valves designed to isolate a set of outlets or inlets from the rest of the system are called _____.
    a. isolation valves
    b. area valves
    c. sectional cutoff valves
    d. zone valves

12. The condition of rubber gloves can be checked visually, or with a(n) _____.
    a. tension test
    b. air test
    c. abrasion test
    d. conductivity test

13. When doing hot work, you should know the location of fire extinguishers and _____.
    a. the first-aid kit
    b. what type extinguisher to use on different fires
    c. the date the extinguisher was last inspected
    d. the distance to the nearest telephone

14. Before brazing can be done on copper tubes, a contractor must obtain a(n) _____.
    a. site license
    b. brazing certificate
    c. hot-work permit
    d. OSHA Form 1012

15. NFPA 101 classifies life safety systems according to the _____.
    a. level of hazard posed
    b. number of zones involved
    c. flammability of the gas used
    d. ventilation required

16. For use in medical gas systems, NFPA specifies the use of hard-drawn seamless copper tubing designated as _____.
    a. Type A
    b. Type HD
    c. Type K
    d. Type L

17. In addition to basic plumbing tools, work on medical gas and vacuum systems requires _____.
    a. oxyacetylene cutting tools
    b. adapters for connecting a pressure gauge
    c. a set of metric wrenches
    d. a soldering gun

18. Before medical gas tubing is charged with oil-free, dry nitrogen gas and sealed at the factory, the inside must be _____.
    a. dehydrated
    b. coated with epoxy
    c. deoxidized
    d. irradiated

19. Forgetting to remove internal plugs from medical gas piping can cause damage when the system is _____.
    a. purged
    b. connected
    c. tested
    d. completed

20. On-site cleaning of medical gas and vacuum tubing is generally prohibited because the chemicals used _____.
    a. are toxic
    b. can damage the copper tubes
    c. are very costly
    d. are a potential environmental hazard

21. After tubing is cut, burrs must be removed with a procedure called _____.
    a. surfacing
    b. reaming
    c. grinding
    d. debriding

22. When cutting and reaming copper tube, if the tube becomes deformed, bring the tube back to roundness using a _____.
    a. rounding tool
    b. round or half-round file
    c. sizing tool
    d. deburring tool

23. A medical gas line labeled with white type on a blue background is used for _____.
    a. nitrous oxide
    b. oxygen
    c. nitrogen
    d. carbon dioxide

24. A shutoff valve must be labeled with the line's operating pressure if _____.
    a. the line carries oxygen
    b. the pressure is different from the standard gauge pressure
    c. the pressure is above 14.7 psig
    d. the system has a pressure-relief valve

25. Medical gas and vacuum tubing must be joined with brazed _____.
    a. tapered joints
    b. butt joints
    c. socket joints
    d. compression joints

26. To protect against joint failure in a fire, the brazing alloy must have a melting temperature above _____.
    a. 1,000°F
    b. 1,200°F
    c. 1,600°F
    d. 1,800°F

27. During brazing operations, the joint must be continuously purged with oil free, dry _____.

   a. carbon dioxide
   b. nitrogen
   c. helium
   d. nitrous oxide

28. An initial pressure test is conducted by pressurizing the system to at least _____.

   a. 14.7 psig
   b. 75 psig
   c. 150 psig
   d. 195 psig

29. A standing pressure test conducted by a verifier has a duration of _____.

   a. 10 minutes
   b. 30 minutes
   c. 4 hours
   d. 24 hours

30. In a piping purity test, the moisture concentration at the outlet farthest from the source must not exceed _____.

   a. 100 ppm
   b. 500 ppm
   c. 1,250 ppm
   d. 1,500 ppm

Trade Terms Quiz

Fill in the blank with the correct term that you learned from your study of this module.

1. A(n) _____ is a person who inspects medical gas and vacuum systems and verifies that they meet all applicable standards and codes as well as the manufacturer's specifications.

2. An installation test in which oil-free, dry nitrogen gas is blown through the distribution piping to remove dirt and other particulate matter is called a(n) _____.

3. A(n) _____ is a medical gas system in which equipment or system failure is likely to cause serious injury or even death.

4. If a product meets the specifications for medical purity established by the United States Pharmacopeial Convention, it is said to be _____.

5. A(n) _____ is the equipment used to store oxygen and distribute it to the supply line at system pressure.

6. Medical air that has been certified as pharmaceutical grade according to the standards of the United States Pharmacopeial Convention is called _____.

7. A(n) _____ is a medical gas system in which equipment failure would not affect patient care at all.

8. An installation test performed to ensure that the distribution piping in a medical gas system will not experience excessive pressure loss when under service pressure is called a(n) _____.

9. A(n) _____ is an installation test performed to ensure that there is no pressure loss in the distribution piping.

10. A condition in which all water has been removed is called _____.

11. An installation test performed to ensure that medical gas and vacuum systems are not cross-connected is called a(n) _____.

12. A(n) _____ is an alarm that continually monitors one system and is located near the system being monitored.

13. A special type of medical vacuum system designed to suction away bodily fluids during surgery is called a(n) _____.

14. _____ is work performed on heated metal.

15. An alarm that continually monitors medical gas and vacuum systems that are located in a specific area, such as at a treatment room is called a(n) _____.

16. A(n) _____ is an alarm that continually monitors the gas supply source and mainline pressure for all systems and is directly connected to each of the devices it monitors.

17. An installation test performed to purge the system of any remaining dirt and other particulate matter is called a(n) _____.

18. A(n) _____ is an atmosphere that contains more than 23.5 percent oxygen by volume.

19. Something that is considered essential for ensuring the safety of patients is considered _____.

20. A(n) _____ is a gas that is inhaled to inhibit sensitivity to pain.

21. A medical gas system in which equipment failure is likely to cause minor injury is called a(n) _____.

22. A(n) _____ is a set of equipment and piping that uses negative pressure to remove waste gases, liquids, and solids, along with the associated pumps and compressors, filters and purifiers, valves, alarms, gauges, and controls.

23. A signal that cannot be turned off manually is called _____.

24. _____ is compressed air, not to be confused with medical air USP, that is used to power pneumatic tools and to clean instruments.

25. A medical gas system in which equipment failure is unlikely to cause injury but may cause some discomfort to patients is called a(n) _____.

26. _____ is the medical term for the breathing cycle.

27. A meter that measures and indicates the volumetric flow rate of a gas or a liquid is called a(n) _____.

28. A(n) _____ is an installation test performed to ensure that vacuum system piping in a medical vacuum system will not experience excessive vacuum loss when in use.

29. A set of equipment and piping that supplies nonflammable medical gases under pressure, along with the associated pumps and compressors, filters and purifiers, valves, alarms, gauges, and controls is called a(n) _____.

30. A(n) _____ is an electrical device that converts an analog signal, such as line pressure, into an electrical signal that can be read by an electrical device such as an alarm circuit board.

31. To be filled with an inert gas such as oil-free, dry nitrogen is also called _____.

Trade Terms

Anesthetic	Dehydrated	Life safety
Area alarm	Flowmeter	Local alarm
Bulk oxygen system	Hot work	Master alarm
Category 1 system	Initial blowdown test	Medical air USP
Category 2 system	Initial cross-connection test	Medical gas system
Category 3 system	Initial piping-purge test	Medical-surgical vacuum system
Category 4 system	Initial pressure test	
Charged	Instrument air	

Trade Terms Introduced in This Module

Anesthetic: A gas that is inhaled to inhibit sensitivity to pain.

Area alarm: An alarm that continually monitors medical gas and vacuum systems that are located in a specific area, such as at a treatment room.

Bulk oxygen system: The equipment used to store oxygen and distribute it to the supply line at system pressure.

Category 1 system: A medical gas system in which equipment or system failure is likely to cause serious injury or even death.

Category 2 system: A medical gas system in which equipment failure is likely to cause minor injury.

Category 3 system: A medical gas system in which equipment failure is unlikely to cause injury but may cause some discomfort to patients.

Category 4 system: A medical gas system in which equipment failure would not affect patient care at all.

Charged: Filled with an inert gas such as oil-free, dry nitrogen.

Dehydrated: A condition in which all water has been removed.

Flowmeter: A meter that measures and indicates the volumetric flow rate of a gas or a liquid.

Hot work: Work performed on heated metal.

Initial blowdown test: An installation test in which oil-free, dry nitrogen gas is blown through the distribution piping to remove dirt and other particulate matter.

Initial cross-connection test: An installation test performed to ensure that medical gas and vacuum systems are not cross-connected.

Initial piping-purge test: An installation test performed to purge the system of any remaining dirt and other particulate matter.

Initial pressure test: An installation test performed to ensure that there is no pressure loss in the distribution piping.

Instrument air: Compressed air that is used to power pneumatic tools and to clean instruments; not to be confused with medical air USP.

Life safety: Essential for ensuring the safety of patients.

Local alarm: An alarm that continually monitors one system and is located near the system being monitored.

Master alarm: An alarm that continually monitors the gas supply source and mainline pressure for all systems and is directly connected to each of the devices it monitors.

Medical air USP: Medical air that has been certified as pharmaceutical grade according to the standards of the United States Pharmacopeial Convention.

Medical gas system: Equipment and piping that supplies nonflammable medical gases under pressure, along with the associated pumps and compressors, filters and purifiers, valves, alarms, gauges, and controls.

Medical-surgical vacuum system: A special type of medical vacuum system designed to suction away bodily fluids during surgery.

Medical vacuum system: Equipment and piping that uses negative pressure to remove waste gases, liquids, and solids, along with the associated pumps and compressors, filters and purifiers, valves, alarms, gauges, and controls.

Noncancelable: A signal that cannot be turned off manually.

Oxygen-enriched atmosphere: An atmosphere that contains more than 23.5 percent oxygen by volume

Pharmaceutical grade: Meeting the specifications for medical purity established by the United States Pharmacopeial Convention.

Respiration: The breathing cycle.

Standing pressure test: An installation test performed to ensure that the distribution piping in a medical gas system will not experience excessive pressure loss when under service pressure.

Standing vacuum test: An installation test performed to ensure that vacuum system piping in a medical vacuum system will not experience excessive vacuum loss when in use.

Transducer: An electrical device that converts an analog signal, such as line pressure, into an electrical signal that can be read by an electrical device such as an alarm circuit board.

Verifier: A person who inspects medical gas and vacuum systems and verifies that they meet all applicable standards and codes as well as the manufacturer's specifications.

Additional Resources

This module presents thorough resources for task training. The following resource material is suggested for further study.

ASSE/IAPMO/ANSI Standard #6010, *Medical Gas Systems Installers*, Latest Edition. Washington, DC: American National Standards Institute.

ASSE/IAPMO/ANSI Standard #6020, *Medical Gas Systems Inspector*, Latest Edition. Washington, DC: American National Standards Institute.

ASSE/IAPMO/ANSI Standard #6030, *Medical Gas Systems Verifiers*, Latest Edition. Washington, DC: American National Standards Institute.

ASSE/IAPMO/ANSI Standard #6040, *Medical Gas Systems Maintenance Personnel*, Latest Edition. Washington, DC: American National Standards Institute.

ASTM B88, *Standard Specification for Seamless Copper Water Tube*, Latest Edition. West Conshohocken, PA: ASTM International.

ASTM B280, *Standard Specification for Seamless Copper Tube for Air Conditioning and Refrigeration Field Service*, Latest Edition. West Conshohocken, PA: ASTM International.

ASTM B819, *Standard Specification for Seamless Copper Tube for Medical Gas Systems*, Latest Edition. West Conshohocken, PA: ASTM International.

Brazing, Second edition. Mel M. Schwartz. 2003. Novelty, OH: ASM International.

CGA M-1, *Guide for Medical Supply Systems at Consumer Sites*, Latest Edition. Chantilly, VA: Compressed Gas Association.

The Copper Tube Handbook. 2010. New York: Copper Development Association.

NFPA 55, *Compressed Gases and Cryogenic Fluids Code*, Latest Edition. Quincy, MA: National Fire Protection Association.

NFPA 99, *Health Care Facilities Code*, Latest Edition. Quincy, MA: National Fire Protection Association.

NFPA 101, *Life Safety Code*, Latest Edition. Quincy, MA: National Fire Protection Association.

United States Pharmacopeia, Latest Edition. Rockville, MD: United States Pharmacopeial Convention, Inc.

Figure Credits

Answer	Section Reference	Objective
Section One		
1. a	1.1.1	1a
2. d	1.2.3	1b
Section Two		
1. d	2.1.4	2a
2. b	2.2.2	2b
3. b	2.3.1	2c
Section Three		
1. a	3.1.0	3a
2. d	3.2.0	3b
3. c	3.3.0	3c
4. d	3.4.0	3d
Section Four		
1. c	4.0.0	4
2. a	4.2.1	4b
Section Five		
1. b	5.1.0	5a
2. c	5.2.0	5b
Section Six		
1. c	6.1.5, 6.1.6	6a
2. d	6.2.9	6b

NCCER CURRICULA — USER UPDATE

NCCER makes every effort to keep its textbooks up-to-date and free of technical errors. We appreciate your help in this process. If you find an error, a typographical mistake, or an inaccuracy in NCCER's curricula, please fill out this form (or a photocopy), or complete the online form at **www.nccer.org/olf**. Be sure to include the exact module ID number, page number, a detailed description, and your recommended correction. Your input will be brought to the attention of the Authoring Team. Thank you for your assistance.

Instructors – If you have an idea for improving this textbook, or have found that additional materials were necessary to teach this module effectively, please let us know so that we may present your suggestions to the Authoring Team.

NCCER Product Development and Revision
13614 Progress Blvd., Alachua, FL 32615

Email: curriculum@nccer.org
Online: www.nccer.org/olf

❏ Trainee Guide   ❏ Lesson Plans   ❏ Exam   ❏ PowerPoints   Other _____

Craft / Level: _____ Copyright Date: _____

Module ID Number / Title: _____

Section Number(s): _____

Description: _____

_____

_____

_____

_____

Recommended Correction: _____

_____

_____

_____

_____

Your Name: _____

Address: _____

_____

Email: _____ Phone: _____

Glossary

Accounting: The process of compiling and analyzing financial records.

Accounts receivable: Money that is owed to a business but has not been paid.

Active system: A solar heating system that uses a pump and controls to circulate hot water through the system.

Aeration system: A type of soil absorption system that uses an aeration tank to clarify effluent.

Aeration tank: A waste storage and separation tank used in an aeration system. In an aeration tank, anaerobic bacteria decompose sludge and then aerobic bacteria clarify the effluent.

Aerobic bacteria: Bacteria that need oxygen to live. They are used in aeration tanks to clarify effluent.

Air break: A backflow preventer in which a smaller pipe drains into a larger pipe.

Air gap: A simple backflow preventer that consists of a space between the indirect waste system and the building waste system. The gap must be two times the diameter of the indirect waste disposal pipe and not less than 1 inch.

Airtrol valve: In a hydronic system, a valve that purges air from the system into the expansion tank.

Anaerobic bacteria: Bacteria that do not need oxygen to live. They are used in septic tanks and aeration tanks to decompose solid wastes.

Anesthetic: A gas that is inhaled to inhibit sensitivity to pain.

Anodic inhibitor: A chemical applied to a pipe to prevent galvanic corrosion.

Aquifer: An underground reservoir containing water-saturated soil, sand, and rock.

Area alarm: An alarm that continually monitors medical gas and vacuum systems that are located in a specific area, such as at a treatment room.

Artesian aquifer: A deep aquifer that does not lose water through runoff. Also called a confined aquifer.

Asset: Any type of property owned by a business.

Automatic trap primer: A device that feeds water into a low-use trap, allowing the trap to maintain a constant seal. Automatic trap primers may be designed to close automatically (also called automatic dosing systems), or to operate electronically or under the pressure of gravity.

Average flow: The flow rate in a water pressure booster system under average operating conditions, used to calculate off-peak loads.

Backwater valve: A valve that prevents the backflow of sewage into a pool drain.

Baffle: Partition inside the body of an interceptor. Baffles are designed to reduce turbulence caused by incoming wastewater and to block wastes from escaping through the outlet.

Balance: The condition of an account when assets and liabilities are equal.

Balance sheet: A form used to track assets and liabilities.

Balancing valves: Valves in a recirculation system that ensure steady water flow.

Batch system: A solar heating system in which water is both heated and stored in the collector. The system uses the principle of convection to circulate water.

Bathing load: The maximum number of people allowed in a pool per hour, based on capacity and turnover rate.

Bid: A formal offer to do work according to an estimate.

Biochemical oxygen demand (BOD): A measure of life-sustaining oxygen content in water.

Boiler: In a hydronic system, a device used to heat or boil water. Boilers differ from typical water heaters in that they have a greater capacity, operate at higher temperatures and pressures, and have a greater heat output.

Bond: A financial guarantee that work will be completed according to the terms of a contract.

Bored well: A type of well drilled by an auger and lined with a well casing.

Bulk oxygen system: The equipment used to store oxygen and distribute it to the supply line at system pressure.

Capacity: The volume of a pool in gallons.

Catch basin: A reservoir installed before the sanitary piping. It allows sediment to settle out of wastewater.

Category 1 system: A medical gas system in which equipment or system failure is likely to cause serious injury or even death.

Category 2 system: A medical gas system in which equipment failure is likely to cause minor injury.

Category 3 system: A medical gas system in which equipment failure is unlikely to cause injury but may cause some discomfort to patients.

Category 4 system: A medical gas system in which equipment failure would not affect patient care at all.

Cesspool: A lined, covered pit for storing solid and liquid wastes. Cesspools allow effluent to seep into the soil.

Change order: An agreement to perform work in addition to what was originally agreed to in a contract.

Charged: Filled with an inert gas such as oil-free, dry nitrogen.

Chlorinate: To disinfect a well with a chlorine compound such as calcium hypochlorite or liquid bleach.

Circuit Setter™: A type of balancing valve used to regulate the flow on the return of an entire circuit, with a dial permitting fine adjustments.

Cistern: A water storage tank used to supply some private water supply well systems with potable water.

Closed loop system: Another name for an indirect system.

Closed system: A private waste-disposal system that stores wastes until they can be disposed of.

Code: A legal document enacted to protect the public and property, establishing the minimum standards for materials, practices, and installations.

Collector: A device used in a solar heating system to receive heat from sunlight and transfer it to water or another liquid.

Combined upfeed and downfeed system: A recirculation system in which hot water is supplied in the upward and downward flow through the system.

Comprehensive Consensus Codes®: A collection of ANSI-compliant codes and standards developed jointly by the International Association of Plumbing and Mechanical Officials (IAPMO) and the National Fire Protection Association (NFPA).

Constant-speed system: A water pressure booster system in which water circulates continuously through the system and pumps replacement water from the main.

Continuous demand: Demand for water caused by outlets, pumps, and other devices with a relatively constant flow.

Contract: A legal agreement between a contractor and a client for work to be done.

Convection: Circulation caused by the sinking of dense cold water and the rising of hot water.

Cradle-to-grave responsibility: A contractor's legal responsibility for the management of hazardous wastes from creation to disposal.

Critical activities: Tasks in the critical path management process that could cause significant delay if not completed on time.

Critical path management (CPM): A task-planning process in which diagrams show the relationships between tasks and estimated completion times.

Daily work report: A report written by a job superintendent describing activities at a job site.

Debt: Money owed by a business for payment of direct costs and indirect costs.

Deep well: A well that is drilled deeper than 25 feet.

Deep-well jet pump: A pump used in a deep well that uses a nozzle and venturi to accelerate the water flow.

Dehydrated: A condition in which all water has been removed.

Differential thermostat: In a solar heating system, a control that activates a pump when water in the collector reaches a specified temperature above that of the water in the storage tank.

Direct cost: An expense for an activity or an item related directly to a project.

Direct return system: A two-pipe system in which the return water flows in the opposite direction from the flow in the supply pipe.

Direct system: A solar heating system that uses a pump and temperature controls to control the flow of hot water. It is also called an open loop system.

Distribution box: A watertight container that directs effluent to different parts of a leach field.

Diverter tee: In a one-pipe system, a tee that diverts the water flow by using an internal baffle to create a pressure drop.

Double-suction impeller: A pump impeller with two cavities, one each on opposite sides of the impeller. It is used to channel water in high-flow conditions in water pressure booster systems.

Downfeed system: A recirculation system that supplies hot water as it travels down the system.

Drain-down system: A direct system in which water drains out of the collector when the pump shuts off.

Draw-off hose: A hose used to siphon grease and oil from interceptors.

Drilled well: A type of well created by chisels or rotary drills.

Driven well: A type of well drilled by a pointed well head driven by a hammer.

Drop pipe: Another name for the discharge pipe that connects the well to the water storage tank.

Drywell: A covered pit filled with loose sand and gravel that collects roof and basement drainage and allows it to percolate into the surrounding soil.

Dug well: A type of well excavated by shovel and backhoe and lined with brick, stone, or concrete.

Effluent: Liquid waste that separates from solid waste. In soil absorption systems and cesspools, effluent is allowed to percolate into the soil.

Elevated-tank system: The oldest and simplest type of water pressure booster system, featuring a water tank on the roof operated by the weight of the water column.

Equity: The value of the business if all debts are paid.

Estimating: Calculating the costs to complete a specific project, which are submitted as part of a bid.

Ethylene glycol: An alcohol-based liquid used in an indirect system to transfer heat from a collector to a heat exchanger.

Expansion joint: A mechanical device installed on a pipe that allows the pipe to expand and flex as hot water flows through it.

Expansion tank: A tank attached to the supply pipes in a recirculation system that allows water to cool and slow down as it flows.

Floater: A device that releases chlorine through vents while floating in a pool.

Flow control fitting: A device that regulates the flow of wastewater into grease and oil interceptors.

Flowmeter: A meter that measures and indicates the volumetric flow rate of a gas or a liquid.

Foot valve: A valve used to prime water-lubricated pumps in wells deeper than 32 feet.

Forced circulation system: A recirculation system that uses a pump to circulate hot water through the system.

Forced heat system: A type of hydronic system that uses fans and ducts to circulate air heated by passing it over coils containing hot water.

Gravity draw-off system: A system that automatically drains oil from an interceptor into a storage tank.

Gravity return system: A recirculation system in which the force of gravity circulates hot water throughout the system.

Gross income: All money earned before expenses are subtracted.

Header pipe: A pipe used to relieve hydrostatic pressure. The pipe runs underneath the pool to a pump connection on the pool deck.

Heat exchanger: A device used in an indirect system to transfer heat from one liquid to another in such a way that the two liquids do not come into contact with each other.

High-limit aquastat: A safety device that deactivates a hot water boiler when the water reaches the maximum temperature.

Hot work: Work performed on heated metal.

Humus: In an organic system, the dried solid waste that results from decomposition. Humus can be used as fertilizer after a time.

Hydrojet: A nozzle that sends a high-pressure stream of air and water into a spa.

Hydrologic cycle: The process through which water cycles through the environment as both a liquid and a vapor.

Hydronic system: A plumbing system that uses water or steam to heat a building. It is also called a radiant system.

Hydropneumatic tank system: A water pressure booster system that uses air pressure to circulate the water.

Hydrostatic pressure: The force applied to the bottom of a pool, spa, or hot tub by high groundwater levels beneath it.

Indirect cost: An expense for an activity or an item that supports the administrative and overhead activities of a business.

Indirect system: An active system that uses a liquid other than water to distribute heat from the collector. It is also called a closed loop system.

Indirect waste system: A drainpipe attached to a fixture or appliance that is separated from the regular drainage system by a backflow preventer.

Indirect waste: Wastewater that flows through an indirect waste pipe.

Initial blowdown test: An installation test in which oil-free, dry nitrogen gas is blown through the distribution piping to remove dirt and other particulate matter.

Initial cross-connection test: An installation test performed to ensure that medical gas and vacuum systems are not cross-connected.

Initial piping-purge test: An installation test performed to purge the system of any remaining dirt and other particulate matter.

Initial pressure test: An installation test performed to ensure that there is no pressure loss in the distribution piping.

Instrument air: Compressed air that is used to power pneumatic tools and to clean instruments. Not to be confused with medical air USP.

Interceptor: A device that separates special wastes before they can enter the sewer, septic, or stormwater system.

Intermittent demand: Demand for water caused by fixtures that are used no more than about five minutes at a time.

International Plumbing Code® (IPC): The model plumbing code of the International Code Council (ICC). It was first issued in 1995.

Invoice: An itemized list of materials for purchase or rent.

Leach: To percolate into soil and be absorbed.

Leach field: In a soil absorption system, an area of soil that is designed to accept effluent from distribution pipes. Leach fields are also sometimes called absorption fields or drain fields.

Liability: In accounting, money owed by a business. In insurance, a legal responsibility.

Life safety: Essential for ensuring the safety of patients.

Limit switch: In a solar heating system, a switch that activates when the water in the collector reaches a preset temperature.

Local alarm: An alarm that continually monitors one system and is located near the system being monitored.

Low-water cutoff device: A device that deactivates a pump when the water level in a well falls to a preset low point. See also *tail pipe* assembly.

Manual draw-off system: A system that drains oil out of an interceptor through a valve-operated draw-off hose.

Master alarm: An alarm that continually monitors the gas supply source and mainline pressure for all systems and is directly connected to each of the devices it monitors.

Maximum flow: The theoretical total flow in a water pressure booster system if all outlets are simultaneously opened.

Maximum probable flow: The flow in a water pressure booster system during periods of peak demand.

Mechanical grease interceptor: A grease interceptor fitted with a timer-activated self-cleaning capability.

Medical air USP: Medical air that has been certified as pharmaceutical grade according to the standards of the United States Pharmacopeial Convention.

Medical gas system: Equipment and piping that supplies nonflammable medical gases under pressure, along with the associated pumps and compressors, filters and purifiers, valves, alarms, gauges, and controls.

Medical vacuum system: Equipment and piping that uses negative pressure to remove waste gases, liquids, and solids, along with the associated pumps and compressors, filters and purifiers, valves, alarms, gauges, and controls.

Medical-surgical vacuum system: A special type of medical vacuum system designed to suction away bodily fluids during surgery.

Mobilization: The final stage of preplanning, during which all items required to start construction are moved to the job site, arrangements are made with the other trades to coordinate job-site activities, and all job-site details are organized in preparation for construction.

Model code: A set of comprehensive, general guidelines that establish and define acceptable plumbing practices and materials and list prohibited installations.

Modified waste line: A vertical pipe at the terminal of a waste line that allows indirect waste to mix with air.

Net income: Another term for profit.

Noncancelable: A signal that cannot be turned off manually.

One-pipe system: A hydronic system in which water or steam is fed to the heating units and returned to the boiler through a single run of pipe.

Open loop system: Another name for a direct system.

Open-hub waste receptor: A waste drainpipe or pipe hub that extends above the floor line.

Organic system: A private waste disposal system that allows waste to decompose through exposure to aerobic bacteria. The resulting dried waste is called humus.

Overhead: The costs of maintaining a business that are not applicable to a specific job. Office and administrative expenses are common examples of overhead.

Oxygen-enriched atmosphere: An atmosphere that contains more than 23.5 percent oxygen by volume

Packer-type ejector: For jet pumps, an alternative discharge method to the two-pipe system in which the space between the water supply pipe and the well casing acts as a pressure pipe.

Parallel: A one-pipe system in which each heating unit draws only the amount of water it requires from the supply line.

Passive system: A solar heating system that relies on convection to circulate water.

Perched water table: A small water table that is cut off from a larger reservoir.

pH value: A measure of the hydrogen ions in water. The term is an abbreviation for "potential hydrogen."

Pharmaceutical grade: Meeting the specifications for medical purity established by the United States Pharmacopeial Convention.

Pitless adapter: A device that permits a discharge pipe to exit a well and connect to a water storage tank. Pitless adapters protect well water from contamination and freezing.

Premium: What the contractor pays the insurance company to ensure that the project is covered.

Preplanning: The phase immediately following the awarding of a contract but before the start of construction, during which subcontractors are selected, purchase orders are issued, construction schedules are developed, and estimates are reviewed in depth.

Pressure control valve: A valve that diverts a portion of a pump's discharge through the ejector, thereby maintaining adequate pressure in the ejector.

Pressure distribution system: A soil absorption system in which effluent is pumped through the distribution pipes under pressure.

Pressurized air-bladder storage tank: A tank used in a water pressure booster system that uses compressed air to maintain water pressure.

Primary-secondary system: A one-pipe system with a parallel arrangement.

Production figure: An estimate of the time required for a crew to perform a task.

Profit: Money left over after expenses are subtracted from income.

Profit and loss statement: A list of all income and expenses over a given period of time.

Purchase order: An order form submitted to a supplier.

Radiant loop: A heating unit that consists of a network of hot water pipes in the floor or ceiling.

Radiant system: Another name for a hydronic system.

Radiation: The transfer of heat between bodies through space.

Radiator: A heating unit that emits heat from water via a series of metal fins.

Recapitulation sheet: A summary of material, labor, and equipment costs for each task in a project. Also called a recap sheet.

Receptor: A basin designed to hold liquids.

Recirculation system: A plumbing installation that circulates hot water within a building, providing customers with hot water on demand.

Respiration: The breathing cycle.

Reverse return system: A two-pipe system in which the return water flows in the same direction as the flow in the supply pipe.

Reverse thermosiphoning: In a thermosiphon system, a condition in which water flows backward through the system.

Rework: To redo a previously completed task.

Sediment bucket: A removable basin in a drain that prevents solid wastes from entering a sanitary system.

Septic system: A soil absorption system that uses a septic tank to separate out sludge.

Septic tank: A tank used in a soil absorption system to settle out sludge and scum and to decompose them using anaerobic bacteria.

Series: A one-pipe system in which all the water in the supply loop goes through each heating unit.

Seven-minute peak demand period: An estimate of the greatest possible water demand in a private water supply well system.

Shallow well: A well that is no more than 25 feet deep.

Shallow-well jet pump: A pump used in a shallow well that uses a nozzle and venturi to accelerate the water flow.

Sidewall drain: A drain installed on a swimming pool wall below the waterline.

Single-suction impeller: A type of impeller used in centrifugal pumps that have a single-suction cavity. Single-suction impellers are used in low-flow water pressure booster systems.

Skimmer: A device that traps pool debris in a removable basket.

Soil absorption system: A private waste-disposal system that uses tanks to separate out effluent, then allows it to drain into a leach field.

Soil mottle: Spots and streaks in soil that indicate water saturation.

Soil profile description: A report that specifies the depths and thicknesses of soil layers in a proposed leach field.

Solar heating system: A hydronic system that uses the sun to heat water.

Sole source: A type of bid in which a project is offered to a contractor of the client's choice.

Special waste: Waste that must be removed, diluted, or neutralized before it can be discharged into a sanitary system.

Specific heat: A measure of how much heat a liquid can hold per pound, measured in British thermal units (Btus).

Stage: In a vertical turbine pump, the arrangement of a single impeller and its water passage.

Standing pressure test: An installation test performed to ensure that the distribution piping in a medical gas system will not experience excessive pressure loss when under service pressure.

Standing vacuum test: An installation test performed to ensure that vacuum system piping in a medical vacuum system will not experience excessive vacuum loss when in use.

Steam dome: In a steam hydronic system, the place in the top of the boiler where steam collects.

Submersible pump: A pump consisting of a multistage centrifugal pump and a submersible electric motor. Submersible pumps are usually installed in the bottom of deep wells.

Tail pipe assembly: A device that matches a pump's discharge rate to a well's recovery rate. See also *low-water cutoff device*.

Task planning: The process of organizing activities and materials so that tasks are completed in the right order and on time.

Tempering mixing valve: A valve that adds cold water to the hot water flow to control water temperature.

Thermal purge valve: A valve that allows heated water to drain from the recirculation system when no-flow conditions exist.

Thermistor: In a solar heating system, a control that adjusts a pump's speed according to changes in the water temperature.

Thermosiphon system: A passive system in which hot water is stored in a tank located higher than the solar collector.

Thermosiphoning: In a thermosiphon system, the process whereby cold water flows back to the collector to be reheated.

Transducer: An electrical device that converts an analog signal, such as line pressure, into an electrical signal that can be read by an electrical device such as an alarm circuit board.

Turbidity: A measure of suspended solids in well water.

Turnover rate: The time required for all the water in a pool to cycle through the filter.

Two-pipe system: A hydronic system that uses separate supply and return lines; for jet pumps, an alternative to the packer-type ejector discharge method. One pipe draws well water through an ejector, and another pipe feeds the pressurized water to the jet.

Uniform Plumbing Code® (UPC): The model code of the International Association of Plumbing and Mechanical Officials (IAPMO). It has been published since 1945.

Unit cost: In estimating materials used in a job, the total material cost divided by the number of units used.

Upfeed system: A type of recirculation system that supplies hot water as it travels up the system.

Vacuum fitting: A pool drain that serves as a connection for cleaning hoses.

Variable-capacity system: A type of water pressure booster system that uses two or more pumps to provide water in response to peaks and lows in demand.

Venturi: A device that accelerates water and lowers its pressure by passing it through a narrow pipe opening. It is also called a diffuser.

Verifier: A person who inspects medical gas and vacuum systems and verifies that they meet all applicable standards and codes as well as the manufacturer's specifications.

Vertical turbine pump: A pump used in water pressure booster systems where high-pressure and low-flow conditions exist.

Vibration isolators: Devices such as rubber pads and flexible line connectors that reduce the effects of pump vibration.

Volute: In centrifugal pumps, a geometrically curved outlet path.

Water pressure booster system: A plumbing installation that increases water pressure in the fresh water supply system.

Water supply fitting: A fixture that allows water to flow from a recirculation system back into a swimming pool.

Water table: The top of an aquifer, below which rock, sand, and soil are completely saturated with groundwater.

Water table aquifer: An aquifer in which water flows toward bodies of water. Also called an unconfined aquifer.

Well casing: A pipe section inserted into a well that maintains the shape of the well and prevents contamination.

Well head: A pointed rod used to create a driven well by being forced into the ground with a hammer.

Well-casing adapter: A well cover that protects the well casing from air leaks.

Well: A hole in the ground that connects with an aquifer. Wells are equipped with pumps to lift water from the aquifer into water storage tanks.

Zone control valve: In a hydronic system, a valve that allows the temperature in various parts of the building to be preset.

Index

Cartridge filters, (02410):5–6
Cast-Iron Soil Pipe Institute (CISPI), (02406):10, 22
Catch basins, special waste systems, (02404):6, 13, 19
Category 1 medical gas and vacuum systems, (02412):1, 2, 12, 49
Category 2 medical gas and vacuum systems, (02412):1, 2, 12, 49
Category 3 medical gas and vacuum systems, (02412):1, 2, 10, 13, 49
Category 4 medical gas and vacuum systems, (02412):1, 2, 49
Cellular telephones, (46101):3
Centrifugal pumps
    components, (02403):7
    double-suction impeller, (02403):1, 5, 29
    hot water circulation, (02405):4
    single-suction impeller, (02403):1, 5, 29
    swimming pool recirculation systems, (02410):4
    water pressure boosters, (02403):4–7
    water-sourced heat pumps, (02405):1
Cesspool, (02409):1, 8, 24
Change, motivating with, (46101):20
Change orders, (02401):1, 11, 14, 28
Charged, (02412):29, 30, 49
Chart of accounts, (02401):3
Check valves
    hydronic system, (02405):16–18
    recirculation systems, (02403):20
    solar heating system, (02405):21
    water supply systems, mobile home parks, (02411):4
Chemical waste systems, (02404):6, 7, 15
Chlorinate, (02408):1, 5, 24
Chlorinator, (02410):9–10
Circuit setter™, (02403):14, 21, 29
Circulating pumps, (02405):4
CISPI. See Cast-Iron Soil Pipe Institute (CISPI)
Cistern, (02408):1, 24
Clay distribution pipe, (02409):16, 17
Clear-water installations, (02404):4
Closed heating systems, (02409):1, 24
Closed loop heating system, (02405):7, 12, 29
Closed waste disposal systems
    cesspools, (02409):8
    components, (02409):9
    defined, (02409):1, 24
    drywells, (02409):8
    process, (02409):7
Code changes, typical, (02406):15–16
Code-development organizations, (02406):1–2
Code of Hammurabi, (02406):2
Code revision process
    International Code Council (ICC), (02406):12–13
    Uniform Plumbing Code® (UPC) (IAMPO), (02406):10, 12
Codes. See also Model codes; Plumbing codes
    defined, (02406):1, 2, 20
    energy conservation and efficiency, (02406):16
    fittings, (02406):16
    historically, (02406):3
    online, (02406):3
Cold weather protection
    pitless adapter, (02408):14
    water supply systems, mobile home parks, (02411):4, 5
Collector, (02405):1, 29
Combined upfeed and downfeed circulation system, (02403):14, 17, 29
Commander leadership style, (46101):13
Communication
    defined, (46101):14

in diverse workforces, (46101):4, 17
electronic, (46101):3, 4, 16–17
feedback in training, (46101):2
on-the-job, (02401):17
leadership and, (46101):12
listening
    active, (46101):15
    effective, (46101):15–16
non-verbal, (46101):16
organizational structure for, (46101):7
in project planning, (46101):56
tailoring, (46101):17
verbal, (46101):14–15, 15
visual messaging, (46101):16–17
websites for, (02406):1
written, (46101):16–17
Communication styles, gender differences in, (46101):4
Community-corporate partnerships, (46101):1
Company vehicles, (02401):9
Composting tanks, (02409):7
Comprehensive Consensus Codes® (C3)
    defined, (02406):1, 20
    International Association of Plumbing and Mechanical Officials (IAMPO)
        Uniform Mechanical Code™ (IAMPO), (02406):6
        Uniform Plumbing Code®, (02406):6
    National Fire Protection Association
        NFPA 1, Fire Code, (02406):6
        NFPA 3A, Code for Motor Fuel Dispensing Facilities and Repair Garages, (02406):6
        NFPA 30, Flammable and Combustible Liquids Code, (02406):6
        NFPA 54, National Fuel Gas Code, (02406):6
        NFPA 58, Liquefied Petroleum Gas Code, (02406):6
        NFPA 70, National Electrical Code®, (02406):6
        NFPA 101, Life Safety Code®, (02406):6
        NFPA 900, Building Energy Code, (02406):6
Compressors, medical air systems, (02412):9
Confined-space safety procedures, (02409):19
Conflict
    with crew workers, avoiding, (46101):11
    cultural, resolving, (46101):4–5
    employee inability to work with others, (46101):26
Constant-speed system
    defined, (02403):1, 29
    highlighted in text, (02403):2
    water pressure boosters, (02403):3, 4
Construction cost control
    cost performance assessment and, (46101):64
    crew leader's role, (46101):65
    field reporting system for, (46101):64–65
Construction cost control, preplanning for
    change orders, (02401):11, 14
    company vehicles, (02401):9
    cost control systems, establishing, (02401):7
    equipment records, (02401):8, 11
    equipment requirements, (02401):6–7
    estimates
        reviewing, (02401):6
        software for, (02401):13
    materials handling
        deliveries, (02401):7, 8
        invoices, (02401):8, 10
        ordering, (02401):8
        securing, (02401):8
        storage, (02401):7–8
    mobilization, (02401):1, 4, 7, 28
    purchase orders, (02401):5, 6